Fascinating Life Sciences

This interdisciplinary series brings together the most essential and captivating topics in the life sciences. They range from the plant sciences to zoology, from the microbiome to macrobiome, and from basic biology to biotechnology. The series not only highlights fascinating research; it also discusses major challenges associated with the life sciences and related disciplines and outlines future research directions. Individual volumes provide in-depth information, are richly illustrated with photographs, illustrations, and maps, and feature suggestions for further reading or glossaries where appropriate.

Interested researchers in all areas of the life sciences, as well as biology enthusiasts, will find the series' interdisciplinary focus and highly readable volumes especially appealing.

More information about this series at http://www.springer.com/series/15408

Dieter Thomas Tietze
Editor

Bird Species

How They Arise, Modify and Vanish

Springer Open

Editor
Dieter Thomas Tietze
Natural History Museum Basel
Basel, Switzerland

ISSN 2509-6745 ISSN 2509-6753 (electronic)
Fascinating Life Sciences
ISBN 978-3-319-91688-0 ISBN 978-3-319-91689-7 (eBook)
https://doi.org/10.1007/978-3-319-91689-7

Library of Congress Control Number: 2018948152

Cover illustration: Coal Tit *Periparus ater melanolophus* (current status) or Spot-winged Tit *Periparus melanolophus* (common sense for many decades in the past)? Marker-gene sequences place it in the trans-Eurasian Coal Tit assemblage, its song fits in as well, but its plumage is much more colorful than that of other Coal Tit subspecies. Photograph taken near Rakchham, Himachal Pradesh, India, on 10 October 2012 by Gunjan Arora.

This Springer imprint is published by the registered company Springer Nature Switzerland AG
The registered company address is: Gewerbestrasse 11, 6330 Cham, Switzerland

In this book, ornithologists from around the world elaborate on various aspects on how the fascinating diversity of birds found on Earth has formed, how bird species change in time and space, and how they get lost—for natural reasons and under increasing pressure from human activities.

Foreword

According to the compilation used in this book, 10828 species of birds fall into 40 orders. One of these 40 orders, the Passeriformes, arose relatively recently, yet contains 60% of all bird species. The order Passeriformes comprises three suborders. One suborder, the oscines, contains 45% of all recognized bird species, which are found on all continents. The second, the suboscines, contains 15% of all species, mostly confined to South America. The third, the New Zealand wrens, contains just two extant species. What accounts for the large differences in diversity and distribution across the bird portion of the tree of life? They ultimately trace back to differences in speciation rates, extinction rates, and dispersal. Research on speciation, and even research on *bird* speciation, has grown tremendously over the past 15 years. Research on extinction has also grown tremendously, but for birds this has been largely concerned with present-day extinctions and not what happened before humans. Research on bird dispersal is undergoing a revolution, not only through the use of phylogenetic reconstruction and advanced analytical techniques, but through real-time tracking. Up-to-date reviews of the burgeoning literature are needed to keep us informed. Several fine ones are included in this compilation.

One issue that is being revolutionized by new methods for quantifying species traits, including the use of genomic data, is in the species definition itself. Clearly if we talk about speciation, extinction, and dispersal, we need to know what is supposed to be speciating, what is going extinct, and what is dispersing. George Sangster in Chapter 2 of this book gives a good introduction to this contentious issue. As someone who generally employs the biological species concept, I take this opportunity to expand upon some of his remarks.

The introduction of molecular dating in the late twentieth century led to surprisingly old dates, generally millions of years, separating even the youngest species that breed in the same place. These species, if crossed, are likely to produce infertile hybrids, as a result of genetic incompatibilities, and the failure to interbreed maintains their distinctive features. Now, a broader survey and the application of new genomic analyses have led to the discovery of several examples of very young co-occurring species, including the parasitic *Vidua* finches of Africa, Darwin's ground finches, and the *Sporophila* finches of South America. These may truly have recently formed. Other species came together, before they were completely reproductively isolated and whose co-occurrence has led to ongoing hybridization, gene flow, and genetic homogenization across most of the genome. Whatever the reason, they provide the focal point for the debate over species concepts. Young species are recognized as species, because they co-occur, but remain distinct. They are good *biological* species, because they rarely hybridize (premating isolation) and also because hybrids have low fitness generally for ecological reasons (postmating isolation). They are defined by the reproductive isolating mechanisms that keep them apart, which thereby maintain lineage distinctiveness among sympatric forms, and the study of biological speciation is the study of how such mechanisms arose. But according to various phylogenetic concepts, such young species should not be considered species at all, because the vast majority of the genetic variation present is not restricted to such species; hence, lineages are not (yet) distinct. This view has been most strongly promoted by Robert Zink, e.g., in his discussions of Darwin's finches (Zink 2002).

While in the previous paragraph I gave examples of young biological species in sympatry, much older allopatric populations are often classified as belonging to the same biological species and given the status of subspecies. They are considered to be subspecies and not species, because their inferred reproductive isolation is incomplete and the two taxa would interbreed and collapse back into one, if they came into sympatry. These subspecies may be reciprocally monophyletic in many genes, i.e., already be quite genetically distinctive. Such allopatric forms are "phylogenetic species" harboring substantial amounts of unique genetic variation. Hence, in the literature we have biological species that are not phylogenetic species (young co-occurring forms) and phylogenetic species that are not biological species (older allopatric forms). At present, most classifications and studies of speciation use the biological species concept, relatively easily applied to sympatric forms, but often requiring subjective inference on whether related allopatric taxa would interbreed, if they were to come together. Perhaps we need to move to a clear statement of the species definition being used.

These considerations highlight a major issue associated with biological speciation. Birds are particularly suitable for investigating the way by which reproductive isolation is generated as a result of divergent selection pressures (so-called ecological speciation). Studies of habitat choice, food choice, host choice in parasitic birds, migration differences, timing of breeding, and urban–rural differentiation are all nicely summarized in this book. Sometimes, these selection pressures lead to rapid divergence in ecological traits that enable co-occurrence. Hence, biological speciation can be rapid. But the co-occurrence of such young species, which have no intrinsic isolating barriers, creates the opportunity for interbreeding, hybridization, and species collapse as environments change. So perhaps communities of old co-occurring species were formed after a longer time in allopatry, with first the formation of phylogenetic species, possibly with strong reproductive barriers. In fact, it seems likely that both periods in allopatry and ecologically divergent selection pressures are involved in most biological speciation events. This book sets the stage for what will surely be an important research direction in the next decades: integrating the role of ecologically divergent selection pressures with an understanding of the origin of those genes that drive genetic incompatibilities, i.e., those genes that ensure species are permanently reproductively isolated from each other. They will undoubtedly show that speciation is often a long and protracted process.

In this light, I now return to the question posed at the beginning of this foreword and use our current state of knowledge about speciation, extinction, and dispersal to generate a viable hypothesis for why the oscines are so species rich and why the suboscines are largely restricted to South America. We infer from both fossils and phylogenetic reconstructions that the ancestor to the songbirds arose in the Australian region. It then seems that a great cooling some 34 million years ago (the Eocene/Oligocene boundary) was the time when descendants of that species set off to world domination, spreading north and multiplying as they moved through Asia and into Africa and later over the Bering Strait into North America. (One species that crossed over the Bering Strait perhaps 20 million years ago gave rise to more than 800 species, including tanagers, New World warblers, and New World blackbirds.) South America was the last major land mass to be reached and entry from the north is still going on: Two migratory swallow species have recently started to breed in Argentina. Dispersal through the world in a cooling climate was correlated with extinctions in other orders, as inferred from the fossil record. Extinctions likely affected the suboscines as well, and the late entry of oscines into South America may be one reason why so many suboscines persist there. The reasons for the great success of the oscines are still debated, but are likely to include elements of speciation, extinction, and dispersal. Several authors of chapters in this book emphasize song learning as a possible accelerator of speciation. But other researchers, including Luis Baptista and Pepper Trail (1992) and Storrs Olson (2001), have noted the possible importance of large brains and base intelligence; nests can be constructed craftily and several oscine species store food, very useful in harsh climates. Perhaps attributes such as these promoted successful dispersal and persistence in new locations, thereby accelerating speciation by giving time for reproductive isolation to develop and lowering extinction rates. I find the intelligence argument quite compelling.

Now, millions of years later, we see an intelligent organism moving toward world domination, albeit on a much grander scale than the oscines. This book summarizes aspects of bird speciation, but also pressing issues associated with the other side of the diversification coin, extinction. Timescales of extinction in the past are poorly understood for birds, and extinction addressed here is contemporary. The emergent conclusion is that the time it has taken to produce the diversity of species we now see is completely discordant from current times to extinction. As the fossil record shows for other groups, it will take many millions of years to recover from the extinction wave now in process. The more we learn about contributions of speciation and extinction to diversity, the more obvious it becomes that we need to lower extinction rates. This book is an important summary by showing where we currently are, especially in our understanding of speciation, and what we need to do next.

University of Chicago
Chicago, IL, USA

Trevor Price

References

Baptista LF, PW Trail (1992) The role of song in the evolution of passerine diversity. Syst Biol 41:242–247

Olson SL (2001) Why so many kinds of passerine birds? Bioscience 51:268–269

Zink RM (2002) A new perspective on the evolutionary history of Darwin's finches. Auk 119:864–871

Contents

Chapter 1
Introduction: Studying Birds in Time and Space

Dieter Thomas Tietze

Abstract Birds are of high public interest and of great value as indicators of the state of the environment. Some 11,000 species are a number relatively well to handle. From a scientific point of view, it is not easily answerable what a species is, since speciation and extinction are ongoing evolutionary processes and differentiation among species works on various traits. Contemporary systematics attempts to take into account as many criteria as possible to delimit species. The currently most influential approach is the use of genomic sequences, be it as a neutral marker or to discover the underpinnings of functional traits. The study of the outer appearance of birds nevertheless remains fundamental, since that is the interface between a bird and its biotic and abiotic environment. For the majority of bird species, acquired traits of vocal communication add to this complex. Birds can also vary the timing of important behavior such as breeding or molting. Most fascinating among circannual behavior are the long-distance movements that can quite fast evolve and have genetic bases. Despite such dispersal ability for many bird species, geographic barriers play a large role for distribution and speciation in birds. Extant, former, and potential future ranges of a species can be modeled based on the abiotic niches individuals of this species have. Within a species' range, genetic and phenotypic traits vary and promote to process toward species splits. Beside geographic frameworks, ecological circumstances play a major role and contribute to natural selection but also trigger individual responses such as phenotypic plasticity, modification of the environment, and habitat selection. Anthropogenic global impacts such as climate and land-use changes (e.g., urbanization) force extant species to accelerated modifications or population splits or let them vanish forever. Only if humans leave more room and time to birds and other organisms can we expect to maintain such a number of diverse bird species, although they will keep modifying, splitting, and becoming extinct—but for natural reasons.

D. T. Tietze (✉)
Natural History Museum Basel, Basel, Switzerland
e-mail: thomas.tietze@bs.ch

D. T. Tietze (ed.), *Bird Species*, Fascinating Life Sciences,
https://doi.org/10.1007/978-3-319-91689-7_1

Keywords Speciation · Integrative taxonomy · Genomics · Morphology · Bird song · Allochrony · Bird migration · Biogeography · Niche modeling · Phylogeography · Hybridization · Ecological speciation · Global change · Climate change · Urbanization

1.1 Why and How to Study Bird Species

For many people around the world, birds are among the most fascinating fellow animals on Earth. Ornithology, the scientific study on birds, is one of the oldest organized scientific disciplines (Birkhead et al. 2014). And birds are among the best studied organisms (del Hoyo et al. 1992–2013). So why are we presenting yet another book on birds and especially bird speciation?

What we nowadays recognize as a bird is the relatively small-sized feathered survivor of the dinosaur assemblage of reptiles. These warm-blooded vertebrates diversified into tens of thousands of species within the last 150 and especially 65 million years. And while this diversification process keeps going, species have become extinct—for natural and increasingly for non-natural reasons. While there are many ways how we humans are letting the numbers of bird individuals and species diminish, there is also a lot we could and should do to halt this trend and preserve avian diversity.

To that end—and for many scientific reasons—it is important to understand what a bird species is and how it arises, is modified, and vanishes. It is far less easy than it appears to circumscribe a bird species. This is mainly due to the transient nature of species. Being one descendant of another species, a species can slowly become another species (anagenesis), die out, or split into two or more daughter species (cladogenesis).

In the 2000s, Newton (2003) and Price (2008) summarized the state of knowledge on this last aspect, which is so fundamentally important for the generation of bird diversity or biodiversity in general. A lot of advancement has been achieved since then, be it in the field of genomic foundations of species, bird distributions and their modeling, or (macro)ecological insights. And on the downside, human impact through land-use and climate changes has challenged more and more bird species and also reduces the population sizes of hitherto abundant species.

That is why publisher and editor both wished to compile an update on the topic of bird speciation but also to widen the audience from specialists to general bird enthusiasts and conservationists. A variety of experts—from PhD students to senior researchers—elaborate on various aspects in the following chapters.

George Sangster starts with the timely approach on how to circumscribe or even define a bird species (Chap. 2). Instead of using a fixed degree of differentiation in a certain genetic marker or some other scoring, he favors an integrative synopsis of as many lines of evidence available for the birds in question such as morphology, genetics, distribution, and behavior. This does not make things easier but proved to be most adequate.

We wish to keep your confusion as little as possible when diving deeper into bird species mysteries and follow the taxonomy and nomenclature of birds according to the IOC World Bird List version 7.1 (Gill and Donsker 2017, worldbirdnames.org) throughout. In contrast to other such comprehensive bird species lists, this global endeavor has scientific reasoning as first priority and works on a peer-reviewing basis. Since this meets the spirit of the time most, scientific societies such as the British Ornithologists' Union (BOU) and the German Ornithologists' Society (DO-G) decided to trust this expert panel and save the human resources in their own ranks from evaluating updates in bird systematics.

Although species delimitation should not rely on any single method alone, the study of the molecules of life, especially DNA, let research in this field make a huge leap forward three decades ago (Sibley et al. 1988). Progress in this field accelerated, and whole genomes have now become the focus (Ellegren 2013). The B10K project (https://b10k.genomics.cn) is ambitiously attempting to assemble genomic data for all known bird species. It kicked off at the level of bird orders (Jarvis et al. 2014).

Most chapters will cite work that has at least incorporated research on molecular markers. Daronja Trense and I thus provide some basic knowledge about how the genetic information of birds is organized, used by the birds themselves and by researchers who want to understand evolutionary changes and differences between individuals as well as between taxa on higher levels (Chap. 3). This chapter is mainly written for readers without training in biology, but due to the fast development in molecular biology, it might offer interesting updates for biologists who graduated more than few years ago.

1.2 Physical and Behavioral Aspects of Birds

Phylogenies based on genome sequences might provide the "backbone" for further studies, but variability at the molecular level is not necessarily detectable in the individual birds and their phenotypes. So whatever we observe as outer appearance or behavior (e.g., vocalizations, temporal changes, migration) are the characteristics that are also perceived by conspecifics and other animals as well. They constitute the phenotype on which selection works, and thus they are also very important to gain understanding for species and speciation.

Till Töpfer shows that the morphological variability within and among bird species can have many reasons such as sexual dimorphism, changes during an individual lifespan, adaptation to local environmental conditions or rather short-term reactions to environmental influences (phenotypic plasticity) (Chap. 4). He also presents several methods to describe structural or measure two- or three-dimensional features of a bird and emphasizes the importance of public bird collections that allow for robust comparative studies.

While morphological traits are mostly inherited, be it under the interplay of several genes, bird behavior, specifically song, can have some learning component. This adds cultural evolution as another driving force for bird species diversity and makes birds also models for the study of the evolution of human language. Martin Päckert introduces the three major bird clades that are capable of learning their song which plays an important role in species recognition and sexual selection (Chap. 5). They together comprise 54% of the 10,828 bird species recognized in this book (Gill and Donsker 2017). Similar to the advancement in genetic studies, bioacoustic methods such as sound recording in the field and sonagraphic analysis on screen have allowed for new insights in bird behavior and species delimitation.

Singing and other vocal behavior are controlled neuronally as well as hormonally and occur at specific times of the day and of the year. Barbara Helm and Robyn Womack describe some examples for timing in birds and demonstrate how different timing of breeding alone can lead to species divergence (Chap. 6). If breeding time is an inherited trait, then all descendants of an "innovative" breeding pair will keep to this alternative time and not form pairs with "traditional" conspecifics.

One of the most fascinating phenomena we observe in birds is their migratory behavior, i.e., seasonal movements, mostly between breeding areas and some sort of maintenance range(s) outside the breeding season. This is another example of an annual rhythm, with additional aspects to consider including navigation and the buildup of fat deposits. The when and where to of bird migration has to be adapted or adjusted to climatic changes and food availability. Such behavioral traits are thus quite labile. Nevertheless, they can be studied on both micro- and macroevolutionary levels. Miriam Liedvogel and Kira Delmore provide an overview on important aspects of bird migration within the scope of our book, report state-of-the-art genomic studies to understand the regulation of this behavior, and show how migration might play a role in the speciation process (Chap. 7).

1.3 The Spatial Component

For the diversification of birds, geography has been most important. Manuel Schweizer and Yang Liu demonstrate for major basal lineages of birds how they apparently arose before the background of continental drift (Chap. 8). The majority of bird species are able to fly and manage to disperse to hitherto uninhabited areas, before they become extinct in the current range. Such exchange between two areas can become impossible over time when barriers such as mountains or rivers grow. If the separated populations remain on their own for sufficient time (in allopatry), they might become reproductively incompatible and do not interbreed upon secondary contact. And if they also diversify in ecological respect so that their ecological niches do not overlap too much, they can also live in the same area (in sympatry) and their ranges overlap instead. For historical reasons as well as due to gradients in environmental resources across the Earth, the number of bird species varies a lot across space.

Within a certain timeframe, the ecological niche of a bird species remains quite stable and might also be inherited by daughter species (niche conservatism). Nevertheless, niches of sister species have to diverge to allow range overlap. Darius Stiels and Kathrin Schidelko explain how avian niches can be characterized based on the abiotic features under which we find representatives of extant species (Chap. 9). This approach allows researchers to model the (potential) distribution of a taxon now, in the (recent) past, and the—very near!—future (under various climate change scenarios or for cases of species introductions).

Finally, Leo Joseph elaborates from not only, but a more Australian perspective how intraspecific genetic differences lead to observable patterns and potentially to allospecies, geographically separated (young) species (phylogeography) (Chap. 10). Upon secondary contact, these taxa—no longer subspecies, not yet full species—may still exchange genetic information and even produce an intermediate taxon by such hybridization. Various genetic markers and phenotypic traits may or may not resolve the same pattern and elucidate or rather blur the current stage of the respective populations on the speciation trail. In consequence, we need to accept that any bird population is rather a continually evolving evolutionary lineage than a permanent species or subspecies.

1.4 Ecology Matters: Bird Species in the Anthropocene

In contrast to allopatric speciation emphasized so far, it has in the recent past become clearer and clearer that also in birds various ways of sympatric speciation can be found. Pim Edelaar elaborates on various modes of ecological speciation in birds: natural selection, phenotypic plasticity, adjustments of the environment, and habitat selection (Chap. 11). Foremost, individuals in a given population that have more suitable trait states for the area and habitat they live in right now have higher chances to survive and to produce a higher number of offspring (i.e., being fitter in the evolutionary sense = Darwinian natural selection). On a smaller temporal scale, individuals react to challenges within their lifetime within the framework of their inherited constitution (phenotypic plasticity), and this may nevertheless lead to segregation within the population. Regarding the bird's environment, they can actively make it more suitable for their phenotype by either modifying their surroundings or moving to areas better for foraging or nesting than the current one. Again, if only part of the population chooses to react that way, it might split in the long run.

The two final chapters deal with two major anthropogenic threats to bird-species diversity: climate and land-use change. Neither is completely "new" to birds, because both climate and surface structure have changed tremendously over the history of the Earth. But in the Anthropocene, these changes have occurred (almost) globally and so quickly that many species cannot maintain their population sizes, range extents, and/or genetic diversity.

We are well aware of the fact that increased output of certain gases by human activities (CO_2 emissions increased by 62% from 1992 until 2016 alone; Ripple et al. 2017) has intensified the generally necessary greenhouse effect and not only raised the average temperature on Earth (by approx. 0.5 °C from 1992 until 2016 alone; Ripple et al. 2017) but also consequently triggered various other climatic effects. A vast body of literature has accumulated that describes, models, and interprets the impact of these climatic changes on birds (and other organisms). And indeed, birds are affected in many ways: timing of migration and breeding have been adjusted, population sizes changed, ranges shifted, population and community compositions altered. Sven Trautmann summarizes approaches and findings as well as potential consequences for bird species conservation (Chap. 12).

Among the many changes of land use humans perform, the building of our own homes in relatively high density is a very striking example with severe consequences for birds. Caroline Isaksson provides numerous examples for urbanization impacts on birds (Chap. 13). While some species (urban exploiters) even profit from the environmental conditions humans create in cities, urban adapters are somehow able to live in human settlements but suffer from certain restrictions and might exhibit lower survival rates and fecundities than their rural conspecifics. Finally, species that are too shy or too specialized on habitats not found in cities (urban avoiders) get their ranges reduced in overall size, fragmented, and disconnected. While the latter has population-genetic effects and might lead to allopatric speciation in the long run, the coexistence of urban and rural populations of a single species has at least vast phenotypic consequences at the extent of what has naturally occurred over much longer time.

As a consequence of the insights into (natural) speciation processes and (anthropogenic) constraints on bird life these days, we need to reflect on traditional bird conservation strategies and ask if they sufficiently take into account the pervasiveness of human impact. Consequences of human action can be found globally and do not spare nature reserves (protected or even managed with a lot of effort) on the one hand and that birds on the other hand—as most living beings—adhere to intrinsic and extrinsic evolutionary "rules" that make it impossible to preserve a given "species" forever.

(Bird) species arise, modify, and vanish. If we want to protect "nature"—as a cultural ethic value—in its diversity and the birds we love for various reasons, we need to prevent their extinction and allow them to change, but in their way and speed. Therefore, we have to provide them sufficient space and time. Space: Restrict human settlements to far less than 100% of the continental surface (and marine space) and abandon developmental goals that promise optimal (urban or even metropolitan) living conditions to each human and every remote village. Time: Slow down or even reverse environmental impacts such as greenhouse effect, pollution, and land-use changes.

Acknowledgments Ommo Hüppop, secretary-general of the German Ornithologists' Society (DO-G) organized a symposium on evolution and speciation at the annual convention in Stralsund in 2016. This stimulated Springer editor Verena Penning to have a book on this topic in Springer's "Fascinating Life Sciences" series. She invited me as editor, as I also chaired the symposium.

I had gained some overview of the topic during my doctoral studies with Jochen Martens and postdoctoral research with Trevor Price and Michael Wink.

I also thank Trevor Price for his influential book (Price 2008), commenting on chapter drafts, and the foreword.

Lars Koerner took over from Verena Penning at Springer and together with Daniel Ignatius Jagadisan and his production team brought the project to a successful end.

However, most grateful I am to the other authors who reflected on bird species and research on their generation, alteration, and disappearance from various angles. The Stiftung zur Förderung des Naturhistorischen Museums Basel made open-access publishing possible.

References

Birkhead TR, Wimpenny J, Montgomerie B (2014) Ten thousand birds. Ornithology since Darwin. Princeton University Press, Princeton

del Hoyo J, Elliott A, Sargatal J (eds) (vol 1–7); Christie DA (eds) (vol 8–16) (1992–2013) Handbook of the birds of the world. Lynx Edicions

Ellegren H (2013) The evolutionary genomics of birds. Annu Rev Ecol Evol Syst 44:239–259

Gill F, Donsker D (eds) (2017) IOC world bird list (v 7.1). doi:https://doi.org/10.14344/IOC.ML.7.1

Jarvis ED, Mirarab S, Aberer AJ, Li B, Houde P, Li C, Ho SYW, Faircloth BC, Nabholz B, Howard JT, Suh A, Weber CC, Fonseca d, Rute R, Li J, Zhang F, Li H, Zhou L, Narula N, Liu L, Ganapathy G, Boussau B, Bayzid MS, Zavidovych V, Subramanian S, Gabaldon T, Capella-Gutierrez S, Huerta-Cepas J, Rekepalli B, Munch K, Schierup M, Lindow B, Warren WC, Ray D, Green RE, Bruford MW, Zhan X, Dixon A, Li S, Li N, Huang Y, Derryberry EP, Bertelsen MF, Sheldon FH, Brumfield RT, Mello CV, Lovell PV, Wirthlin M, Schneider C, Paula M, Prosdocimi F, Samaniego JA, Velazquez V, Missael A, Alfaro-Nunez A, Campos PF, Petersen B, Sicheritz-Ponten T, Pas A, Bailey T, Scofield P, Bunce M, Lambert DM, Zhou Q, Perelman P, Driskell AC, Shapiro B, Xiong Z, Zeng Y, Liu S, Li Z, Liu B, Wu K, Xiao J, Yinqi X, Zheng Q, Zhang Y, Yang H, Wang J, Smeds L, Rheindt FE, Braun M, Fjeldsa J, Orlando L, Barker FK, Jonsson KA, Johnson W, Koepfli K-P, O'Brien S, Haussler D, Ryder OA, Rahbek C, Willerslev E, Graves GR, Glenn TC, McCormack J, Burt D, Ellegren H, Alstrom P, Edwards SV, Stamatakis A, Mindell DP, Cracraft J, Braun EL, Warnow T, Jun W, Gilbert MTP, Zhang G (2014) Whole-genome analyses resolve early branches in the tree of life of modern birds. Science 346:1320–1331

Newton I (2003) The speciation and biogeography of birds. Academic, London

Price T (2008) Speciation in birds. Roberts and Company, Greenwood Village, CO

Ripple WJ et al (Ripple WJ, Wolf C, Newsome TM, Galetti M, Alamgir M, Crist E, Mahmoud MI, Laurance WF, and 15,364 scientist signatories from 184 countries) (2017) World scientists' warning to humanity: a second notice. BioScience 67:1026–1028

Sibley CG, Ahlquist JE, Monroe BL (1988) A classification of the living birds of the world based on DNA-DNA hybridization studies. Auk 105:409–423

Chapter 2
Integrative Taxonomy of Birds: The Nature and Delimitation of Species

George Sangster

Abstract Species are the basic currency in biodiversity studies, but what constitutes a species has long been controversial. Since the late 1990s, debates over species have shifted from theoretical questions (e.g., What is a species? Which species concept is best?) to empirical questions (How can we document species both efficiently and accurately?). A growing number of taxonomists agree that species can be discovered and documented in many different ways, preferably by employing and combining multiple types of evidence ("integrative taxonomy"). This chapter examines how and why avian taxonomy has become integrative, how species hypotheses are documented and falsified, and how the growth of taxonomic knowledge provides new and valuable insights into the speciation process, biogeography, and conservation biology.

Keywords Aves · Integrative taxonomy · Pluralism · Speciation · Species criteria · Species limits

2.1 The Centrality of Species

The concept of species is one of several key concepts in biology (Keller and Lloyd 1992; Pigliucci and Kaplan 2006; Sober 2006). While the discovery, description, and naming of species are strictly the responsibility of taxonomists, species are important in many other disciplines in society, including conservation (IUCN list of endangered species), health care (e.g., infectious diseases), and law (CITES, ESA; Geist 1992). Species are crucial for various concepts in other branches of biology, and society in general, as reflected by such terms as "keystone species" (ecology), "flagship species" (conservation), "speciation" (evolutionary biology), and "species-ism" (ethics). In biology the concept of species is shared across various hierarchies

---

G. Sangster (✉)
Department of Bioinformatics and Genetics, Swedish Museum of Natural History, Stockholm, Sweden

Naturalis Biodiversity Center, Leiden, The Netherlands

D. T. Tietze (ed.), *Bird Species*, Fascinating Life Sciences,
https://doi.org/10.1007/978-3-319-91689-7_2

Table 2.1 Species are important in and shared among various hierarchies

Ecology	Taxonomy[a]	Genealogy	Conservation
	Domain		
	Kingdom		
Ecosphere	Phylum		
Biome	Class		
Regional ecosystem	Order		Landscape
Local ecosystem	Family	Tree of life	Ecosystem (service)
Community	Genus	Monophyletic group	Habitat
Species	**Species**	**Species**	**Species**
Population	Subspecies	Deme	ESU
Organism		Organism	Management unit
Organ system		Genome (germ line)	Population/breed
Organ		Gene	Genetic diversity
Tissue			Adaptation
Cell			Gene

[a]See Dubois (2006) for a much more extensive set of categories

(Table 2.1). The broad relevance of species does not mean that the term species means the same thing to different biologists or that species taxa are uniform across groups. Far from it.

2.2 Why Is There a Species Problem?

There are numerous reasons why many professional biologists and philosophers have different opinions about "species." It is impossible to do justice to this rich debate in only a few paragraphs, so I will only briefly discuss the main points of contention. For the purpose of showing the diversity of opinion, I contrast the most divergent views, but it should be noted that there are numerous intermediate viewpoints.

2.2.1 Monism vs. Pluralism

A major issue among philosophers of science and more philosophically inclined systematists is the question whether species possess a single, unique quality that differentiates them from all other categories. Pluralists believe that there is no single unified way to delimit species; nature can be viewed from a multitude of perspectives which are equally valid. Most philosophers of science seem to agree that this is extremely likely and that pluralism is the only realistic approach. There are at least two brands of pluralism. One suggests that there are many permissible ways to classify species in any given situation, depending on the needs of the particular biologist (Kitcher 1984). Thus, ecologically defined species of ducks are just as acceptable as phylogenetically or morphologically defined species of ducks. The other suggests that in each group there is only one optimal classification, but that the optimal criteria to classify species may differ among groups (Mishler and Donoghue 1982).

The opposite view is monism, which holds that there is a single optimal way to divide nature into species. This is the view preferred by most biologists, because they would like to find *the* answer to the species problem and preferably have this derived from first principles in evolutionary biology or systematics (Ruse 1969; Pigliucci and Kaplan 2006).

2.2.2 Realism vs. Anti-realism

The issue of whether there is one or more than one valid way to classify species is closely related to the question of whether species taxa are real. Not surprisingly, monists (i.e., most biologists) tend to accept that species exist independent of our ability to recognize or find them. For instance, Mayr (1982) noted that variation is often discontinuous and that nature consists mostly of discrete units. Diamond (1966) demonstrated that members of "primitive" human tribes in New Guinea classify local birds in almost exactly the same way as professional taxonomists. Modern taxonomists may further point out that in many cases, different taxonomic methods identify the same species units, indicating that such species are not a fabric of the human mind.

Opponents of this view are not convinced by such observations and have argued that discrete groups are sometimes elusive, that all species are connected vertically through their ancestry, or that there is simply nothing special about species. The view that species are not real, i.e., that they have no objective reality outside of the human mind, is also called "nominalism" (Stamos 2003). Modern-day adherents to this view typically come from a background in either botany (where species limits tend to be more blurred by gene flow than in zoology) or paleontology (where inferring species limits through time has always been considered difficult). Botanist Van Valen (1976), for instance, noted that the classification of oaks is so problematic as a result of gene flow that it is impossible to find any discrete species. He asked "why can't there be taxa without species?" In his presidential address to the Paleontological Society, Shaw (1969) argued that species are not real (and should be abandoned), because drawing species limits in time is completely arbitrary. Mishler (1999) emphasized that all taxonomic ranks are arbitrary, but that clades are real. He argued that in a truly phylogenetic system, there is no place for ranks and as a consequence, no place for species. In his view, species are phylogenetic groups just like higher taxa, and all other putative properties of the species rank that have been proposed by various taxonomists (phenetic groupings, interbreeding, reticulation) can also be found at other ranks or are inconsistent with phylogenetic groups.

2.2.3 *Theoretical vs. Operational*

Philosophers of science have proposed several theoretical "solutions" to the species problem. These include the idea that "species are individuals" (Ghiselin 1974) or that "species are relations" (Stamos 2003). However, these do not connect with taxonomic practice. The Hennigian species concept, which proposes that species can be delimited in time by branching events (nodes) and extinction events (Hennig 1966), has also been proposed as a solution (Ridley 1989) but fails for the same reason. A formal description of this species concept (Kornet 1993) is mathematically sound, is universally applicable, and is able to assign every individual organism to a species. Yet, this method is hopelessly unrealistic because one needs to know the position of every individual within the genealogy.

At the other end of the spectrum is the phenetic school, which is completely theory-free (Sokal and Crovello 1970). However, without any theoretical underpinnings, it is impossible to determine which of many methods to group individuals into "species" is the best. The phylogenetic species concept defines species as "the smallest diagnosable cluster of individual organisms within which there is a parental pattern of ancestry and descent" (Cracraft 1983) but does not tell what those species are beyond the somewhat vague "evolutionary units."

It is safe to say that theoretical concepts are usually not very strong operationally and that operational concepts do not tend to be strong theoretically (Hull 1997, 1999). Both aspects are connected, because in order to know how one should find species, one must first know what to look for. Whereas many biologists would agree with Cracraft (2000) that a species concept "should work," no species concept is entirely satisfactory without a solid theoretical basis.

2.2.4 *Pattern vs. Process*

Some species concepts emphasize evolutionary mechanisms and processes, such as mate choice, species recognition, interbreeding, hybridization, and gene flow, whereas others focus on the results and patterns of evolution (such as diagnosability, monophyly, and congruency). The biological and recognition concepts are examples of process-based species concepts, whereas phylogenetic, monophyletic, and internodal concepts are pattern-based approaches. Evolutionary biologists and population geneticists tend to prefer process-based species concepts, whereas taxonomists and phylogeneticists tend to prefer pattern-based approaches.

2.2.5 *Prospective vs. Historical*

Under the biological species concept, taxa are recognized as species, if they remain "reproductively isolated," in the sense that they do not fuse into a single population (Mayr 1982, 1996). The BSC, therefore, is prospective (O'Hara 1993, 1994; Maddison 1997); only future events will show whether currently recognized taxa remain reproductively isolated or fuse into each other.

The alternative viewpoint is that species-level taxa should be delimited on the basis of historical subdivisions (i.e., historical patterns), rather than present-day or possible future interactions and processes (Lidén and Oxelman 1989), such as hybridization and gene flow. Maddison (1997) argued that species concepts which are prospective and which require speculations about the future are not helpful in biology; since all of our data are of the present and past, the units by which we interpret these data must also be strictly historical. There are two major kinds of historical units: species as clades and species as lineages. The species-as-clades view was developed by Donoghue (1985) and de Queiroz and Donoghue (1988, 1990) and considers phylogenetic species to be the smallest *monophyletic* groups of organisms supported by autapomorphies (unique derived character states). The species-as-lineages view regards species as (parts of) branches in a phylogeny (Mayden 1997; de Queiroz 1998, 1999).

2.2.6 *Concerns by End Users*

Some end users of species have been very vocal about how taxonomists should delimit species. For instance, some conservationists and biodiversity scientists have complained that if pattern-based species concepts were to be implemented widely, the number of species would become "too high" (e.g., Agapow et al. 2004) and result in "pseudoreplication" of species (Isaac et al. 2004). Other complaints made by some end users of taxonomy are (1) that taxonomists have recognized species on the basis of "trivial" morphological character differences or molecular differences (Collar 1997; Tobias et al. 2010), (2) that taxonomy has become "unstable" due to the use of phylogenetic species concepts (Isaac et al. 2004; Mace 2004), and (3) that species designations no longer correspond to genetically incompatible populations (affecting captive breeding and reintroductions; Frankham et al. 2012). These are not scientific (i.e., empirical) concerns, but they do illustrate what some end users have come to expect from species.

All these viewpoints underscore the many dimensions of the species problem. An important insight was that the species problem actually comprises several distinct problems (e.g., Frost and Kluge 1994; Reydon 2004): What are species? How do species originate? How can, may, or should one delimit species in practice? It seems obvious that one should have a good sense of what species are before one can tell how species originate or how to delimit species. Thus, the concept of species is the primary problem. This set the stage for a major breakthrough in the late 1990s.

2.3 The Lineage Concept

The lineage concept of species was first articulated by Simpson (1951) and further developed by Hennig (1966), Wiley (1978), and Wilson (1995). These authors viewed species as branches in the tree of life. This view was long known as the "evolutionary species concept" or "Hennigian species concept." Because these views were not accompanied by, and did not easily translate into, operational criteria, these have long been ignored by taxonomists (but see Frost and Hillis 1990). The idea that species are best seen as lineages gained strong support in the late 1990s due to three now-classic publications by Richard Mayden and Kevin de Queiroz (Mayden 1997; de Queiroz 1998, 1999). Their case rested on two important insights.

First, there is a fundamental distinction between the *species concept* (primary species concept, Mayden 1997), which tells us what species are, and *species criteria* (secondary species concepts, Mayden 1997), which describe how species taxa may be delimited in practice. Both Mayden and de Queiroz observed that most species concepts are actually little more than prescriptions of how species should be delimited (i.e., species criteria). Thus, many "species concepts" have confounded operational and theoretical aspects of species and have defined the items to be discovered by the discovery process (Frost and Kluge 1994).

Second, most views on species and species criteria are consistent with the idea that *species are segments of population lineages* (de Queiroz 1998, 1999), i.e., branches or parts of branches in the tree of life. Although some criteria can only be applied at one time-slice (e.g., reproductive isolation), it does not follow that species are only real at that point in time, and this does not preclude one from seeing such species as lineages.

2.4 Corollaries of the Lineage Concept

The lineage concept does not by itself dictate how species should be delimited, but it provides a few important hints. First, because different species may have different properties, all traditional species criteria are potentially relevant for finding species. Importantly, different species criteria highlight different aspects of species. Thus, it makes sense to apply as many criteria as possible, not only to avoid overlooking species but also to describe and better understand the nature of these species taxa (Sangster 2014).

Second, there is no reason to presume that all the properties identified by traditional species concepts evolve in all lineages, let alone all at the same time. Thus, a species may exist even if it does not possess one's preferred species criterion

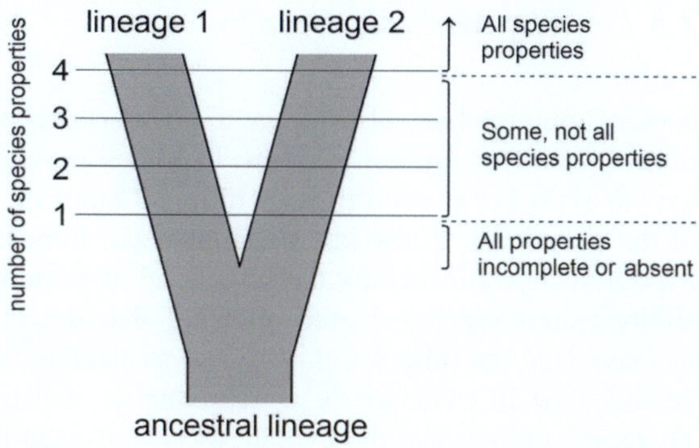

Fig. 2.1 Different species properties (e.g., reproductive isolation, diagnosability, monophyly, ecological niches) may evolve at different times during divergence. The absence of a particular property (e.g., property 4) does not falsify species rank, if there is other evidence that it is a separate lineage (e.g., properties 1–3)

(Fig. 2.1). The absence of a particular species property does not mean that the lineage itself does not exist. This is an important observation, because from this it follows that *no property can be considered "defining" for species* (de Queiroz 1999, 2007). This explicitly rejects the equality of species; some may possess all properties, others may not; some may be highly distinctive, others are cryptic; some may be monophyletic in a gene tree, whereas others are not; some may be reproductively isolated, but others may not.

Third, if species are lineages, then the species delimited by taxonomists are *hypotheses of lineages*. Thus, species are not simply described ("facts"), but their existence is a testable hypothesis. Like any other scientific statement, these should be documented as thoroughly as possible.

2.5 Integrative Taxonomy

Integrative taxonomy may be considered the operational counterpart of the lineage concept. The term "integrative taxonomy" was introduced almost simultaneously in two papers which appeared in 2005 (Dayrat 2005; Will et al. 2005). Both papers emphasized that it is important to combine and integrate multiple approaches to species delimitation. Will et al.'s (2005) motivation for their paper was to provide a critique of barcoding, which is the use of a single line of molecular evidence to identify and discover species. Ironically, it later became apparent that barcoding may also be a valid part of integrative taxonomy, because it may produce new hypotheses of species taxa, which may be tested, refined, falsified, or corroborated by other evidence (Padial and de la Riva 2007).

2.5.1 *Why Multiple Data?*

Seeking multiple lines of evidence to document species taxa is a logical consequence of the view that species taxa are hypotheses. Indeed, it is common sense that a hypothesis is better confirmed by different kinds of evidence than by more evidence of the same kind. It also can be rationalized from Bayesian principles. Each time a hypothesis is confirmed by the same kind of evidence (e.g., morphology), the probability expressing the degree of belief that it will do so in the future gradually increases. On the other hand, the prior probability that a hypothesis is confirmed by another type of evidence (e.g., reproductive isolation) may initially be quite low. Therefore, finding that the hypothesis is confirmed by a new line of evidence represents a significant increase in the probability that the hypothesis is true (Chalmers 1999). The validity of questionable species taxa based on small numbers of specimens or small numbers of characters is often addressed with other datasets, typically molecular data (e.g., Bensch and Pearson 2002; Sangster et al. 2016a; Pérez-Emán et al. 2018). Another reason why it is best to seek multiple kinds of evidence is that all taxonomic criteria have both strengths and weaknesses (Table 2.2).

2.5.2 *Why Integrate?*

Although the use of multiple lines of evidence in taxonomy is not new, the emphasis on integration of evidence is novel. Collecting multiple kinds of evidence is not enough, for several reasons. As noted above, various species criteria are complementary, because these highlight different aspects of lineage divergence. Sometimes the interpretation of one type of evidence depends on information on other properties. For instance, to demonstrate that two putative sympatric species are reproductively isolated, it is often necessary to demonstrate that these species are diagnosable or reciprocally monophyletic. Evidence for species may be scattered across various papers and thus should be collected and assessed together. Thus, integrative taxonomy may be defined as *the theory and practice of documenting and evaluating hypotheses about the taxonomic status of putative species by integrating multiple lines of evidence.*

2.6 Strengths of Integrative Taxonomy

By combining and integrating taxonomic evidence in a single approach, integrative taxonomy has several qualities that distinguish it from traditional approaches.

First, if one views different kinds of taxonomic data as complementary accounts of species divergence, it becomes apparent that many taxonomic conflicts are not based on incompatible interpretations of species limits but simply stem from

Table 2.2 Strengths and drawbacks of various species criteria

Criterion	Strengths	Drawbacks
Diagnosability (fixed difference in heritable character)	Easy to apply Cost-effective Applicable to a broad range of taxa Applicable to a broad range of data types Can be expressed in statistical terms (e.g., Wiens and Servedio 2000)	Heritability often unknown Sample size dependent Nonheritable differences may also identify evolutionary units (e.g., song, biometric differences) Difficult to apply to sequence data
Degree of difference (phenotypic/genetic divergence)	Phenotypic divergence: appeals to the wish for "good species," popular among end-users of taxonomy Genetic divergence: easily obtained as by-product of phylogenetic analysis	Requires a better supported reference taxonomy (Sangster 2000) Cutoff point typically arbitrary Difficult to apply in complex taxonomic problems May conflict with phylogenetic relationships
Reproductive isolation (intrinsic barriers to gene flow)	Often point of no return (permanent lineage divergence) Mechanism of maintenance of species boundaries	Limited scope: cannot be applied directly in allopatry, at different time slices, or in paleontology Requires prediction of future interactions Ability to interbreed is plesiomorphic: reproductive compatibility may conflict with phylogenetic relationships Arbitrary: many species hybridize; when are species (not) reproductively isolated? Degree of gene flow/isolation may differ among genes
Prezygotic isolation	Biological meaning ("species recognition")	Recognition or cues may be dependent on environment, phase of breeding season Density dependance: degree of interbreeding may depend on the availability of homospecific partners (Hubbs' rule)
Postzygotic isolation	Biological meaning ("genetic incompatibility")	Extremely hard to document Application beyond expertise of most trained taxonomists Fitness of hybrids and/or parental species may vary with local circumstances (e.g., bounded hybrid superiority; Moore 1977)
Non-monophyly	Removes nonhistorical units	Only a suitable basis for splitting if two taxa are not sisters (other evidence is required if the two "species" are nested)

(continued)

Table 2.2 (continued)

Criterion	Strengths	Drawbacks
Reciprocal monophyly (of gene trees)	Demonstration of historical isolation Easy to apply	Cannot be used in isolation (monophyletic groups exist at many levels, so further evidence is required) Evolves slowly, thus prone to type II errors Choice of loci is arbitrary Monophyly may be based on chance, rather than historical isolation (Irwin 2002)

differences in the sensitivity of taxonomic criteria. For instance, the finding that two diagnosable species do not form reciprocally monophyletic groups does not represent a major taxonomic dilemma but is best interpreted as a case where one property (diagnosability of morphological character states) simply became fixed before the other (exclusive coalescence of gene trees). Thus, integrative taxonomy may lay numerous taxonomic controversies to rest.

Second, integrative taxonomy is an efficient approach to taxonomy. Because evidence for species rank is not limited to one ("defining") property, integrative taxonomy may be informed by all kinds of available data.

Third, the integration of different kinds of evidence may help to discover additional species. Evidence which is not very relevant or informative in isolation may be crucial when combined with other evidence. For instance, the finding that two migratory subspecies of Bean Goose *Anser fabalis* have overlapping wintering ranges in Western Europe is not remarkable nor is their return to different breeding areas in Siberia (Burgers et al. 1991). However, when combined with the knowledge that high-latitude species of waterfowl find their mates on the wintering grounds, these findings become very significant for taxonomy. If pair formation takes place on the wintering grounds, then both forms have the opportunity to form mixed pairs. If no birds ringed in winter return in the breeding area of the other form, then this is prima facie evidence that the two are not forming mixed pairs and remain reproductively isolated (Sangster and Oreel 1996). This hypothesis received some further support from the findings that both groups formed separate groups in a molecular phylogeographic analysis (Ruokonen et al. 2008) and may not even be each other's closest relatives (Ottenburghs et al. 2016).

Fourth, examples such as these highlight the fundamentally analytic and iterative nature of integrative taxonomy and underscore that species delimitation is a science (Yeates et al. 2011; Sluys 2013). Thus, modern taxonomy is more than a mere baptismal act or a one-time "description." Hypotheses of species rank require documentation and reassessment in the light of new evidence, just like any other scientific hypothesis.

Finally, by including all species criteria under a single unified approach, integrative taxonomy shifts away attention from the yes-or-no question of whether or not

Table 2.3 Examples of falsifications of species limits based on defining criteria

Species pair	Original hypothesis of species distinctiveness based on:	Proposed falsification of species status based on:
Ducks: *Anas carolinensis*—*A. crecca*	Non-sister group relationship (paraphyly with respect to *A. flavirostris*) High level of sequence divergence (mtDNA) Diagnostic differences in male plumage (Sangster et al. 2001)	Inferred gene flow of nuclear DNA (Peters et al. 2012)
Giant petrels: *Macronectes halli*—*M. giganteus*	Sympatric (allochronic) breeding Plumage and bare parts (diagnostic differences; Bourne and Warham 1966)	Low genetic divergence (<1%; Penhallurick and Wink 2004)
Kites: *Milvus fasciicauda*—*M. milvus*	Plumage (diagnostic differences; Hazevoet 1995)	Non-monophyly of cytochrome b sequences (Johnson et al. 2005)
Owls: *Otus madagascariensis*—*O. rutilus*	Plumage and vocalizations (diagnostic differences, Rasmussen et al. 2000)	Non-monophyly of ND2 sequences (Fuchs et al. 2007)

species A meets the criterion of species concept X but instead focuses on the information content of the total body of evidence.

2.7 What Is Not Integrative Taxonomy?

Many types of data, analyses, and criteria can be valid parts of integrative taxonomy, but this is not to say that anything goes in integrative taxonomy. At least two approaches are inconsistent with integrative taxonomy: (1) the falsification of species rank by invoking a "defining" criterion and (2) the use of a set of "standard" criteria that restricts the scope of taxonomic fact-finding.

2.7.1 Falsification by a "Defining" Species Criterion

Species taxa may show many different combinations of species properties, and not every property may be present in each species (Fig. 2.1). Thus, it is important to keep in mind that *while any particular species property may help to document the existence of a species, its absence in other species does not falsify their species rank* (de Queiroz 2007). From this, it follows that a hypothesis about the validity of a species taxon can only be falsified if all lines of evidence supporting that taxon are falsified. Unfortunately, this is sometimes overlooked. Table 2.3 lists several recent instances where a "defining" criterion was used to falsify species that had been recognized on the basis of other evidence. In each of these cases, the authors have added new data but have inappropriately placed that evidence above all previous evidence. Some authors have used this strategy to draw far-reaching but highly dubious conclusions. For instance, Johnson et al. (2005) suggested that the Cape Verde Kite *Milvus fasciicauda* "does not exist," because cytochrome *b* sequences of that species were nested phylogenetically within those of the Red Kite *M. milvus*. This ignores not only that these species are still diagnosable by plumage characters (Hazevoet 1995) but also that non-monophyly may simply result from incomplete lineage sorting due to recent speciation—a common phenomenon in bird species (reviewed by Funk and Omland 2003; McKay and Zink 2010) of which new examples are still being discovered (Päckert et al. 2012; Luna et al. 2017). Appeals to defining criteria will no doubt continue to be made in the taxonomic literature but should be resisted, because it results in unnecessary confusion and instability. This is especially true, if defining criteria differ from paper to paper (as suggested by Table 2.3).

2.7.2 *Standardization of Species Criteria*

A recent proposal attempted to standardize the assessment of species limits by limiting this process to a small number of data classes and a few characters per class (Tobias et al. 2010). This set of criteria explicitly excluded entire data classes that are commonly used in taxonomy (molecular data, phylogenetic data, multivariate analyses, and detailed evidence from hybrid zones). The approach put forward by Tobias et al. (2010) seems to be rooted in what philosophers of science call "anti-realism": the merit of the criteria is defined in terms of their generality and simplicity, not in terms of whether these result in the delimitation of real evolutionary units. This is underscored by the taxonomic revisions that resulted from application of these criteria (del Hoyo and Collar 2014, 2016), which include several species that modern molecular phylogenetic studies have documented as paraphyletic assemblages of multiple distinct taxa (e.g., Golden-fronted Woodpecker *Melanerpes aurifrons*, Red-crowned Parakeet *Cyanoramphus novaezelandiae*, Mourning Wheatear *Oenanthe lugens*; see Kearvell et al. 2003; Schweizer and Shirihai 2013; García-Trejo et al. 2009; Rawlence et al. 2015). The criteria put forward by Tobias et al. (2010) are inappropriate, because the lineages that we call "species" are things that we discover, not create. Taxonomy is a fact-finding endeavor; any practice that limits, distorts, or interferes with this endeavor should be abandoned.

Attempts to falsify species taxa by appeal to a particular defining criterion and pleas to standardize species taxonomy by limiting its scope can be viewed as attempts to prescribe how species *must* be found, so that any species not found by that method is not a species. The flaws of this approach may be illustrated by analogy to a criminal court case. Taxonomy and the criminal justice process share many similarities owing to the fact that both strongly rely on empirical data (Table 2.4). Like a criminal court case, the task of species taxonomy is to assess objectively what

Table 2.4 Similarities between the empirical aspects of species-level taxonomy (evaluating species limits) and the criminal justice process (evaluating guilt)

Aspect	General principle	Taxonomy (species limits)	Criminal justice (guilt)
Aim	Assess objectively what has happened	Speciation (*has speciation occurred*?)	Crime (*did this suspect commit this crime*?)
Hypotheses	The null hypothesis should be accepted unless sufficient evidence shows this to be false	H_0: no lineage separation has occurred (taxa must not be recognized as species) H_1: speciation has occurred (taxa are best treated as species)	H_0: defendant is innocent (and must be acquitted) H_1: defendant committed the crime
Strategy	*Document* the best explanation for the observed evidence; rule out alternatives	(1) Collect as much evidence and as many types of evidence as possible; (2) integrate evidence; (3) define scenarios; (4) evaluate and compare the evidence for all plausible scenarios; repeat, if necessary	Idem
Evidence	Reliance on *empirical* data	*Examples*: morphology, genetics, bioacoustics, behavior	*Examples*: eye witness testimony, DNA fingerprinting, examination of physical objects, written confession
Quality assessment	Assessment of quality (meaning) of the evidence	*Examples*: rule out misidentification (of specimens), contamination (of DNA samples), inappropriate methodology, effects of small sample size or poor reference data, comparison among non-homological character states	*Examples*: rule out misidentification (of people or objects), contamination (of DNA samples), inappropriate methodology, effects of small sample size or poor reference data
Complementarity of evidence	Multiple evidence is good; multiple *independent* lines of evidence is even better	*Example*: finding a diagnostic bioacoustic character state *plus* reciprocal monophyly of DNA is more convincing evidence for lineage separation than finding two diagnostic bioacoustic character states	*Example*: two credible witnesses saying the same thing is good, but a single credible witness *plus* DNA evidence is even better
Integration of evidence	Different lines of evidence highlight	*Example*: bioacoustic data may show a pattern	*Example*: an eye witness may describe what

(continued)

Table 2.4 (continued)

Aspect	General principle	Taxonomy (species limits)	Criminal justice (guilt)
	different aspects; must be *integrated*	(consistent differences in songs), playback tests may highlight lack of mate recognition, DNA data may show the result (lack of gene flow) (see also Table 2.2)	happened, whereas DNA may tie the defendant to the crime scene
Uniqueness	Each species/criminal court case is unique	No single type of evidence (or combination thereof) can be assumed to "work" in every case	Idem

has happened at some time in the past. In both cases, a hypothesis must be documented, and alternative hypotheses must be excluded. Those who assess and evaluate the hypothesis did not witness the relevant events and thus have to rely on *evidence*. Needless to say, the more evidence the better. Having multiple types of evidence is better than having more of the same type of evidence. If multiple kinds of evidence tell the same story, this increases the plausibility of the hypothesis. Speciation (like a criminal act) may have been a complex, protracted process, and therefore multiple types of evidence may be useful to highlight various aspects of this process. In many cases, these lines of evidence are *complementary* and must be *integrated*. In addition, each pair of species (just like each crime) is unique in one way or another, and there is no a priori reason to suspect that a single type of evidence works in all cases. Finally, like a criminal court case, taxonomy is an *iterative* process. When important new evidence emerges, the original hypothesis must be evaluated again within the light of all available evidence. To argue which species criterion should be adopted as the defining criterion in taxonomy is just as futile as arguing which single type of evidence should be used to "define" guilt in all criminal court cases. Any attempt to a priori limit the evidence used to evaluate a taxonomic case study interferes with the fundamentally empirical nature of taxonomy.

2.8 The Dynamics of Taxonomic Change

After the official start of zoological nomenclature in 1758, the number of bird species recognized by taxonomists increased steadily until reaching a peak in the early twentieth century (Fig. 2.2). During the 1840s, zoologists Carl Sundevall and Herman Schlegel began to treat poorly defined species as geographical races

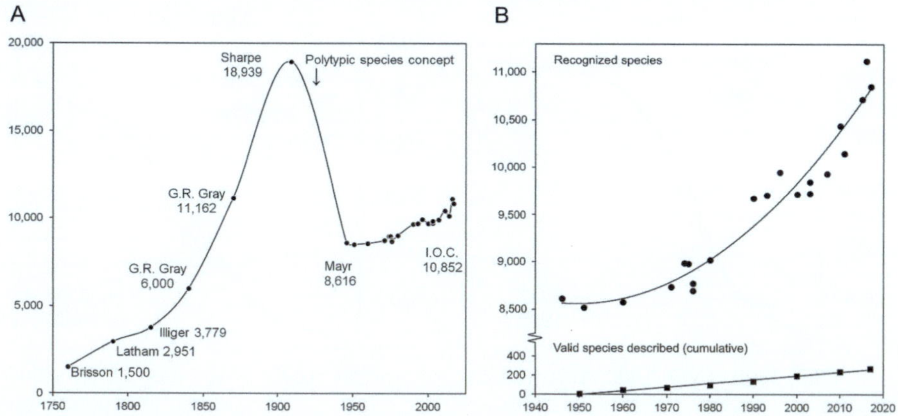

Fig. 2.2 (**a**) Trends of the total number of recognized species of recent birds from 1760 to 2017. Note the dramatic (>50%) reduction in the number of recognized species between 1909 (Sharpe) and 1946 (Mayr) and the *relatively* slow increase since 1946 (*Updated from Haffer* 1992). (**b**) Trends of the total number of recognized species of recent birds (upper panel, based on various estimates and classifications, as indicated by the filled circles) and newly described species of birds (lower panel) from 1946 to 2016. Note that the number of recognized species increases much faster (c. 92 species per year in 2007–2017) than the number of newly described valid species (c. 5.3 per year). A second-order trend line was added to illustrate the trend (*Updated from Sangster and Luksenburg* 2015)

(subspecies). In subsequent decades, increased exploration and collection of specimens led to improved understanding of the geographic and morphological continuity of previously recognized species. In the second half of the nineteenth century, a growing number of zoological works started ranking taxa as subspecies (Stejneger 1884). During this period, intergradation became the cornerstone of subspecific taxonomy (Allen 1871). According to this view, subspecies are distinguishable forms that intergrade, whereas species do not intergrade (Stejneger 1884). Because the criterion of intergradation can only be applied to geographically contiguous (i.e., continental) populations, taxonomists used an arbitrary assessment of the degree of difference to rank island populations (Stone 1899).

By the early twentieth century, the recognition of subspecies had degraded to the principle of "geographic representation." Many thousands of distinctive and less distinctive taxa were combined in large polytypic species simply based on the non-overlap of their breeding ranges. The inclusion of subspecies into polytypic species was seen as a simplification of taxonomy, because it reduced the number of recognized species (Stejneger 1884; Mayr 1942). The effect on the number of bird species was substantial: within about 40 years, the number of recognized species had dropped from 18,939 (Sharpe 1909) to 8590 (Mayr and Amadon 1951).

By the mid-twentieth century, the number of bird species started to increase again, and this trend is accelerating (Fig. 2.2b). Increases of species numbers have also been documented in other vertebrates, but the rate of the increase differs among groups (Fig. 2.3). For instance, the number of extant bird species increased by 0.9%

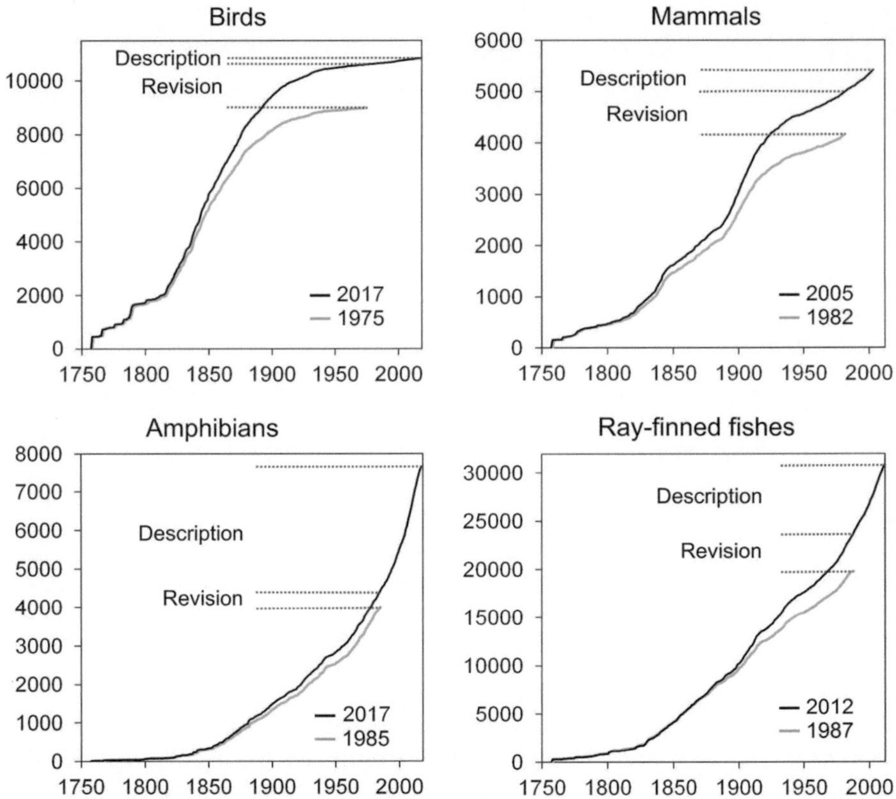

Fig. 2.3 Cumulative numbers of species of birds, mammals, amphibians, and ray-finned fishes recognized in classifications published in 1975–1987 (gray lines) and 2005–2012 (black lines). The relative contributions of taxonomic description (new species) and taxonomic revision (subspecies upgraded to species rank) differ greatly among vertebrate groups, with revisions predominating in birds and mammals and descriptions predominating in ray-finned Fishes and especially amphibians

per year during 2007–2016, whereas in amphibians the increase was a whopping 2.4% per year.

2.9 The Drivers of Taxonomic Change

Why does the number of bird species keep rising even after more than 250 years of taxonomic research? Three major reasons have been proposed for the increasing number of vertebrates recognized by specialists: (1) new discoveries, (2) revisions based on new empirical evidence (e.g., molecular data), and (3) revisions based on changes of the way species are delimited. Hanken (1999) attributed the increase of the number of amphibian species to all three of these, whereas Isaac et al. (2004)

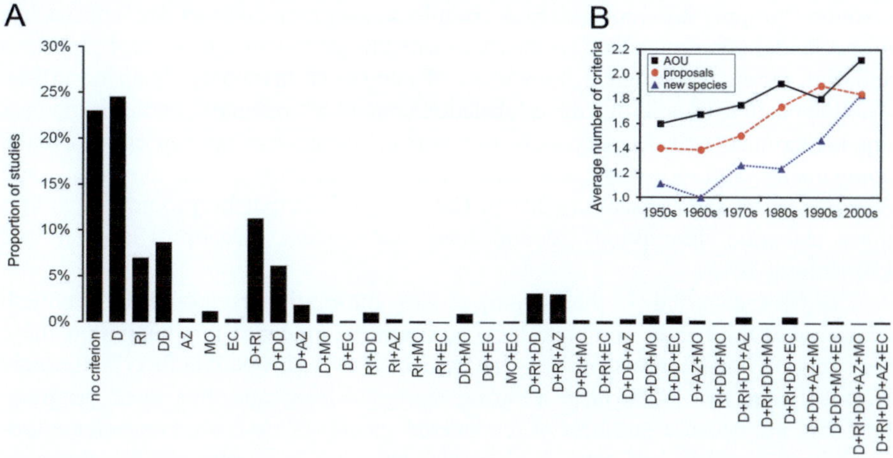

Fig. 2.4 (**a**) Species criteria used in a dataset of 1249 taxonomic studies (new descriptions, revisions, and recommendations made by the AOU Committee on Classification and Nomenclature) published between 1950 and 2009 (Sangster 2014). Note the large number of different combinations of criteria, illustrating the eclectic nature of species-level taxonomy in birds. *D* diagnosability, *RI* reproductive isolation, *DD* degree of difference, *AZ* adaptive zone, *MO* monophyly, *EC* exclusive coalescence of gene trees. (**b**) Changes in the average number of criteria per taxonomic hypothesis, illustrating the increasingly pluralistic nature of species delimitation (Modified from Sangster 2014)

incorrectly attributed the rising number of recognized species of birds and mammals to a shift toward the phylogenetic species concept (see below).

The relative role of descriptions and revisions in the increasing number of species is easily inferred by comparing plots of the cumulative number of species described over time of a recent classification with that of a less-recent classification (Fig. 2.3). These show that in birds, revisions play a much greater role in the increase than descriptions, whereas the opposite is true in amphibians (Fig. 2.3). More specifically, of the c. 92 species of birds that on average were added each year during 2007–2016, only about five per year were newly described species (Fig. 2.2b).

Analyses of the taxonomic literature have shown that the rising number of bird species cannot be attributed to a shift toward the phylogenetic species concept (Sangster 2009). The trend to recognize more species of birds started at least two decades before the introduction of the phylogenetic species concept, and most newly recognized species were not based exclusively on criteria associated with that concept (Sangster 2009). In a set of 747 taxonomic proposals published in major ornithological journals, about 85% of the newly proposed species were supported by new taxonomic data. Thus, in birds taxonomic revision is the predominant proximate driver of the growing number of species, and revisions are very often based on new empirical data.

Taxonomic practice is pluralistic (employing multiple types of evidence and criteria) and increasingly eclectic (employing many different combinations of criteria). Using a dataset of >1200 studies, a recent study found that six taxonomic criteria were applied in >30 unique combinations (Fig. 2.4; Sangster 2014). Avian taxonomists apply different criteria as complementary rather than as rival approaches to species delimitation, which is consistent with the general lineage concept of species and with the methodological framework of integrative taxonomy (Sangster 2014). Another positive trend is that the information content of taxonomic studies is increasing; taxonomists write longer papers, use more characters and more specimens, make comparisons with more taxa, and make greater use of illustrations, sonagrams, and maps (Sangster and Luksenburg 2015). There is evidence that this promotes stability: more elaborate descriptions require fewer subsequent revisions (Sangster and Luksenburg 2015).

The great magnitude of the lumping process during the first half of the twentieth century and the nearly complete lack of empirical evidence that supported these revisions are important reasons why the taxonomy of birds is still in flux. Taxonomic changes since the 1950s have a strong empirical basis and thus mark progress. Furthermore, because numbers of recognized species of birds are increasing exponentially (Fig. 2.2b) and many bird species have not been revised since the early twentieth century, species inventories of birds are still far from complete.

2.10 Benefits of Integrative Taxonomy to Other Fields

2.10.1 Speciation Studies

The processes by which new evolutionary lineages diversify may be highly complex and may differ among species. Identifying the main processes involved in speciation typically requires in-depth study from multiple angles. Research questions asked in speciation research are often identical to those that taxonomists ask (e.g., In which ways do the populations differ from each other? What is their population structure? When did the populations diverge? Do members of these populations recognize each other as potential mates/competitors? Is there any hybridization and does this lead to gene flow?). Modern taxonomic studies are crucial for speciation research, because one must first document patterns of variation before one can identify and understand the processes responsible for these patterns. Indeed, many interesting case studies in speciation research recently have been identified from integrative taxonomic research, as the following examples illustrate.

The black Carrion Crow *Corvus corone* and gray-and-black Hooded Crow *C. cornix* have long been treated as subspecies, because they meet and hybridize along a narrow contact zone in central Europe and Scotland. A series of modern studies in their hybrid zone showed that the crows actually behave as species (reviewed by Parkin et al. 2003). Further research on mitochondrial and nuclear DNA divergence found almost no divergence at the DNA level (Haring et al. 2007, 2012; Wolf et al. 2010), indicating either very recent speciation or widespread gene flow or both. Building on this work, full-genome studies demonstrated that the Carrion and Hooded Crows indeed show widespread gene flow in large parts of their genome, but not in the genes responsible for mate choice (preference for "black" vs. "gray-and-black" phenotypes) and plumage type. Both sets of genes are located rather close to each other and occupy a <2-megabase stretch of DNA (Poelstra et al. 2014). This shows that speciation may result from changes in a relatively small and local part of the genome. In turn, this finding is relevant for taxonomy, because it shows that gene flow does not necessarily falsify species rank and that traditional molecular methods (mtDNA and a few ncDNA markers) may not suffice to find all lineages.

Similarly, the discovery that two populations of *Oceanodroma* storm petrels in the Azores (Portugal) most likely originated through sympatric speciation by allochrony (Friesen et al. 2007) was preceded by a series of taxonomic studies which documented various aspects of their divergence. Storm petrels breeding during the summer (hot season) and those breeding during winter (cold season) were first found to show differences in morphology and timing of molt (Monteiro and Furness 1998). This was followed by studies which showed that these populations also differ by vocalizations, response to playback, and mtDNA and microsatellites but are more closely related to each other than to any other storm petrel (Bolton 2007; Friesen et al. 2007; Smith et al. 2007). The combined data provided compelling evidence that hot- and cold-season storm petrel populations arose sympatrically (Friesen et al. 2007). This represented the first evidence for sympatric speciation by allochrony in tetrapods.

Hybrid speciation is another rare form of speciation. Based on the descriptive taxonomic works of Meise (1936) and Johnston (1969), Italian Sparrow *Passer italiae* had long been suspected to be a hybrid species resulting from past interbreeding between Spanish *P. hispaniolensis* and House Sparrows *P. domesticus*. This was corroborated by recent work on morphology, microsatellites, and mitochondrial and nuclear DNA sequences (Elgvin et al. 2011; Hermansen et al. 2011, 2014; Trier et al. 2014). Conversely, these studies have helped clarify the taxonomic status of Italian Sparrow, which is now treated as a full species (Sangster et al. 2015).

2.10.2 Biogeography

Biogeography, the science that attempts to document and understand spatial patterns in biology, is one of the primary components of systematic research. Taxonomy (including phylogenetic study) provides the scientific basis for historical biogeography. Refinement of taxonomic designations results in increased precision in biogeographic studies. For instance, modern phylogeographic and taxonomic studies (Ribas et al. 2011; Fernandes et al. 2014) have not only corroborated the existence of multiple areas of endemism in the Amazon region of South America (Cracraft 1985) but also contributed to the designation of a new area of endemism northwest of Manaus, Brazil (Borges and da Silva 2012). Similarly, the importance of the white-sand forests of the northern Peruvian Amazon was only recently recognized due to the discovery of several new species confined to this habitat (Whitney and Alvarez Alonso 1998, 2005; Alvarez Alonso and Whitney 2001; Isler et al. 2002).

Contact zones have long been of interest for biogeographers, because these may help identify which ecological and historical factors have influenced the past and present distributions of populations and their divergence (Swenson 2010). Recent taxonomic studies have revealed several previously overlooked contact zones between morphologically similar taxa. Examples include the geese *Branta canadensis/hutchinsii* in Arctic North America (Paxinos et al. 2002), the pittas *Erythropitta macklotii/habenichti* in New Guinea (Irestedt et al. 2013), and the grey shrikes *Lanius excubitor/borealis* in Western Siberia (Olsson et al. 2010).

The processes responsible for high tropical species richness have attracted much debate, but with little consensus so far. For instance, it has been proposed that speciation rates in the tropics are higher or that extinction rates are lower or that clades originate in the tropics but only recently spread to higher latitudes leaving little time for species richness to accumulate in temperate zones (Pianka 1966; Rahbek and Graves 2001; Willig et al. 2003; Weir and Schluter 2007). Adequate, unbiased inventories of the diversity at different latitudes are an important first step in distinguishing among these and many other explanations. Recent phylogenetic and taxonomic research suggests that species diversity has been underestimated especially in the tropics (Lohman et al. 2010; Milá et al. 2012; Voelker et al. 2013), and this has bearing on some of the explanations that have been proposed for latitudinal diversity gradients (e.g., Tobias et al. 2008; Irestedt et al. 2013). Recognition of more species in the tropics results in steeper latitudinal gradients, but also in shorter branch lengths, suggesting more recent speciation (Irestedt et al. 2013).

2.10.3 Conservation

Taxonomy provides the scientific underpinnings of biodiversity conservation (Wheeler and Cracraft 1997; Cotterill et al. 2017). There are at least three major ways in which conservation work benefits from continued taxonomic research. First, the discovery and documentation of species lineages represents a crucial first step in biodiversity conservation, because species are often considered the basic currency in conservation (e.g., IUCN Red List). By documenting previously overlooked species and removing invalid species, taxonomic research helps conservationists prioritize their work. Threatened species which had remained hidden due to incorrect taxonomy are still being documented even in supposedly well-known parts of the world (Fig. 2.5), underscoring the urgency of taxonomic work.

Second, taxonomic research helps to delimit Endemic Bird Areas which are an efficient way to identify conservation priorities for restricted-range species (Stattersfield et al. 1998). An Endemic Bird Area is an area which encompasses the overlapping breeding range of at least two species that have a range of $<50{,}000\ km^2$. Two examples of Endemic Bird Areas recently identified by taxonomic research are the Balearic Islands and Corsica. Until the mid-1990s, only one endemic species was recognized (Corsican Nuthatch *Sitta whiteheadi*), but subsequent taxonomic study has added two endemic species to the Balearic Islands (Balearic Shearwater *Puffinus*

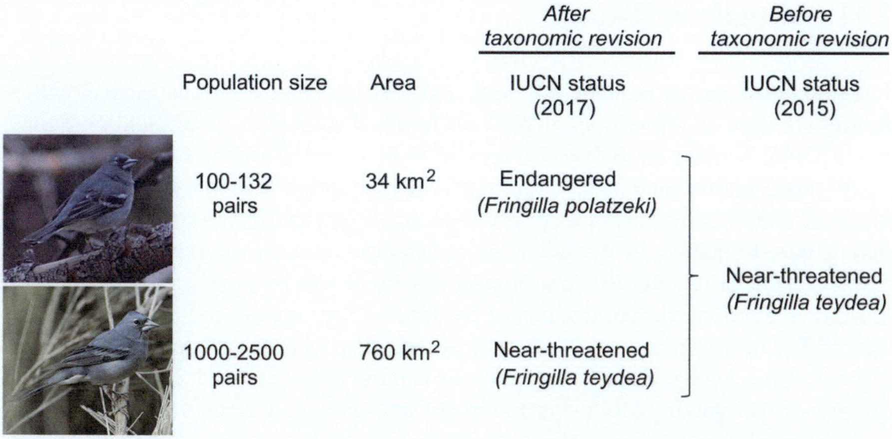

Fig. 2.5 Taxonomic revision affects the recognition of threatened species. The Blue Chaffinch *Fringilla teydea* was long considered a polytypic species with two subspecies, *F. t. polatzeki* on Gran Canaria and *F. t. teydea* on Tenerife. Until 2016, Blue Chaffinch was classified as a "near-threatened" species, because the range size and population numbers of the Tenerife taxon (now *F. teydea*) overwhelmed those of the much rarer Gran Canaria taxon (now *F. polatzeki*). After two studies showed that the two taxa differed in plumage, morphometrics, vocalizations, response to playback, sperm morphology, and nuclear DNA (Lifjeld et al. 2016; Sangster et al. 2016b), *F. polatzeki* was classified as an endangered species. Photographs courtesy of Jolanda A. Luksenburg (upper) and Ian Merrill (lower)

mauretanicus and Balearic Warbler *Sylvia balearica*), one to Corsica (Corsican Finch *Carduelis corsicana*), and two which are shared between the two areas and parts of mainland Italy (Moltoni's Warbler *Sylvia subalpina* and Mediterranean Flycatcher *Muscicapa tyrrhenica*) (Sangster 2000; Shirihai et al. 2001; Sangster et al. 2002; Brambilla et al. 2008; Förschler et al. 2009; Pons et al. 2016).

Finally, taxonomic revisions help identify the lineages that contain most evolutionary history, which is an important measure for prioritizing conservation efforts (e.g., Vane-Wright et al. 1991; Mooers and Atkins 2003). Using metrics that capture the evolutionary distinctiveness and conservation status of species, Jetz et al. (2014) identified the top-ranking bird species for phylogenetic conservation measures. Several of these represent species that have only recently been revealed by taxonomic study, including North Island Brown Kiwi *Apteryx mantelli* (Burbidge et al. 2003), Udzungwa Forest Partridge *Xenoperdix udzungwensis* (Dinesen et al. 1994), Australian Painted-snipe *Rostratula australis* (Baker et al. 2007), and New Zealand Storm Petrel *Fregetta maoriana* (Robertson et al. 2011).

2.11 Remaining Issues

Integrative taxonomy is an active field, and it is likely that various aspects will be further clarified and developed in the near future (Padial et al. 2010; Schlick-Steiner et al. 2010). Among the issues that remain to be clarified are the following:

As noted before, one of the arguments for integrative taxonomy is that different kinds of evidence highlight different aspects of species divergence: the tempo of divergence, frequency of hybridization, adaptation, and the geography and mechanisms of speciation. But this also means that there will be more and more cases of discordance among different kinds of evidence. How should one deal with discordance? The most sensitive methods may not be the most accurate methods. On which grounds does one decide which hypothesis is most likely correct? Should one err on the side of recognizing too few or too many species (e.g., Carstens et al. 2013)?

Continued progress in taxonomy depends largely on the collection of new empirical data (including specimens, sound recordings, and DNA sequences). Obtaining sufficient data may be problematic in several cases, such as (1) recent speciation when many markers may not be sensitive enough to detect divergence, (2) in paleontology when data are typically limited to bone material, and (3) when species are known from very few specimens (e.g., singletons; Lim et al. 2012). Finding efficient ways to maximize confidence in taxonomic hypotheses will remain a major challenge.

Acknowledgments I am grateful to Jolanda Luksenburg and Ian Merrill for the permission to use their photographs of Blue Chaffinches. I was supported by grants from Mark Constantine and the Swedish Research Council (grant 2015-06455).

References

Agapow P-M, Bininda-Emonds ORP, Crandall KA, Gittleman JL, Mace GM, Marshall JC, Purvis A (2004) The impact of species concept on biodiversity studies. Q Rev Biol 79:161–179. https://doi.org/10.1086/383542

Allen JA (1871) On the mammals and winter birds of East Florida. Bull Mus Comp Zool II (3):161–450

Alvarez Alonso J, Whitney BM (2001) A new *Zimmerius* tyrannulet (Aves: Tyrannidae) from white sand forests of northern Amazonian Peru. Wilson Bull 113:1–9. https://doi.org/10.1676/0043-5643(2001)113[0001:ANZTAT]2.0.CO;2

Baker AJ, Pereira SL, Rogers DI, Elbourne R, Hassell CJ (2007) Mitochondrial-DNA evidence shows the Australian Painted Snipe is a full species, *Rostratula australis*. Emu 107:185–189. https://doi.org/10.1071/MU07024

Bensch S, Pearson D (2002) The large-billed Reed Warbler *Acrocephalus orinus* revisited. Ibis 144:259–267. https://doi.org/10.1046/j.1474-919X.2002.00036.x

Bolton M (2007) Playback experiments indicate absence of vocal recognition among temporally and geographically separated populations of Madeiran Storm-petrels *Oceanodroma castro*. Ibis 149:255–265. https://doi.org/10.1111/j.1474-919X.2006.00624.x

Borges SH, da Silva JMC (2012) A new area of endemism for Amazonian birds in the Rio Negro Basin. Wilson J Ornithol 124:15–23. https://doi.org/10.1676/07-103.1

Bourne WRP, Warham J (1966) Geographical variation in the giant petrels of the genus *Macronectes*. Ardea 54:45–67

Brambilla M, Vitulano S, Spina F, Baccetti N, Gargallo G, Fabbri E, Guidali F, Randi E (2008) A molecular phylogeny of the *Sylvia cantillans* complex: cryptic species within the Mediterranean basin. Mol Phylogenet Evol 48:461–472. https://doi.org/10.1016/j.ympev.2008.05.013

Burbidge ML, Colbourne RG, Robertson HA, Baker AJ (2003) Molecular and other biological evidence supports the recognition of at least three species of brown kiwi. Conserv Genet 4:167–177. https://doi.org/10.1023/A:1023386506067

Burgers J, Smit JJ, van der Voet H (1991) Origins and systematics of two types of the Bean Goose *Anser fabalis* (Latham, 1787) wintering in the Netherlands. Ardea 79:307–316

Carstens BC, Pelletier TA, Reid NM, Satler JD (2013) How to fail at species delimitation. Mol Ecol 22:4369–4383. https://doi.org/10.1111/mec.12413

Chalmers AF (1999) What is this thing called science? 3rd edn. Open University Press, Maidenhead

Collar NJ (1997) Taxonomy and conservation: chicken and egg. Bull Br Ornithol Club 117:122–136

Cotterill FP, Groves CP, Taylor PJ (2017) Taxonomy: refine rather than stabilize. Nature 547:162. https://doi.org/10.1038/547162d

Cracraft J (1983) Species concepts and speciation analysis. Curr Ornithol 3:159–187

Cracraft J (1985) Historical biogeography and patterns of differentiation within the South American avifauna: areas of endemism. Ornithol Monogr (36):49–84

Cracraft J (2000) Species concepts in theoretical and applied biology: a systematic debate with consequences. In: Wheeler QD, Meier R (eds) Species concepts and phylogenetic theory: a debate. Columbia University Press, New York, pp 3–14

Dayrat B (2005) Towards integrative taxonomy. Biol J Linn Soc 85:407–415. https://doi.org/10.1111/j.1095-8312.2005.00503.x

de Queiroz K (1998) The general lineage concept of species, species criteria, and the process of speciation: a conceptual unification and terminological recommendations. In: Howard DJ, Berlocher SH (eds) Endless forms: species and speciation. Oxford University Press, Oxford, pp 57–75

de Queiroz K (1999) The general lineage concept of species and the defining properties of the species category. In: Wilson RA (ed) Species: new interdisciplinary essays. MIT Press, Cambridge, pp 49–89 [a highly influential description of the general lineage concept of species]

de Queiroz K (2007) Species concepts and species delimitation. Syst Biol 56:879–886. https://doi.org/10.1080/10635150701701083

de Queiroz K, Donoghue MJ (1988) Phylogenetic systematics and the species problem. Cladistics 4:317–338. https://doi.org/10.1111/j.1096-0031.1988.tb00518.x

de Queiroz K, Donoghue MJ (1990) Phylogenetic systematics and species revisited. Cladistics 6:83–90. https://doi.org/10.1111/j.1096-0031.1990.tb00527.x

del Hoyo J, Collar NJ (2014) Illustrated checklist of the birds of the world, vol 1. Lynx Edicions, Barcelona

del Hoyo J, Collar NJ (2016) Illustrated checklist of the birds of the world, vol 2. Lynx Edicions, Barcelona

Diamond J (1966) Zoological classification system of a primitive people. Science 151:1102–1104. https://doi.org/10.1126/science.151.3714.1102

Dinesen L, Lehmberg T, Svendsen JO, Hansen LA, Fjeldså J (1994) A new genus and species of perdicine bird (Phasianidae, Perdicini) from Tanzania; a relict form with Indo-Malayan affinities. Ibis 136:2–11. https://doi.org/10.1111/j.1474-919X.1994.tb08125.x

Donoghue MJ (1985) A critique of the biological species concept and recommendations for a phylogenetic alternative. Bryologist 88:172–181. https://doi.org/10.2307/3243026

Dubois A (2006) Proposed rules for the incorporation of nomina of higher-ranked zoological taxa in the International Code of Zoological Nomenclature. 2. The proposed rules and their rationale. Zoosystema 28:165–258

Elgvin TO, Hermansen JS, Fijarczyk A, Bonnet T, Borge T, Sæther SA, Voje KL, Sætre G-P (2011) Hybrid speciation in sparrows II: a role for sex chromosomes? Mol Ecol 20:3823–3837. https://doi.org/10.1111/j.1365-294X.2011.05182.x

Fernandes AM, Wink M, Sardelli CH, Aleixo A (2014) Multiple speciation across the Andes and throughout Amazonia: the case of the spot-backed antbird species complex (*Hylophylax naevius/Hylophylax naevioides*). J Biogeogr 41:1094–1104. https://doi.org/10.1111/jbi.12277

Förschler MI, Senar JC, Perret P, Björklund M (2009) The species status of the Corsican finch *Carduelis corsicana* assessed by three genetic markers with different rates of evolution. Mol Phylogenet Evol 52:234–240. https://doi.org/10.1016/j.ympev.2009.02.014

Frankham R, Ballou JD, Dudash MR, Eldridge MDB, Fenster CB, Lacy RC, Mendelson JR, Porton IJ, Ralls K, Ryder AO (2012) Implications of different species concepts for conserving biodiversity. Biol Conserv 153:25–31. https://doi.org/10.1016/j.biocon.2012.04.034

Friesen VL, Smith AL, Gómez-Díaz E, Bolton M, Furness RW, González-Solís J, Monteiro LR (2007) Sympatric speciation by allochrony in a seabird. Proc Natl Acad Sci 104:18589–18594. https://doi.org/10.1073/pnas.0700446104

Frost DR, Hillis DM (1990) Species in concept and practice. Herpetologica 46:87–104

Frost DR, Kluge AG (1994) A consideration of epistemiology in systematic biology, with special reference to species. Cladistics 10:259–294. https://doi.org/10.1111/j.1096-0031.1994.tb00178.x

Fuchs J, Pons J-M, Pasquet E, Raherilalao MJ, Goodman SM (2007) Geographical structure of genetic variation in the Malagasy Scops-owl inferred from mitochondrial sequence data. Condor 109:408–418. https://doi.org/10.1650/0010-5422(2007)109[408:GSOGVI]2.0.CO;2

Funk DJ, Omland KE (2003) Species-level paraphyly and polyphyly: frequency, causes, and consequences, with insights from animal mitochondrial DNA. Annu Rev Ecol Evol Syst 34:397–423. https://doi.org/10.1146/annurev.ecolsys.34.011802.132421

García-Trejo EA, Espinosa De Los Monteros A, Del Coro Arizmendi M, Navarro-Siüenza AG (2009) Molecular systematics of the Red-bellied and Golden-fronted Woodpeckers. Condor 111:442–452. https://doi.org/10.1525/cond.2009.080017

Geist V (1992) Endangered species and the law. Nature 357:274–276

Ghiselin MT (1974) A radical solution to the species problem. Syst Zool 23:536–544. https://doi.org/10.1093/sysbio/23.4.536

Haffer J (1992) The history of species concepts and species limits in ornithology. Bull Br Ornithol Club Centenary Suppl 112A:107–158

Hanken J (1999) Why are there so many new amphibian species when amphibians are declining. Trends Ecol Evol 14:7–8. https://doi.org/10.1016/S0169-5347(98)01534-1

Haring E, Gamauf A, Kryukov A (2007) Phylogeographic patterns in widespread corvid birds. Mol Phylogenet Evol 45:840–862. https://doi.org/10.1016/j.ympev.2007.06.016

Haring E, Däubl B, Pinsker W, Kryukov A, Gamauf A (2012) Genetic divergences and intraspecific variation in corvids of the genus *Corvus* (Aves: Passeriformes: Corvidae) – a first survey based on museum specimens. J Zool Syst Evol Res 50:230–246. https://doi.org/10.1111/j.1439-0469.2012.00664.x

Hazevoet CJ (1995) The birds of the Cape Verde Islands. BOU Check-list No. 13. BOU, Tring

Hennig W (1966) Phylogenetic systematics. University Illinois Press, Urbana

Hermansen JS, Sæther SA, Elgvin TO, Borge T, Hjelle E, Sætre G-P (2011) Hybrid speciation in sparrows I: phenotypic intermediacy, genetic admixture and barriers to gene flow. Mol Ecol 20:3812–3822. https://doi.org/10.1111/j.1365-294X.2011.05183.x

Hermansen JS, Haas F, Trier CN, Bailey RI, Nederbragt AJ, Marzal A, Sætre G-P (2014) Hybrid speciation through sorting of parental incompatibilities in Italian sparrows. Mol Ecol 23:5831–5842. https://doi.org/10.1111/mec.12910

Hull DL (1997) The ideal species concept-and why we can't get it. In: Claridge MF, Dawah HA, Wilson MR (eds) Species, the units of biodiversity. Systematics Association Special Volume 54. Chapman and Hall, London, pp 357–380

Hull DL (1999) On the plurality of species: questioning the party line. In: Wilson RA (ed) Species: new interdisciplinary essays. MIT Press, Cambridge, pp 23–48

Irestedt M, Fabre P-H, Batalha-Filho H, Jønsson KA, Roselaar CS, Sangster G, Ericson PGP (2013) The spatio-temporal colonization and diversification across the Indo-Pacific by a 'great speciator' (Aves, *Erythropitta erythrogaster*). Proc R Soc Lond B 280:20130309. https://doi.org/10.1098/rspb.2013.0309

Irwin DE (2002) Phylogeographic breaks without geographic barriers to gene flow. Evolution 56:2383–2394. https://doi.org/10.1554/0014-3820(2002)056[2383:PBWGBT]2.0.CO;2

Isaac NJB, Mallet J, Mace GM (2004) Taxonomic inflation: its influence on macroecology and conservation. Trends Ecol Evol 19:464–469. https://doi.org/10.1016/j.tree.2004.06.004

Isler ML, Alvarez Alonso J, Isler PR, Valqui T, Begazo A, Whitney BM (2002) Rediscovery of a cryptic species and description of a new subspecies in the *Myrmeciza hemimelaena* complex (Thamnophilidae) of the Neotropics. Auk 119:362–378. https://doi.org/10.1642/0004-8038(2002)119[0362:ROACSA]2.0.CO;2

Jetz W, Thomas GH, Joy JB, Redding DW, Hartmann K, Mooers AO (2014) Global distribution and conservation of evolutionary distinctness in birds. Curr Biol 24:919–930. https://doi.org/10.1016/j.cub.2014.03.011

Johnson JA, Watson RT, Mindell DP (2005) Prioritizing species conservation: does the Cape Verde kite exist? Proc R Soc Lond B 272:1365–1371. https://doi.org/10.1098/rspb.2005.3098

Johnston RF (1969) Taxonomy of house sparrows in the Mediterranean basin. Condor 71:129–139. https://doi.org/10.2307/1366074

Kearvell JC, Grant AD, Boon WM (2003) The orange-fronted parakeet (*Cyanoramphus malherbi*) is a distinct species: a review of recent research into its taxonomy and systematic relationship within the genus *Cyanoramphus*. Notornis 50:27–35

Keller EF, Lloyd EA (eds) (1992) Keywords in evolutionary biology. Harvard University Press, Cambridge

Kitcher P (1984) Species. Philos Sci 51:308–333. https://doi.org/10.1086/289182

Kornet DJ (1993) Permanent splits as speciation events: a formal reconstruction of the internodal species concept. J Theor Biol 164:407–435. https://doi.org/10.1006/jtbi.1993.1164

Lidén M, Oxelman B (1989) Species–pattern or process? Taxon 38:228–232. https://doi.org/10.2307/1220837

Lifjeld JT, Anmarkrud JA, Calabuig P, Cooper JE, Johannessen LE, Johnsen A, Kearns AM, Lachlan RF, Laskemoen T, Marthinsen G, Stensrud E, Garcia-del-Rey E (2016) Species-level divergences in multiple functional traits between the two endemic subspecies of Blue Chaffinches *Fringilla teydea* in Canary Islands. BMC Zool 1(1):4. https://doi.org/10.1186/s40850-016-0008-4

Lim GS, Balke M, Meier R (2012) Determining species boundaries in a world full of rarity: singletons, species delimitation methods. Syst Biol 61:165–169. https://doi.org/10.1093/sysbio/syr030

Lohman DJ, Ingram KG, Prawiradilaga DM, Winker K, Sheldon FH, Moyle RG, Ng PKL, Ong PS, Wang LK, Braile TM, Astuti D, Meier R (2010) Cryptic genetic diversity in "widespread" Southeast Asian bird species suggests that Philippine avian endemism is gravely underestimated. Biol Conserv 143:1885–1890. https://doi.org/10.1016/j.biocon.2010.04.042

Luna LW, Souza TO, Carneiro LS, Silva WAG, Schneider H, Sampaio I, Araripe J, Rêgo PSD (2017) Molecular data and distribution dynamics indicate a recent and incomplete separation of manakins species of the genus Antilophia (Aves: Pipridae) in response to Holocene climate change. J Avian Biol 48:1177–1188. https://doi.org/10.1111/jav.01378

Mace GM (2004) The role of taxonomy in species conservation. Philos Trans R Soc Lond B 359:711–719. https://doi.org/10.1098/rstb.2003.1454

Maddison WP (1997) Gene trees in species trees. Syst Biol 46:523–536. https://doi.org/10.1093/sysbio/46.3.523

Mayden RL (1997) A hierarchy of species concepts: the denouement in the saga of the species problem. In: Claridge MF, Dawah HA, Wilson MR (eds) Species, the units of biodiversity. Systematics Association Special Volume 54. Chapman and Hall, London, pp 381–424

Mayr E (1942) Systematics and the origin of species. Columbia University Press, New York

Mayr E (1982) The growth of biological thought. Harvard University Press, Cambridge

Mayr E (1996) What is a species, and what is not? Philos Sci 63:262–277. https://doi.org/10.1086/289912

Mayr E, Amadon D (1951) A classification of recent birds. Am Mus Novit 1496:1–42
McKay BD, Zink RM (2010) The causes of mitochondrial DNA gene tree paraphyly in birds. Mol Phylogenet Evol 54:647–650. https://doi.org/10.1016/j.ympev.2009.08.024
Meise W (1936) Zur Systematik und Verbreitungsgeschichte der Haus- und Weidensperlinge, *Passer domesticus* (L.) und *hispaniolensis* (T.). J Ornithol 84:631–672. https://doi.org/10.1007/BF01951015
Milá B, Tavares ES, Muñoz Saldaña A, Karubian J, Smith TB, Baker AJ (2012) A trans-Amazonian screening of mtDNA reveals deep intraspecific divergence in forest birds and suggests a vast underestimation of species diversity. PLoS One 7(7):e40541. https://doi.org/10.1371/journal.pone.0040541
Mishler BD (1999) Getting rid of species? In: Wilson RA (ed) Species: new interdisciplinary essays. MIT Press, Cambridge, pp 307–315
Mishler BD, Donoghue MJ (1982) Species concepts: a case for pluralism. Syst Zool 31:491–503. https://doi.org/10.2307/2413371
Monteiro LR, Furness RW (1998) Speciation through temporal segregation of Madeiran Storm-petrel (*Oceanodroma castro*) populations in the Azores? Philos Trans R Soc Lond B 353:945–953. https://doi.org/10.1098/rstb.1998.0259
Mooers AØ, Atkins RA (2003) Indonesia's threatened birds: over 500 million years of evolutionary heritage at risk. Anim Conserv 6:183–188. https://doi.org/10.1017/S1367943003003226
Moore WS (1977) An evaluation of narrow hybrid zones in vertebrates. Q Rev Biol 52:263–277. https://doi.org/10.1086/409995
O'Hara RJ (1993) Systematic generalization, historical fate and the species problem. Syst Biol 42:231–246. https://doi.org/10.1093/sysbio/42.3.231
O'Hara RJ (1994) Evolutionary history and the species problem. Am Zool 34:12–22. https://doi.org/10.1093/icb/34.1.12
Olsson U, Alström P, Svensson L, Aliabadian M, Sundberg P (2010) The *Lanius excubitor* (Aves, Passeriformes) conundrum—Taxonomic dilemma when molecular and non-molecular data tell different stories. Mol Phylogenet Evol 55:347–357. https://doi.org/10.1016/j.ympev.2009.11.010
Ottenburghs J, Megens H-J, Kraus RHS, Madsen O, van Hooft P, van Wieren SE, Crooijmans RPMA, Ydenberg RC, Groenen MAM, Prins HHT (2016) A tree of geese: a phylogenomic perspective on the evolutionary history of true geese. Mol Phylogenet Evol 101:303–313. https://doi.org/10.1016/j.ympev.2016.05.021
Päckert M, Martens J, Wink M, Feigl A, Tietze DT (2012) Molecular phylogeny of Old World swifts (Aves: Apodiformes, Apodidae, *Apus* and *Tachymarptis*) based on mitochondrial and nuclear markers. Mol Phylogenet Evol 63:606–616. https://doi.org/10.1016/j.ympev.2012.02.002
Padial JM, de la Riva I (2007) Integrative taxonomists should use and produce DNA barcodes. Zootaxa 1586:67–68
Padial JM, Miralles A, De la Riva I, Vences M (2010) The integrative future of taxonomy. Front Zool 7:16. https://doi.org/10.1186/1742-9994-7-16 [a good review of integrative taxonomy]
Parkin DT, Collinson M, Helbig AJ, Knox AG, Sangster G (2003) The taxonomic status of Carrion and Hooded Crows. Br Birds 96:274–290
Paxinos EE, James HF, Olson SL, Sorenson MD, Jackson J, Fleischer RC (2002) mtDNA from fossils reveals a radiation of Hawaiian geese recently derived from the Canada Goose (*Branta canadensis*). Proc Natl Acad Sci U S A 99:1399–1404. https://doi.org/10.1073/pnas.032166399
Penhallurick J, Wink M (2004) Analysis of the taxonomy and nomenclature of the Procellariiformes based on complete nucleotide sequences of the mitochondrial cytochrome b gene. Emu 104:125–147
Pérez-Emán JL, Ferreira JP, Gutiérrez-Pinto N, Cuervo AM, Céspedes LN, Witt CC, Cadena CD (2018) An extinct hummingbird species that never was: a cautionary tale about sampling issues in molecular phylogenetics. Zootaxa 4442:491–497
Peters JL, McCracken KG, Pruett CL, Rohwer S, Drovetski SV, Zhuravlev YN, Kulikova I, Gibson DD, Winker K (2012) A parapatric propensity for breeding precludes the completion of speciation in common teal (*Anas crecca*, sensu lato). Mol Ecol 21:4563–4577. https://doi.org/10.1111/j.1365-294X.2012.05711.x

Pianka ER (1966) Latitudinal gradients in species diversity: a review of concepts. Am Nat 100:33–46. https://doi.org/10.1086/282398
Pigliucci M, Kaplan J (2006) Making sense of evolution: the conceptual foundations of evolutionary biology. University Chicago Press, Chicago
Poelstra JW, Vijay N, Bossu CM, Lantz H, Ryll B, Müller I, Baglione V, Unneberg P, Wikelski M, Grabherr MG, Wolf JBW (2014) The genomic landscape underlying phenotypic integrity in the face of gene flow in crows. Science 344:1410–1414. https://doi.org/10.1126/science.1253226
Pons J-M, Thibault J-C, Aymí R, Grussu M, Muntaner J, Olioso G, Sunyer JR, Touihri M, Fuchs J (2016) The role of western Mediterranean islands in the evolutionary diversification of the Spotted Flycatcher (*Muscicapa striata*), a long-distance migratory passerine species. J Avian Biol 47:386–398. https://doi.org/10.1111/jav.00859
Rahbek C, Graves GR (2001) Multiscale assessment of patterns of avian species richness. Proc Natl Acad Sci U S A 98:4534–4539. https://doi.org/10.1073/pnas.071034898
Rasmussen PC, Schulenberg TS, Hawkins F, Voninavoko R (2000) Geographic variation in the Malagasy Scops-Owl (*Otus rutilus* auct.): the existence of an unrecognised species on Madagascar and the taxonomy of other Indian Ocean taxa. Bull Br Ornithol Club 120:75–102
Rawlence NJ, Kennedy M, Scofield RP, Tennyson AJD, Boussès P, Chambers GK (2015) Taxonomic status of *Cyanoramphus* parakeets on the Auckland Islands and implications for the validity of the orange-fronted parakeet as a discrete species. J R Soc N Z 45:197–211. https://doi.org/10.1080/03036758.2015.1084337
Reydon TAC (2004) Why does the species problem still persist? BioEssays 26:300–305. https://doi.org/10.1002/bies.10406
Ribas CC, Aleixo A, Nogueira ACR, Miyaki CY, Cracraft J (2011) A palaeobiogeographic model for biotic diversification within Amazonia over the past three million years. Proc R Soc B 279:681–689. https://doi.org/10.1098/rspb.2011.1120
Ridley M (1989) The cladistic solution to the species problem. Biol Philos 4:1–16. https://doi.org/10.1007/BF00144036
Robertson BC, Stephenson BM, Goldstien SJ (2011) When rediscovery is not enough: taxonomic uncertainty hinders conservation of a critically endangered bird. Mol Phylogenet Evol 61:949–952. https://doi.org/10.1016/j.ympev.2011.08.001
Ruokonen M, Litvin K, Aarvak T (2008) Taxonomy of the bean goose–pink-footed goose. Mol Phylogenet Evol 48:554–562. https://doi.org/10.1016/j.ympev.2008.04.038
Ruse M (1969) Definitions of species in biology. Br J Philos Sci 20:97–119
Sangster G (2000) Genetic distance as a test of species boundaries in the Citril Finch *Serinus citrinella*: a critique and taxonomic reinterpretation. Ibis 142:487–490. https://doi.org/10.1111/j.1474-919X.2000.tb04447.x
Sangster G (2009) Increasing numbers of bird species result from taxonomic progress, not taxonomic inflation. Proc R Soc B 276:3185–3191. https://doi.org/10.1098/rspb.2009.0582
Sangster G (2014) The application of species criteria in avian taxonomy and its implications for the debate over species concepts. Biol Rev 89:199–214. https://doi.org/10.1111/brv.12051 [a detailed analysis of how bird species are delimited in practice]
Sangster G, Luksenburg JA (2015) Declining rates of species described per taxonomist: slowdown of progress or a side-effect of improved quality in taxonomy? Syst Biol 64:144–151. https://doi.org/10.1093/sysbio/syu069
Sangster G, Oreel GJ (1996) Progress in taxonomy of Taiga and Tundra Bean Geese. Dutch Birding 18:310–316
Sangster G, Collinson JM, Helbig AJ, Knox AG, Parkin DT, Prater T (2001) The taxonomic status of Green–winged Teal *Anas carolinensis*. Br Birds 94:218–224
Sangster G, Collinson JM, Helbig AJ, Knox AG, Parkin DT (2002) The specific status of Balearic and Yelkouan Shearwaters. Br Birds 95:636–639
Sangster G, Collinson JM, Crochet P-A, Kirwan GM, Knox AG, Parkin DT, Votier SC (2015) Taxonomic recommendations for Western Palaearctic birds: 10th report. Ibis 157:193–200. https://doi.org/10.1111/ibi.12221

Sangster G, Roselaar CS, Irestedt M, Ericson PGP (2016a) Sillem's Mountain Finch *Leucosticte sillemi* is a valid species of rosefinch (*Carpodacus*, Fringillidae). Ibis 158:184–189. https://doi.org/10.1111/ibi.12323

Sangster G, Rodríguez-Godoy F, Roselaar CS, Robb MA, Luksenburg JA (2016b) Integrative taxonomy reveals Europe's rarest songbird species, the Gran Canaria Blue Chaffinch *Fringilla polatzeki*. J Avian Biol 47:159–166. https://doi.org/10.1111/jav.00825

Schlick-Steiner BC, Steiner FM, Seifert B, Stauffer C, Christian E, Crozier RH (2010) Integrative taxonomy: a multisource approach to exploring biodiversity. Annu Rev Entomol 55:421–438. https://doi.org/10.1146/annurev-ento-112408-085432

Schweizer M, Shirihai H (2013) Phylogeny of the *Oenanthe lugens* complex (Aves, Muscicapidae: Saxicolinae): paraphyly of a morphologically cohesive group within a recent radiation of open-habitat chats. Mol Phylogenet Evol 69:450–461. https://doi.org/10.1016/j.ympev.2013.08.010

Sharpe RB (1909) A hand-list of the genera and species of birds, vol 5. British Museum, London

Shaw AB (1969) Adam and Eve, paleontology, and the non-objective arts. J Paleontol 43:1085–1098

Shirihai H, Gargallo G, Helbig AJ, Harris A, Cottridge D (2001) *Sylvia* warblers: identification, taxonomy and phylogeny of the genus *Sylvia*. Helm, London

Simpson GG (1951) The species concept. Evolution 5:285–298. https://doi.org/10.1111/j.1558-5646.1951.tb02788.x

Sluys R (2013) The unappreciated, fundamentally analytical nature of taxonomy and the implications for the inventory of biodiversity. Biodivers Conserv 22:1095–1105. https://doi.org/10.1007/s10531-013-0472-x

Smith AL, Monteiro L, Hasegawa O, Friesen VL (2007) Global phylogeography of the Band-rumped Storm-petrel (*Oceanodroma castro*; Procellariiformes: Hydrobatidae). Mol Phylogenet Evol 43:755–773. https://doi.org/10.1016/j.ympev.2007.02.012

Sober E (ed) (2006) Conceptual issues in evolutionary biology, 3rd edn. MIT Press, Cambridge

Sokal RR, Crovello TJ (1970) The biological species concept: a critical evaluation. Am Nat 104:127–153. https://doi.org/10.1086/282646

Stamos DN (2003) The species problem: biological species, ontology, and the metaphysics of biology. Lexington Books, Lanham

Stattersfield AJ, Crosby MJ, Long AJ, Wege DC (1998) Endemic bird areas of the World. Birdlife Conservation Series 7, Birdlife International, Cambridge

Stejneger L (1884) On the use of trinomials in American ornithology. Proc US Natl Mus 7:70–80

Stone W (1899) The proper function of 'binomials' and 'trinomials'. Auk 16:374–376. https://doi.org/10.2307/4069392

Swenson NG (2010) Mapping the suturing of a continental biota. Mol Ecol 19:5324–5327. https://doi.org/10.1111/j.1365-294x.2010.04880.x

Tobias JA, Bates JM, Hackett SJ, Seddon N (2008) Comment on "The Latitudinal Gradient in Recent Speciation and Extinction Rates of Birds and Mammals". Science 319:901c. https://doi.org/10.1126/science.1150568

Tobias JA, Seddon N, Spottiswoode CN, Pilgrim JD, Fishpool LDC, Collar NJ (2010) Quantitative criteria for species delimitation. Ibis 152:724–746. https://doi.org/10.1111/j.1474-919x.2010.01051.x

Trier CN, Hermansen JS, Sætre G-P, Bailey RI (2014) Evidence for mito-nuclear and sex-linked reproductive barriers between the hybrid Italian Sparrow and its parent species. PLoS Genet 10:e1004075. https://doi.org/10.1371/journal.pgen.1004075

Van Valen L (1976) Ecological species, multispecies, and oaks. Taxon 25:233–239. https://doi.org/10.2307/1219444

Vane-Wright RI, Humphries CJ, Williams PH (1991) What to protect? Systematics and the agony of choice. Biol Conserv 55:235–254. https://doi.org/10.1016/0006-3207(91)90030-D

Voelker G, Marks BD, Kahindo C, A'genonga U, Bapeamoni F, Duffie LE, Huntley JW, Mulotwa E, Rosenbaum SA, Light JE (2013) River barriers and cryptic biodiversity in an evolutionary museum. Ecol Evol 3:536–545. https://doi.org/10.1002/ece3.482

Weir JT, Schluter D (2007) The latitudinal gradient in recent speciation and extinction rates of birds and mammals. Science 315:1574–1576. https://doi.org/10.1126/science.1135590

Wheeler QD, Cracraft J (1997) Taxonomic preparedness: are we ready to meet the biodiversity challenge? In: Reaka-Kudla ML, Wilson DE, Wilson EO (eds) Biodiversity II. Joseph Henry Press, Washington, pp 435–446

Whitney BM, Alvarez Alonso J (1998) A new *Herpsilochmus* antwren (Aves: Thamnophilidae) from northern Amazonian Peru and adjacent Ecuador: the role of edaphic heterogeneity of terra firme forest. Auk 115:559–576. https://doi.org/10.2307/4089406

Whitney BM, Alvarez Alonso J (2005) A new species of gnatcatcher from white-sand forests of northern Amazonian Peru with revision of the *Polioptila guianensis* complex. Wilson Bull 117:113–127. https://doi.org/10.1676/04-064

Wiens JJ, Servedio MR (2000) Species delimitation in systematics: inferring diagnostic differences between species. Proc R Soc Lond B 267:631–636. https://doi.org/10.1098/rspb.2000.1049

Wiley EO (1978) The evolutionary species concept reconsidered. Syst Zool 27:17–26. https://doi.org/10.2307/2412809

Will KW, Mishler BD, Wheeler QD (2005) The perils of DNA barcoding and the need for integrative taxonomy. Syst Biol 54:844–851. https://doi.org/10.1080/10635150500354878

Willig MR, Kaufman DM, Stevens RD (2003) Latitudinal gradients of biodiversity: pattern, process, scale, and synthesis. Annu Rev Ecol Evol Syst 34:273–309. https://doi.org/10.1146/annurev.ecolsys.34.012103.144032

Wilson BE (1995) A (not-so-radical) solution to the species problem. Biol Philos 10:339–356. https://doi.org/10.1007/BF00852472

Wolf JBW, Bayer T, Haubold B, Schilhabel M, Rosenstiel P, Tautz D (2010) Nucleotide divergence vs. gene expression differentiation: comparative transcriptome sequencing in natural isolates from the carrion crow and its hybrid zone with the hooded crow. Mol Ecol 19(Suppl 1):162–175. https://doi.org/10.1111/j.1365-294X.2009.04471.x

Yeates DK, Seago A, Nelson L, Cameron SL, Joseph L, Trueman JW (2011) Integrative taxonomy, or iterative taxonomy? Syst Entomol 36:209–217. https://doi.org/10.1111/j.1365-3113.2010.00558.x

Suggestion for Further Reading

Ereshefsky M (ed) (1992) The units of evolution. MIT Press, Cambridge [a collection of classic papers in the debate over species concepts, written by biologists and philosophers of science]

Chapter 3
Studying Speciation: Genomic Essentials and Approaches

Daronja Trense and Dieter Thomas Tietze

Abstract A genome comprises the entire genetic material of an organism and consists of DNA, which is in turn constructed of hundreds to billions of nucleotides. Nucleotides are organic molecules composed of three subunits: nitrogenous base, sugar (deoxyribose), and phosphate group. The DNA is differentiated into coding (genes) and noncoding regions. A gene is a specific region of DNA that encodes a function. All genes present within an organism represent its genotype. The genotype determines the phenotype, which is, however, additionally affected by the environment and the individual development (ontogenesis). A gene may affect a single or several phenotypic features (pleiotropy). Likewise, a phenotypic feature may be affected by one or several genes, with the latter comprising polygenic traits. In the process of gene expression, the information encoded by a gene is used to generate a product. The expression of genes is regulated by the external (temperature, stress, resource availability) and internal environment (metabolism, cell division cycle), and the gene-specific role in the respective tissue organism. Several processes underlying evolutionary change, e.g., mutation, genetic drift, gene duplication, selection, and migration, may change the genome at the level of single bases through genes to the organism. Such changes may result in population differentiation and eventually speciation. Molecular genetic studies on microevolution and speciation started with single genetic markers, e.g., the COI marker gene. Today, mainly genomic and transcriptomic approaches, making use of a large number of markers such as single nucleotide polymorphisms or microsatellites, are used to compare species, populations, and individuals.

D. Trense
Institute of Pharmacy and Molecular Biotechnology, Heidelberg University, Heidelberg, Germany

D. T. Tietze (✉)
Natural History Museum Basel, Basel, Switzerland
e-mail: thomas.tietze@bs.ch

D. T. Tietze (ed.), *Bird Species*, Fascinating Life Sciences,
https://doi.org/10.1007/978-3-319-91689-7_3

Keywords Molecular markers · Chicken genome · DNA structure · DNA function · Chromosomes · Nuclear genome · Transcriptomics · Genomics

Understanding the evolution of life is one of the most distinguished tasks of biological research. Recent advances in molecular techniques offer unprecedented opportunities to tackle such issues in a diverse array of taxa including birds, comprising our focal group here. Using genomic data requires detailed knowledge on the composition and function of each component of the genome. In the first section of this chapter, we therefore give an overview from the smallest elements including pentose sugars, double-ringed nitrogenous bases, and phosphate groups, via DNA, genes, and chromosomes to the entire avian genome. Furthermore, we cover functional aspects from DNA replication via transcription and translation through features of cells, tissues, and individuals, in the second section. The third section is about the evolution of the genome. We here highlight the mechanisms bringing about variation in individual genomes as well as the genomes of the next generation. In the last section, we will provide an overview about the components of the genomes that were or are used for understanding speciation, molecular systematics, and other research fields.

3.1 What Is an Avian Genome?

3.1.1 Structure of the Genetic Material

The deoxyribonucleic acid (DNA) is constructed of hundreds to billions of nucleotides, which in turn are constructed of nucleosides. A nucleoside consists of a purine or pyrimidine base linked to a pentose sugar, whereby purine is a double-ringed nitrogenous base such as adenine (A) and guanine (G) and pyrimidine a single-ringed nitrogenous base such as cytosine (C) and thymine (T) (Fig. 3.1) or uracil (U). If a nucleoside is linked to a phosphate group on either the 5′ or 3′ carbon on the deoxyribose, it is called a nucleotide. Two pentose sugar molecules each of two different nucleotide monomers are connected through an individual phosphate molecule, resulting in nucleotides being connected to a long chain. Such a chain creates a single strand of DNA with one end of the chain having a free 5′ and the other a free 3′ end. Two antiparallel and complementary strands can be connected by hydrogen bounds between guanine and cytosine or adenine and thymine, respectively (Alberts et al. 2014). If both DNA strands are wound around each other in an opposite direction, this is called the DNA double helix (Fig. 3.1). In eukaryotes, such as plants, mammals, birds, and many more, which are organisms whose cells have a nucleus and other organelles that are enclosed by membranes, the DNA is organized into chromosomes (Fig. 3.1) within the cell nucleus (plus another DNA molecule in each mitochondrion). Functionally, the genome is divided into genes, i.e., sequences of DNA that encode a single type of ribonucleic acid (RNA).

Due to diploidy of eukaryotic organisms, gene loci occur twice in eukaryotic genomes, as one maternally and one paternally inherited copy. An allele is one of

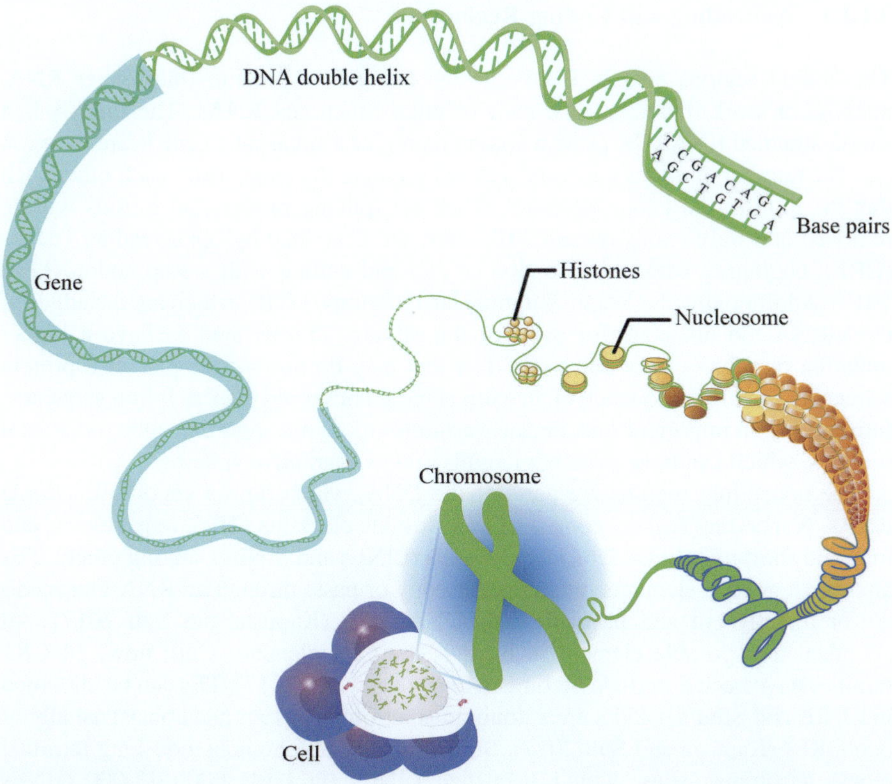

Fig. 3.1 Each cell contains a nucleus with chromosomes. These chromosomes comprise nucleosomes to pack the genetic material in the nucleus. The nucleosome consists of a DNA section, which is wound around the histone. Certain parts of the DNA (genes) carry information for the cell to encode a specific function. The DNA is structured in a double helix and consists of the four bases adenine (A), guanine (G), cytosine (C), and thymine (T) (The graphic was modified for this book chapter and was used from the National Human Genome Research Institute (NHGRI); website www.genome.gov)

several alternative forms of a gene occupying a given locus on a chromosome. All genes of one individual, which were transmitted from its parents, make up the genotype. The genotype produces the phenotype, which is the collection of all observable traits of one organism, e.g., height and eye color (Lesk 2012).

Three bases represent jointly a codon or triplet, and genes include a series of codons that are read sequentially from a starting point on one end to a termination point on the other end. Each triplet codes for a single amino acid in a corresponding protein. There are 64 (4 $\text{bases}^{3\ \text{nucleotides}}$) possible codons but only 20 naturally occurring amino acids. This means that several codons correspond to the same amino acid (Alberts et al. 2014).

In contrast to the DNA, the RNA is evolutionarily older and has a different sugar (ribose) and a different base (uracil), which is replaced in the DNA by base thymine. The ribose makes the RNA less stable than DNA, and the production of uracil is less complex, because uracil is the unmethylated form of thymine (Alberts et al. 2014).

3.1.1.1 Noncoding and Coding Regions

The coding regions encode RNAs, which result in a protein, (messenger RNA, mRNA) or work directly in the case of other functional RNAs. The mRNA is a single-stranded RNA. The protein coding part(s) of a eukaryotic gene is/are the exon (s). The number of exons can vary. Introns separate the exons from each other such that the introns and exons alternate. After the splicing process, all introns will be removed and only exons remain. All exons are described by open reading frames (ORF), beginning with a start codon (ATG) and ending with a stop codon (Lesk 2012). Additionally, the 5′ and 3′ untranslated regions (UTR), which are the edges of the mRNA, do not code for parts of the protein. Introns tend to have a higher mutation rate than exons due to the fact that they do not encode part of a protein sequence. Thus, the sequence of an exon is more conserved than an intron sequence. Introns play an important role, because a single eukaryotic gene can code for several proteins, which can have different lengths due to alternative splicing.

The noncoding regions are parts of the DNA, which do not encode functional RNAs. Noncoding regions consist of transposable elements (TEs), retroviruses, and long and short interspersed nuclear elements (LINEs and SINEs), among others. TEs are selfish genetic elements, which either copy or paste through an RNA intermediate or directly cut and paste in their DNA form (Kapusta and Suh 2017). An abundant transposable element in birds is the CR1 element. Until now, 14 CR1 families have been described in birds (Kapusta and Suh 2017). TEs can be classified in LINEs and SINEs. LINEs are autonomous retrotransposons and consist usually of two ORFs (Kapusta and Suh 2017). SINEs are non-autonomous non-long terminal repeat retrotransposons, which parasitize LINEs (in birds 6000–17,000 SINEs versus 1,500,000 in humans or less than 0.1% of all avian genome sequences) (Kapusta and Suh 2017). Another example for a noncoding region is the retrovirus, an RNA virus, which can convert its sequence into DNA by reverse transcription (explained in Sect. 3.2.2). Endogenous viral elements (EVEs) are retroviruses that rely on obligate integration into the host genome and can be classified as LTR retrotransposons (Kapusta and Suh 2017).

3.1.1.2 Autosomes Versus Sex Chromosomes

A chromosome contains part of the genetic material of a eukaryotic organism and consists of chromatin, which is a complex of DNA and proteins. Most of these proteins are histones (Fig. 3.1), which wrap up the DNA in the nucleus. The number and appearance of chromosomes is called karyotype. Eukaryotic cells can be present either in a diploid or haploid condition. The term haploid means that chromosomes occur in a single set, while diploid cells have a double set of chromosomes. Most eukaryotic organisms have diploid cells; thus, all chromosomes appear twice. However, eukaryotic organisms differ in the number of chromosomes, for example, humans have 46 chromosomes. Furthermore, chromosomes are divided into autosomes and sex chromosomes. Autosomes are pairs of chromosomes in a diploid cell, which have the same form, but each chromosome pair has a specific length. Humans have 44 autosomes and 2 sex chromosomes. Sex chromosomes differ from autosomes in length and function of their genes. They include the sex-determining region Y (SRY) gene on the Y chromosome. Furthermore, in humans, men have two different sex chromosomes, the X and Y chromosomes, while women have two X chromosomes. However, this XY sex-determination system is not present in all eukaryotes, in humans, most other mammals, several insects, some snakes, and a few plants. Another system is the ZW sex-determination system, which can be found in birds, several fishes, crustaceans, some insects, and reptiles. In the ZW sex-determination system, males have two Z chromosomes, while females have a Z and a W chromosome (Scanes 2015). Responsible for the sex determination in birds is probably a gene on the W chromosome, which is similar to the SRY gene on the Y chromosome. The Z and X chromosomes are larger and contain more genes than the W and Y chromosomes. Not only the sex chromosomes can be different in eukaryotic organisms, also the autosomes. In lizards, snakes, turtles, and birds, the autosomes can be divided in micro- and macrochromosomes (Matsubara et al. 2006; Ellegren 2013). Microchromosomes are tiny chromosomes with a length under 20,000,000 bp, while macrochromosomes are larger than 40,000,000 bp and resemble the mammalian autosomes in size. Characteristics of microchromosomes are that they include high rates of meiotic recombination, have high guanine-cytosine (GC) contents, short introns, high densities of genes, and **cytosine-phosphate-guanine (CpG) islands**, low densities of transposable elements and other repeats, but many repetitive sequences (Scanes 2015; Kapusta and Suh 2017). Another important aspect concerning chromosomes is the synteny, which describes the location of genetic loci on the same chromosome within an individual or species or even among species.

3.1.1.3 Nuclear Genome and Mitochondrial Genome

A common feature of eukaryotic organisms (except plants and other photosynthetically active eukaryotes) is that they have two genomes, the nuclear and the mitochondrial genome. The nuclear genome is organized in chromosomes as detailed above, while the mitochondrial genome is circularly or linearly organized and located within the mitochondria (that derived from bacteria). Generally, the nuclear genome is larger and contains many more genes than the mitochondrial genome.

3.1.2 The Chicken Model: History and Overview

An important model organism is the chicken *Gallus gallus domesticus*, because it is the first agricultural animal of which the genome was sequenced and has a relatively recent common ancestor with mammals. The ancestor of mammals and birds diverged 310 million years ago according to mitochondrial findings (Griffin et al.

2007). Furthermore, the chicken is the main laboratory model for over 10,828 extant bird species (Gill and Donsker 2017). The chicken genome has a size of 1.2 gigabases (Gb) (Lesk 2012), while avian genomes average around 1.35 Gb from the smallest, the Black-chinned Hummingbird *Archilochus alexandri* with 0.9 Gb, to the largest, the Common Ostrich *Struthio camelus* with 2.1 Gb (Scanes 2015; Kapusta and Suh 2017). The genome of the chicken was sequenced for the first time in 2004 with a 6.6 X **coverage**, using whole-genome shotgun reads. The resulting **assembly** of the chicken was composed of 933,000,000 bp and a genome size of 1.05 Gb (Hillier et al. 2004).

In chicken, it is difficult to identify genes on the W chromosome as well as on the microchromosomes due to the high number of repetitive sequences. However, the Z chromosome is well explored and contains nearly the same genes in all birds and is therefore highly conserved among bird species. On the Z chromosome of the chicken are about 1000 genes located, which are absent from the W chromosome. The W chromosome is degraded to different extents in some bird lineages (Marshall Graves 2015). This is why it is smaller, poorer in genes, but richer in repeats in most birds (Scanes 2015).

The avian karyotypes have been unusually stable during evolution, but there are some exceptions with chromosome numbers from 40 to 126 due to numerous microchromosomes (Griffin et al. 2007; Scanes 2015). A typical avian karyotype has a $2n$ of 76–80 (Ellegren 2013), and the chicken's haploid karyotype is defined by 39 chromosomes: chromosomes 1 through 10 are macrochromosomes, chromosomes 11 through 38 are microchromosomes, and the 39th chromosome is the sex chromosome (Hillier et al. 2004; Ellegren 2005).

In comparison to other eukaryotic organisms, the reduction of the avian genome size and transposable elements density, which is about 10% in avian genomes, began after the split of birds and crocodilians 250 million years ago (Griffin et al. 2007; Kapusta and Suh 2017). Thus, avian genomes are compact and were selected due to the evolution of flight (Hughes and Piontkivska 2005; Scanes 2015). In comparison with flightless birds, flying birds have a smaller genome. This might be due to the larger body size and longer generation times of flightless birds (Kapusta and Suh 2017).

Birds expanded their repertoire of keratin genes such as feather and claw keratins, and retained genes for egg production (Scanes 2015). The chicken genome has several genes encoding egg-related proteins, which are not represented in the mammalian genome. These are examples for gene losses in the mammalian lineage. On the other hand, there are some genes in chicken and humans that might have changed their function. In contrast to birds, which excrete uric acid, mammals excrete urea. Concomitantly, it seems that the function of some genes is altered in mammals, because genes encoding the enzymes of the mammalian urea cycle are also found in the chicken genome (Hillier et al. 2004).

The **alignment** of the chicken and human genome shows that at least 70,000,000 bp of sequence are likely to be functional in both species (Hillier et al. 2004). It is estimated that 20,000–23,000 protein-coding genes occur in the chicken genome (Hillier et al. 2004). In the human genome, some 20,000 genes have been detected until now. About 60% of protein-coding genes in chicken have a single human ortholog. From these conserved genes in human and chicken, 72% are also conserved in the Japanese pufferfish *Takifugu rubripes* genome. Thus, these genes are most likely present in most vertebrates (Hillier et al. 2004).

3.2 How Does the Genome "Work"?

3.2.1 Replication of the DNA

DNA replication is the process of copying DNA within a cell of an organism. The process starts with the opening of the DNA double helix by the enzyme DNA helicase at a specific position by breaking the hydrogen bonds. Both DNA strands of the DNA double helix serve as a template for the replication of a new complementary strand. After opening, the enzyme DNA polymerase adds complementary nucleotides one by one to the growing DNA chain. The unwinding and adding of new nucleotides to the growing chain stops, if it reaches a region, which is either already replicated, or if a protein binds to the DNA sequence to stop the replication. Afterward, the new DNA strands will be checked by proofreading to remove the mismatches. The results of the DNA replication comprise two DNA double helices with one old and one new DNA strand. Apart from a very small number of copying errors, the two daughter molecules are identical in sequence with the original DNA molecule.

3.2.2 Transcription: RNA Synthesis

When a cell needs a specific protein, the transcription of the respective gene (copying DNA into RNA) starts, which is followed by translation of the nucleotide sequence into the amino acid sequence. The transcription begins with the opening of a small portion of the DNA double helix and its unwinding to display the bases. The enzyme RNA polymerase performs the transcription and knows its target position through a promotor, which is a specific nucleotide sequence of the DNA (Alberts et al. 2014). The promotor is located before the coding region and regulates the expression of genes. One strand of the DNA double helix acts as a template for the synthesis of the mRNA (Fig. 3.2). The sequence of the mRNA chain is defined by complementary base-pairing between free nucleotides and the DNA template. This DNA template is exactly complementary to the precursor messenger RNA (pre-mRNA). The transcription stops at a terminator, which represents the end of a gene (Alberts et al. 2014). Thus, the pre-mRNA is released. In eukaryotes, the pre-mRNA goes through several steps of processing such as polyadenylation, capping, and splicing. Polyadenylation adds a poly(A) tail to the pre-mRNA. This means that a specific number of adenine bases are added to the pre-mRNA. Capping of the pre-mRNA places a specific nucleotide and associated proteins to the 5′ end to stabilize the mRNA. Splicing removes the introns—intragenic regions—from the pre-mRNA; therefore, only exonic sequences exist in the mature mRNA.

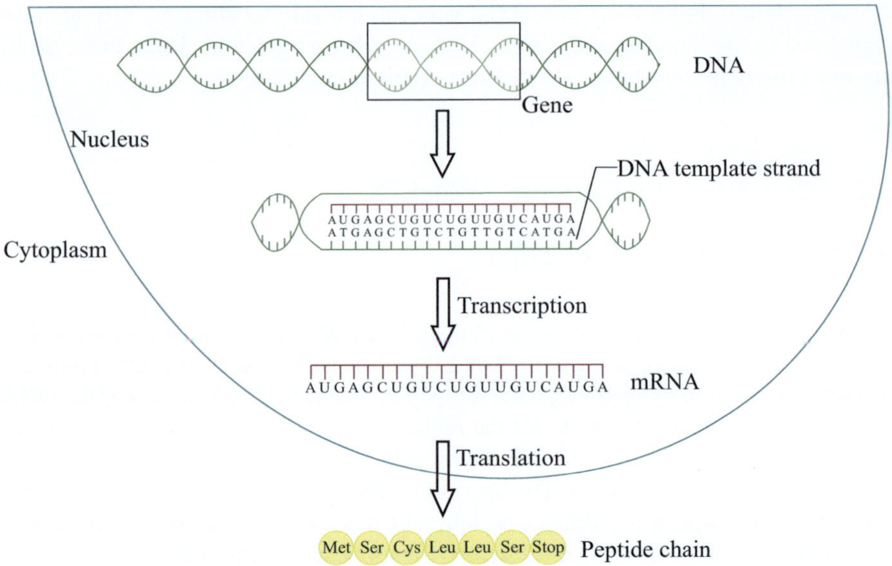

Fig. 3.2 A gene in the DNA provides on the template strand the nucleotide sequence that is transcribed into RNA (change from base thymine to uracil). This synthesis of RNA based on DNA is known as transcription. After the transcription, the mRNA will be released from the nucleus to the cytoplasm. In the cytoplasm, the translation proceeds with synthesizing a peptide chain (protein) based on the nucleotide sequence of the mRNA. In this example, the peptide chain consists of methionine (Met), serine (Ser), cysteine (Cys), leucine (Leu), and a stop codon, which leads to the termination of the translation

3.2.3 *Translation*

After these steps, the translation begins in the cytoplasm on ribosomes, which are complexes of proteins and ribosomal RNA (rRNA). RNA copies are used directly to synthesize the protein (Fig. 3.2) (Alberts et al. 2014). The information of the DNA (or mRNA sequence) comprises the genetic code, which is read by small RNA molecules, the transfer RNA (tRNAs). The tRNA attaches to one end to a specific amino acid and displays at the other end a specific nucleotide triplet, the anticodon. This anticodon recognizes, due to base pairing, a codon in the mRNA. A stop codon is a nucleotide triplet, which has no corresponding tRNA (Alberts et al. 2014); thus, reaching the stop codon in the mRNA terminates translation. Proteins are important for development and functioning: They form parts and build the structure of an organism; perform metabolic reactions, which are necessary for life; participate in regulation as transcription factors and receptors; are key players in signal transduction pathways; and can act as enzymes to catalyze chemical reactions.

3.2.4 One Gene: One Function?

Historically, it has been assumed that each gene encodes a single function. Today though it is well-known that one gene may have different functions. For instance, some genes encode only a subunit of a protein, because several proteins consist of polypeptides encoded by different genes. In other cases, genes do not encode polypeptides, but functional RNA molecules. Furthermore, genes can encode several proteins due to alternative splicing, which is a process following the actual transcription in eukaryotes. During alternative splicing, some exons can be excluded from the pre-mRNA. Thus, different proteins can be coded by the very same gene. This implies that one gene can influence more than one and even unrelated phenotypic features in one individual (pleiotropy). On the other hand, different genes may influence the same (polygenic) phenotypic feature in one individual.

3.2.5 Categorical vs. Quantitative Traits

A trait is defined as a feature of one individual, and this feature can be characterized by an attribute of the physical appearance (e.g., feather color) or a special behavior of the individual (e.g., alarm calls). These traits may be influenced by one or many genes. A categorical trait can be present or absent (e.g., feather crest), depending on the presence or absence of specific genes or alleles. Or in the case of multiple states, trait values can be categorized, for example, as white or black or yellow plumage. In contrast to categorical traits, quantitative traits show no categories, but continuous variation such as beak length.

3.2.6 Phenotypic Plasticity

The phenotypic variation encountered within and among populations may be caused by genetic or environmental factors. Genetically controlled phenotypic variation is caused by genetic polymorphisms, though not all genetic polymorphisms, e.g., at selectively neutral loci, are imperative for phenotypic variation. However, different phenotypes may alternatively persist within a population due to variation under certain environmental conditions (Pigliucci 2001). If such environmentally induced phenotypes result in different morphs, they are referred to as polyphenisms. Polyphenisms are, thus, the result of phenotypic plasticity, which is defined as the ability of a single genotype to produce different phenotypes in different environments (Pigliucci 2001; West-Eberhard 2003). Such changes include modifications of developmental processes as well as in adult phenotypes in response to environmental stimuli. As phenotypic plasticity may quickly change phenotypic traits, it enables an organism to respond to changing environments (Merilä and Hendry 2014). Environmental variation may induce changes in behavior, morphology, or physiology, which may be transient or irreversible. More importantly, phenotypic plasticity may be adaptive or reflect nonadaptive interactions between an organism and its environment (Pigliucci 2001). If adaptive, plasticity alters the fitness of an organism under specific environmental conditions. Consequently, phenotypic plasticity may play an important role in adaptive evolution (Fusco and Minelli 2010).

It may, for instance, shield genotypes from selection, thus slowing down evolutionary rates, or, alternatively, facilitate adaptive evolution through genetic assimilation of environmentally induced phenotypes (Ghalambor et al. 2007). Furthermore, note that effects of genes and the environment may be easily confused. Some environmental conditions may produce phenotypes similar to those produced by genetic factors and vice versa. Finally, both environmental conditions and genetic constitution interact with one another to generate the best adapted phenotype (Fusco and Minelli 2010).

3.3 How Does the Genome Evolve?

3.3.1 Modification of the DNA

DNA methylation is a chemical modification of chromatin. In the methylation process, small molecules (i.e., methyl groups consisting of one carbon atom and three hydrogen atoms) attach to the DNA. If a methyl group is attached to a part of a gene, the gene will be turned off. Modifying the wrong gene or other failures can result in abnormal gene activity or false inactivity of a gene. These errors in the epigenetic processes can lead to, for example, cancer and metabolic disorders. Epigenetics examined why the expression of the gene is activated at a specific time in the development of an organism. Furthermore, it describes an inheritable phenotype, which is created from changing chromosomes without alterations of their DNA sequences (Toraño et al. 2016).

Not only DNA methylation but also acetylation and phosphorylation can result in similar changes. Acetylation is the reaction of an acetyl functional group into a chemical compound, while protein phosphorylation is a modification of a protein by which an amino acid is phosphorylated by the addition of a phosphate group. Both modifications are important in biological regulation such as gene and enzyme regulation.

Another type of modification affects the histones by the attachment of chemical compounds. These chemical compounds can be used by other proteins to decide, if a DNA sequence should be active or ignored within in a specific cell. Covalent histone modifications generate or stabilize the location of specific binding partners to chromatin, while non-covalent mechanisms provide the cell with further tools for introductory changes into the chromatin template. Chromatin remodeling and inclusion of specialized histone variation are examples for non-covalent mechanisms. However, covalent and non-covalent mechanisms can also be combined (Goldberg et al. 2007).

3.3.2 Mutation

Biological diversity would not exist without some degree of error in the hereditary process. Such errors occur from the higher level of karyotype down to the DNA base sequence. The rate at which changes occur in DNA sequences is defined as mutation rate (Alberts et al. 2014). A mutation in which one pyrimidine base is replaced by the other or in which one purine base is replaced by the other is called transition. A transversion is a mutation in which a purine base is replaced by a pyrimidine base or the other way around. A point mutation changes only a single base. A synonymous mutation appears, if the substitution does not change the amino acid sequence of the polypeptide product. This is a type of silent mutation, which is more likely to be fixed by drift. A non-synonymous mutation in a coding region does change the sequence of the amino acid and therefore the polypeptide product. This could result in either the production of a different amino acid or a nonsense or termination codon. Selectively neutral mutations occur, when changes in coding regions have no effects on the phenotype. Further modes of mutation include insertions and deletions of a single base or a sequence of bases. An insertion can be reverted by deletion of the inserted sequence, but a deletion of a sequence cannot be reverted in the absence of some mechanism to restore the lost sequence. There have been very precise repair mechanisms for billions of years, but some mutation rate remains. In most cells, this only affects the actual individual, but in germline cells (egg and sperm cells and their precursors), this leads to genotypic changes in the offspring and potentially their phenotype.

3.3.3 Selection

Selection is a process, which acts on the phenotype and can benefit individuals with a certain feature or genotype. This leads to the spread of traits beneficial to survival and reproduction while eliminating detrimental ones. Individuals with advantageous traits have a higher chance to survive and produce more offspring than individuals with unfavorable traits. A negative or purifying selection eliminates new mutations, because the phenotype is negatively affected by the mutation. If an individual with an advantageous mutation survives, it can produce more fertile offspring than individuals without the mutation (positive selection). Sexual selection is an individual's choice of mates of the other sex from the same species by preferring a presumably advantageous feature. This often led to an arms race (e.g., in plumage coloration, song variability).

3.3.4 Genetic Drift

Genetic drift is a random change in the frequency of a heritably gene variant (allele) in a given population. It may occur in all populations, but its strength is strongly dependent on population size: The smaller the population, the larger the effect of genetic drift. Thus, genetic drift does not depend on specific alleles, either beneficial or harmful. Genetic drift may even lead to the fixation of a harmful allele or the disappearance of beneficial alleles and generally reduces the genetic variation within a population or species. Genetic drift is often associated with founder effects (e.g., settlement on a small island) and population bottlenecks (e.g., glaciation reducing the inhabitable area), owing to the concomitantly reduced population size.

3.3.5 Geographic Variation and Dispersal

The physiological or morphological variation based on genetic features between populations of the same species in its whole range is called geographic variation. Geographic variation may often result from local adaptation, with specific genetic factors of a population being favored by natural selection.

Dispersal means the range expansion of a population by individuals that are adapted to new habitats or places.

3.3.6 Recombination and Migration

The exchange of genetic material either between chromosomes or different regions on the same chromosome is called recombination. Recombination creates new combinations of alleles and genes and gives rise to much of the genetic variability within populations due to different combinations in offspring compared with their parents. In sexually propagating organisms such as birds and humans, it occurs in every generation during the preparation of the germ cells (eggs and sperms). This forms the basis for adaption to changing environmental conditions.

Migration is defined as the change of gene frequency by introducing new allele or more copies of one alleles into a population by a migrant.

3.3.7 Gene Duplication

A duplication of a DNA section, which contains a gene, is defined as gene duplication. Gene duplication can occur during the processes of DNA replication and recombination or when an mRNA is converted back to DNA and new genes integrate into the genome. Gene duplication may allow for the development of a new function. Gene duplication may affect a phenotype, e.g., copies of a gene can lead to a surplus of the gene-specific protein, because the amount of a synthesized protein is regularly proportional to the present number of gene copies (Clancy and Shaw 2008).

Another type of duplication is the duplication of whole chromosomes. This process can occur during cell division when the chromosomes do not separate correctly between the two cells.

3.4 How to Study Speciation Using Genomic Features?

The first molecular markers for species delimitation and taxonomy were isozymes and allozymes. Isozymes describe different molecular forms of an enzyme, which are encoded by different loci. In contrast, allozymes characterize different molecular forms of an enzyme, produced by different alleles at the same locus (Duminil and Di Michele 2009). The term locus refers to a specific position of a gene, while the term gene is related to a DNA section, which contains the information to produce an RNA molecule. The principle approach when using allozymes or isoenzymes is to identify the variation of an enzyme among individuals using electrophoresis. However, nowadays almost exclusively DNA markers instead of protein markers are used for speciation studies because of low resolution due to synonymous mutations.

3.4.1 PCR-Based Molecular Markers

DNA markers can be **codominant** or dominant such as amplified fragment length polymorphisms (AFLP), restriction fragment length polymorphism (RFLP), and random amplified **polymorphic** DNA (RAPD) (Duminil and Di Michele 2009). AFLP studies use restriction enzymes, which digest genomic DNA, followed by the **ligation** of adapters to the **sticky ends** of the restriction fragments. A selection of the restriction fragments will be amplified with polymerase chain reaction (PCR) primers, which have a corresponding adaptor and restriction-site specific sequences. Afterward, the amplicons will be separated through electrophoresis on a gel and visualized. RFLP is a technique, which starts with the cutting process of DNA fragments by restriction enzymes, followed by a gel electrophoresis to order the DNA fragments by their length. RAPD is a special method of the PCR, because it uses short primers and the results are random DNA sequences. The gel electrophoresis shows individual patterns.

There are **codominant** molecular markers, which can be used for species delimitation and taxonomy. For most of these markers, the PCR method is used to multiply a specific DNA sequence of a sample. The method starts with the denaturation of the double-stranded DNA into single strands, called templates. Short DNA sequences, which are generally 18–20 bp long and are known as primers, bind to the templates. This step is called annealing. The next step is elongation, in which the enzyme DNA polymerase synthesizes a new DNA strand, which is complementary to the template, by adding free nucleotides to the single DNA strand. Afterward, the annealing and elongation are repeated in a definite number of cycles, until enough target DNA sequences are available (Semagn et al. 2006). This method can be used to sequence and analyze different DNA sequences for a variety of scientific questions.

3.4.1.1 Ribosomal Genes

The nuclear rDNA encodes rRNA, and both contain highly conserved and variable domains, which is a good condition for analyzing phylogenetic relationships (Hwang and Kim 1999; Patwardhan et al. 2014).

The nuclear small subunit (SSU) rDNA is a highly conserved region of the DNA, which has been used for the reconstruction of phylogenetic relationships in kingdoms, phyla, classes, and orders. The nuclear large subunit (LSU) rDNA contains more variation than the SSU rDNA, and the size of its genes varies among phyla. The LSU rDNA is used for studying genetic relationships in orders and families (Hwang and Kim 1999). Further highly conserved regions like the nuclear SSU rDNA are the 12S and 16S rDNA. They encode the ribosomal RNA, which is part of the small ribosomal subunit of a ribosome in a mitochondrion. The 12S rDNA has been used to study the phylogeny of phyla and subphyla, while the 16S rDNA has been used for analyzing the phylogenetic relationships within families and genera, because the 16S rDNA is more variable than the 12S rDNA (Hwang and Kim 1999).

3.4.1.2 Mitochondrial DNA Markers

Due to the fact that the mitochondrial DNA evolves faster than the nuclear genome, mitochondrial protein-coding regions have been used for analyzing the phylogenetic relationships within families, genera, and species (Hwang and Kim 1999). The first mitochondrial marker used was the control region, which is located in the noncoding region and is part of the regulation and initiation of the mitochondrial DNA replication and transcription (Patwardhan et al. 2014). The mitochondrial control region is variable in size and contains many variations also between individuals of the same species. Thus, it is used for studying genetic relationships in species, subspecies, and populations (Hwang and Kim 1999).

The second mitochondrial marker was the cytochrome oxidase I/II (COI/II), which is a well-known protein of an electron transport chain. In the cytochrome *c* oxidase complex, the COI and COII genes code for two polypeptide subunits. Both have been used for phylogenetic relationships among orders, families, subfamilies, genera, and species. The sequence of the COI gene is one of the sequences that can be used as a barcode for the identification of species (Patwardhan et al. 2014). DNA barcoding is a method to identify species by using short sequences.

Further widely used mitochondrial markers to reconstruct the phylogeny among genera and species are the cytochrome *b* (cytb) and NADH dehydrogenase 2 (nd2) genes.

3.4.1.3 Microsatellites

A microsatellite is a specific DNA motif with a length of two to six base pairs (Fig. 3.3). Microsatellites are used to detect the number of repeats of a sequence to identify an individual. Similar to microsatellites are minisatellites, but their repeat motifs are longer. Microsatellites can be amplified by PCR, for which labeled primers are needed, followed by analyzing the length of the fragment (microsatellite). A large advantage is the small amount of DNA needed for the PCR. Microsatellites are locus-specific, **codominant**, and highly **polymorphic**. A disadvantage of microsatellites is their taxon-specificity. Thus microsatellite libraries need to be generated for each species or closely related sister species (Delaney 2014). Microsatellites are currently mainly used for paternity tests and population genetics but hold large potential for speciation studies due to their potential to distinguish lineages within a species. It is necessary to work with more than one microsatellite locus to have reliable results.

3.4.2 Expressed Sequence Tags

Genes must be converted into mRNA, but RNA is unstable outside the cell. Hence, mRNA needs to be converted into complementary DNA (cDNA) by the reverse transcriptase enzyme. The production of cDNA is the reverse process of transcription, because mRNA is used as the template instead of the DNA. cDNA is more stable than mRNA and contains generally only exons due to splicing of the pre-mRNA. This means that cDNA represents an expressed gene or a part of it. When the cDNA has been isolated, various nucleotides can be sequenced to create expressed sequence tags (ESTs) with a length of 100–800 bp. They allow the discovery of unknown genes and a comparison between different species due to high conservation in the coding regions (Semagn et al. 2006). From ESTs it is possible to develop primer pairs for sequencing genes in other species and to detect single nucleotide polymorphisms (SNPs) (Schlötterer 2004; Semagn et al. 2006).

3.4.3 Single Nucleotide Polymorphisms

A single nucleotide polymorphism (SNP) is the change of a single base in the DNA sequence (Fig. 3.3) (Semagn et al. 2006). Generally, two different nucleotides can be found per position, and SNPs mostly occur in noncoding regions (Grover and Sharma 2016). The simplest method to identify SNPs is to screen a high-quality DNA sequence or an EST. The most common methods like restriction-site-associated DNA sequencing (RAD-seq) and genotyping by sequencing (GBS) will be explained in the following two sections. A comprehensive strategy for detecting SNPs in a genome is the generation of shotgun genome sequences. For this method,

	SNP 1 / SNP 2 / Microsatellite 1
Individual 1	AGTCTTGATCGAATCATGACTGACTCAGGTCCTAGTAACTCACACACACACAGTCAACTGA
Individual 2	AGTCTTGATCGAATCATGACTGACTCTGGTCCTAGTAACTCACACACACACAGTCAACTGA
Individual 3	AGTCTTGATCGAATTATGACTGACTCAGGTCCTAGTAACTCACACACACACACAGTCAACTGA
Individual 4	AGTCTTGATCGAATCATGACTGACTCTGGTCCTAGTAACTCACACACACACACAGTCAACTGA
Individual 5	AGTCTTGATCGAATTATGACTGACTCTGGTCCTAGTAACTCACACACACACAGTCAACTGA
Individual 6	AGTCTTGATCGAATTATGACTGACTCAGGTCCTAGTAACTCACACACACAGTCAACTGA
Individual 7	AGTCTTGATCGAATTATGACTGACTCTGGTCCTAGTAACTCACACACACAGTCAACTGA
Individual 8	AGTCTTGATCGAATCATGACTGACTCTGGTCCTAGTAACTCACACACACACAGTCAACTGA
Individual 9	AGTCTTGATCGAATCATGACTGACTCAGGTCCTAGTAACTCACACACACACAGTCAACTGA
Individual 10	AGTCTTGATCGAATTATGACTGACTCAGGTCCTAGTAACTCACACACACACAGTCAACTGA

Fig. 3.3 Ten individuals from one population are represented with fictional sequences. In these sequences, two single nucleotide polymorphisms (SNPs) and one microsatellite occur. The first SNP is a variation of the bases cytosine (C) and thymine (T). The individuals 1, 2, 4, 8, and 9 carry base C, while the other individuals have a T at the same position. In the individuals 1, 3, 6, 9, and 10, an adenine (A) appears as the second base, whereas the individuals 2, 4, 5, 7, and 8 have the base T on this position. The microsatellite in this example is a repetition of two bases, C and A. In the individuals 1, 2, 5, 8, 9, and 10, it is 12 bases long $(CA)_6$. In the individuals 3 and 4, the microsatellite is 14 bp long $(CA)_7$, while it is shorter in the individuals 6 and 7 $(CA)_5$

a pool of DNA from different individuals should be sequenced. A more efficient approach is the shotgun sequencing with a reduced section of the genome, in which the DNA of many different individuals can be sequenced (Schlötterer 2004). Most of these methods are cost- and time-intensive and the information content of one SNP is very low, but they have a low mutation rate (high stability) and high frequency in the genome, and new analytical methods are being developed and open up new opportunities.

SNPs can be used in different research questions, e.g., investigate about natural selection across species (Künstner et al. 2010), examine recent divergence (McCormack et al. 2012), explore the genetic structure of different morphological features in different species (Silva et al. 2017), and investigate hybridization (Manthey et al. 2016).

3.4.4 Restriction-site-associated DNA sequencing

Restriction-site-associated DNA sequencing (RAD-seq) is the genotyping of short DNA fragments, which are adjacent to the cut site of a restriction enzyme (RE). The first step of RAD-seq is the digestion of the genomic DNA with a chosen RE, followed by the ligation of an adapter (P1) to the overhang of the RE (Baird et al. 2008; Davey and Blaxter 2011). This adapter contains a binding site for the forward primer and a barcode for the sample identification. After ligation, the fragments are pooled and size selected (Baird et al. 2008). The DNA fragments are then ligated to a second adapter (P2), which has a reverse primer site and is a Y adapter with divergent ends (Coyne et al. 2004; Baird et al. 2008). The reason for choosing a Y adapter is that all fragments contain the P1 adapter, because the P2 adapter cannot bind to the reverse primer, before the amplification of the P1 adapter has been finished (Baird et al. 2008; Davey and Blaxter 2011). After ligation of the second adapter, a PCR reaction is performed. The PCR-products are used for next-generation sequencing (3.4.7) (Baird et al. 2008). The resulting reads are trimmed, grouped by barcodes, and mapped to a reference genome or, if no reference genome is available, the same reads are aligned for identifying SNPs (Baird et al. 2008; Davey and Blaxter 2011). The challenges of RAD-seq are the high costs of sequencing and the diversity of RAD-seq protocols with different technical details. Nevertheless, one can choose the protocol most suitable for the own study system or research question (Andrews et al. 2016). RAD-seq can identify and generate thousands of genetic markers, reduces the complexity of the genome, and can be used for species with no or limited existing sequence data (Davey and Blaxter 2011). Furthermore, RAD-seq was extended to use two REs instead of one RE to exclude the step of size selection. This method is called double digest RAD-seq (Peterson et al. 2012).

3.4.5 Genotyping by sequencing

Genotyping by sequencing (GBS) is a highly multiplexed approach for constructing reduced representative libraries for the Illumina next-generation sequencing platform to discover a large number of SNPs. This approach can be used for any species at a low per-sample cost and also incorporates restriction enzymes (RE) to reduce genome complexity (Elshire et al. 2011; Chung et al. 2017). The procedure of GBS like RAD-seq starts with the digestion of DNA by an RE. The selected REs should be suitable for the investigated species by containing an overhang of two to three base pairs, and REs do not cut frequently in the major repetitive fraction of the genome. After the digestion, two adapters are ligated to the ends of the digested DNA. The adapters should be complementary to the overhang of the chosen RE, and one adapter contains a barcode for multiplex sequencing. These adapters contain binding sites for appropriate primers, which are added to perform a PCR reaction to increase the amount of DNA fragments. The PCR products are cleaned up and DNA fragments with a specific size result in a library. Libraries are used for sequencing, followed by filtering reads, which match one of the barcodes and the corresponding cut site of the RE, and are not adapter dimers. These sorted reads are separated by their barcode and after separation the barcode is removed. The filtered reads are mapped to the reference genome, consequently reads, which mapped on the same position are aligned and used to identify SNPs (Elshire et al. 2011). GBS is a cost-effective method to discover SNPs, genotype individuals within a population, and detect molecular markers. The disadvantages are the management of big datasets and the fact that the data do not represent the whole genome, which could have a negative effect on constructing genetic maps (Chung et al. 2017).

3.4.6 Transcriptomics

This is a technique to study an organism's transcriptome, which is the total of all its RNA transcripts. The transcriptome is a snapshot at a specific time of all transcripts in one cell or tissue, for a specific developmental stage. These expressed genes of one organism in different cells, tissues, conditions, or time points give details about the function of uncharacterized genes and the biology of organisms. Furthermore, the comparison of transcriptomes allows the identification of genes, which are expressed in different cells; hence, it gives information about gene regulation. There are two techniques to create a transcriptome: microarrays and RNA-Seq. The microarray approach quantifies a set of predefined sequences, while the RNA-Seq technique uses next-generation sequencing to target "all" expressed genes (Wang et al. 2009).

3.4.7 "Whole" Genome Sequencing

Next-generation sequencing (NGS) is a method to produce a large number of reads of short DNA sequences, between 50 and 150 bp long. The read length of NGS is often short with a high error rate, but this is compensated due to a higher **coverage** of the consensus sequence (Scanes 2015). These reads can be combined to continuous sequences (contigs), and contigs can be in turn linked to scaffolds. Indications about the quality of contigs and scaffolds (genome assemblies) can be provided by the N50 value, which represents the minimum length of long sequences that make up half of the **assembly** of contigs or scaffolds (Kapusta and Suh 2017). Contigs and scaffolds can be used to identify genes, but there are sequences which have no genetic information, which are clustered in chromosome Unknown (chrUn). Annotation is the process of linking DNA reads to information available from previous work (on other taxa) (Scanes 2015).

3.4.7.1 Different Strategies for Sequencing Genomes

The traditional Sanger sequencing with 1-kb-long sequence reads and the Roche 454 sequencing with up to 800 bp sequence reads have been largely replaced by short-read technologies such as Illumina HiSeq with 150 bp sequence reads. There are also even newer technologies available such as Pacific Biosciences with up to 5 kb sequence reads or Ion Torrent with about 500 bp sequence reads (Ekblom and Wolf 2014). The technology of 10× genomics uses short reads from Illumina sequencing to link the short reads to long molecules. In the long molecules, variation can be detected to identify which reads belong to the father or mother of the examined individual. Another method uses single molecules by detecting them and sequencing their DNA. This is called single-molecule genomics.

One of the most common strategies for genome sequencing is the shotgun sequencing. First, DNA is cut into small random fragments, whereby the size of the fragments depends on the technology used. These fragments will be assembled to a longer contig. This process is known as de novo **assembly**. It is important that there is enough overlap between the sequence reads for a correct **assembly**, and this implies also a high **coverage**. If there are longer fragments like several hundred base pairs, both ends of the sequence will be sequenced called paired-end sequencing. Afterward, the resulted contigs are connected to longer sequences (scaffolds) (Ekblom and Wolf 2014).

The genome annotation uses the whole genome sequences in combination with relevant information from gene models, functional information, microRNA, or epigenetic modifications. Consequently, a lack of genomic information will result in low annotation rates. Annotation describes the process of using data of other genomes or transcriptomes to detect genes or transcripts on the newly assembled genome (Ekblom and Wolf 2014).

3.4.7.2 Limitations of Analyzing Genomes

Usually, a genome draft represents the complete nucleotide base sequence for all chromosomes in one species. Nevertheless, there is not just one sequence for a species, due to individual genomic variation, differences among cells within individuals due to diploidy. Thus, the assembled reference genome sequence of one individual will only comprise a subset of the total variation present within a species. Typically, one individual is sequenced, but sometimes a genome is based on a consensus of a few individuals (Ekblom and Wolf 2014). Furthermore, it is not possible to sequence and assemble all nucleotides in the genome due to sequencing errors (Scanes 2015), and most genome **assembly** methods fail on repetitive elements, which are typically not included in reference genomes (Hoban et al. 2016). However, repetitive regions may be characterized through the annotation of a comprehensive dataset compounded of a high-**coverage** single molecule real-time sequencing **assembly**, an assembled optical map, and a generated high-**coverage** short-read sequence **assembly** to a repeat library (Weissensteiner et al. 2017).

3.4.8 Epigenome

In almost all cells of an individual, the same DNA sequence can be found, but nevertheless cells may differ as the information content encoded within the DNA may be used differently. Such differences may arise from chemical modifications of the DNA or histone proteins without changing the DNA sequence. The resulting epigenome includes chemical compounds, which have been added to the DNA to regulate gene activity. These chemical compounds are not part of, but fixed to, the DNA. Epigenomic changes occur in individual development and tissue differentiation and may result in cell division, and, in some circumstances, they can be transferred to the next generation. However, the epigenome can also be influenced by environmental conditions, such that the epigenome may vary between individuals. Through epigenetic changes, genes can be turned off or on (expression), thus determining the production of proteins in specific cells. For example, the eye is specialized for light-sensitive proteins and red blood cells for carrying oxygen. Furthermore, epigenetic changes in DNA and histones play a role in regulatory pathways of eukaryotes (Marshall Graves 2015).

3.5 Closing Words

Speciation is one of the main focusses in evolutionary biology and also the starting point to clarify the relationship between species. Morphological traits and reproduction are important for characterizing one species, but over the last decades, genetic tools got more and more influence in the delimitation of species. Therefore, it is necessary to understand the structure and function of the used genetic material. The genetic investigation of speciation began with short sequences and few genes for small sample sizes. Nowadays, more individuals of one species and additionally more species can be examined. Furthermore, SNPs, transcriptomes, and whole genomes are the newest traits to analyze and understand speciation—also in functional respect. However, methods will be further developed to become more cost-effective, faster, and more informative.

Acknowledgment We thank Klaus Fischer for proofreading this chapter.

References

Alberts B, Johnson A, Lewis J, Morgan D, Raff M, Roberts K, Walter P (2014) Molecular biology of the cell. Garland Science, Taylor & Francis Group

Andrews KR, Good JM, Miller MR, Luikart G, Hohenlohe PA (2016) Harnessing the power of RADseq for ecological and evolutionary genomics. Nat Rev Genet 17(2):81–92

Baird NA, Etter PD, Atwood TS, Currey MC, Shiver AL, Lewis ZA, Selker EU, Cresko WA, Johnson EA, Fay JC (2008) Rapid SNP discovery and genetic mapping using sequenced RAD markers. PLoS One 3(10):e3376

Chung YS, Choi SC, Jun T-H, Kim C (2017) Genotyping-by-sequencing: a promising tool for plant genetics research and breeding. Hortic Environ Biotechnol 58(5):425–431

Clancy S, Shaw KM (2008) DNA deletion and duplication and the associated genetic disorders. Nat Educ 1:23

Coyne KJ, Burkholder JM, Feldman RA, Hutchins DA, Cary SC (2004) Modified serial analysis of gene expression method for construction of gene expression profiles of microbial eukaryotic species. Appl Environ Microbiol 70(9):5298–5304

Davey JW, Blaxter ML (2011) RADSeq: next-generation population genetics. Brief Funct Genomics 9(5–6):416–423

Delaney KS (2014) Landscape genetics of urban bird populations. In: Gil D, Brumm H (eds) Avian urban ecology. Oxford University Press

Duminil J, Di Michele M (2009) Plant species delimitation: a comparison of morphological and molecular markers. Plant Biosyst 143:528–542. https://doi.org/10.1080/11263500902722964

Ekblom R, Wolf JBW (2014) A field guide to whole-genome sequencing, assembly and annotation. Evol Appl 7:1026–1042. https://doi.org/10.1111/eva.12178

Ellegren H (2005) The avian genome uncovered. Trends Ecol Evol 20:180–186. https://doi.org/10.1016/j.tree.2005.01.015

Ellegren H (2013) The evolutionary genomics of birds. Annu Rev Ecol Evol Syst 44:239–259. https://doi.org/10.1146/annurev-ecolsys-110411-160327

Elshire RJ, Glaubitz JC, Qi S, Poland JA, Kawamoto K, Buckler ES, Mitchell SE, Orban L (2011) A robust, simple genotyping-by-sequencing (GBS) approach for high diversity species. PLoS One 6(5):e19379

Fusco G, Minelli A (2010) Phenotypic plasticity in development and evolution: facts and concepts. Introduction. Philos Trans R Soc Lond Ser B Biol Sci 365:547–556. https://doi.org/10.1098/rstb.2009.0267

Ghalambor CK, McKay JK, Carroll SP, Reznick DN (2007) Adaptive versus non-adaptive phenotypic plasticity and the potential for contemporary adaptation in new environments. Funct Ecol 21(3):394–407

Gill F, Donsker D (eds) (2017) IOC world bird list (v7.1). https://doi.org/10.14344/IOC.ML.7.1

Goldberg AD, Allis CD, Bernstein E (2007) Epigenetics: a landscape takes shape. Cell 128:635–638. https://doi.org/10.1016/j.cell.2007.02.006

Griffin DK, Robertson LBW, Tempest HG, Skinner BM (2007) The evolution of the avian genome as revealed by comparative molecular cytogenetics. Cytogenet Genome Res 117:64–77. https://doi.org/10.1159/000103166

Grover A, Sharma PC (2016) Development and use of molecular markers: past and present. Crit Rev Biotechnol 36:290–302. https://doi.org/10.3109/07388551.2014.959891

Hillier LW, Miller W, Birney E, Warren W, Hardison RC, Ponting CP, Bork P, Burt DW, Groenen MAM, Delany ME, Dodgson JB, Chinwalla AT, Cliften PF, Clifton SW, Delehaunty KD, Fronick C, Fulton RS, Graves TA, Kremitzki C, Layman D, Magrini V, McPherson JD, Miner TL, Minx P, Nash WE, Nhan MN, Nelson JO, Oddy LG, Pohl CS, Randall-Maher J, Smith SM, Wallis JW, Yang S-P, Romanov MN, Rondelli CM, Paton B, Smith J, Morrice D, Daniels L, Tempest HG, Robertson L, Masabanda JS, Griffin DK, Vignal A, Fillon V, Jacobbson L, Kerje S, Andersson L, Crooijmans RPM, Aerts J, van der Poel JJ, Ellegren H, Caldwell RB, Hubbard SJ, Grafham D V., Kierzek AM, McLaren SR, Overton IM, Arakawa H, Beattie KJ, Bezzubov Y, Boardman PE, Bonfield JK, Croning MDR, Davies RM, Francis MD, Humphray SJ, Scott CE, Taylor RG, Tickle C, Brown WRA, Rogers J, Buerstedde J-M, Wilson SA, Stubbs L, Ovcharenko I, Gordon L, Lucas S, Miller MM, Inoko H, Shiina T, Kaufman J, Salomonsen J, Skjoedt K, Wong GK-S, Wang J, Liu B, Wang J, Yu J, Yang H, Nefedov M, Koriabine M, deJong PJ, Goodstadt L, Webber C, Dickens NJ, Letunic I, Suyama M, Torrents D, von Mering C, Zdobnov EM, Makova K, Nekrutenko A, Elnitski L, Eswara P, King DC, Yang S, Tyekucheva S, Radakrishnan A, Harris RS, Chiaromonte F, Taylor J, He J, Rijnkels M, Griffiths-Jones S, Ureta-Vidal A, Hoffman MM, Severin J, Searle SMJ, Law AS, Speed D, Waddington D, Cheng Z, Tuzun E, Eichler E, Bao Z, Flicek P, Shteynberg DD, Brent MR, Bye JM, Huckle EJ, Chatterji S, Dewey C, Pachter L, Kouranov A, Mourelatos Z, Hatzigeorgiou AG, Paterson AH, Ivarie R, Brandstrom M, Axelsson E, Backstrom N, Berlin S, Webster MT, Pourquie O, Reymond A, Ucla C, Antonarakis SE, Long M, Emerson JJ, Betrán E, Dupanloup I, Kaessmann H, Hinrichs AS, Bejerano G, Furey TS, Harte RA, Raney B, Siepel A, Kent WJ, Haussler D, Eyras E, Castelo R, Abril JF, Castellano S, Camara F, Parra G, Guigo R, Bourque G, Tesler G, Pevzner PA, Smit A, Fulton LA, Mardis ER, Wilson RK. 2004. Sequence and comparative analysis of the chicken genome provide unique perspectives on vertebrate evolution. Nature 432:695–716. DOI: https://doi.org/10.1038/nature03154.

Hoban S, Kelley JL, Lotterhos KE, Antolin MF, Bradburd G, Lowry DB, Poss ML, Reed LK, Storfer A, Whitlock MC (2016) Finding the genomic basis of local adaptation: pitfalls, practical solutions, and future directions. Am Nat 188:379–397. https://doi.org/10.1086/688018

Hughes AL, Piontkivska H (2005) DNA repeat arrays in chicken and human genomes and the adaptive evolution of avian genome size. BMC Evol Biol 5:12. https://doi.org/10.1186/1471-2148-5-12

Hwang U-W, Kim W (1999) General properties and phylogenetic utilities of nuclear ribosomal DNA and mitochondrial DNA commonly used in molecular systematics. Korean J Parasitol 37:215–228. https://doi.org/10.3347/kjp.1999.37.4.215

Kapusta A, Suh A. 2017. Evolution of bird genomes – a transposon's-eye view. 1389:164–185. DOI: https://doi.org/10.1111/nyas.13295.

Künstner A, Wolf JBW, Backström N, Whitney O, Balakrishnan CN, Day L, Edwards SV, Janes DE, Schlinger BA, Wilson RK, Jarvis ED, Warren WC, Ellegren H (2010) Comparative genomics based on massive parallel transcriptome sequencing reveals patterns of substitution and selection across 10 bird species. Mol Ecol 19(Suppl 1):266–276. https://doi.org/10.1111/j.1365-294X.2009.04487.x

Lesk AM (2012) Introduction to genomics. Oxford University Press

Manthey JD, Robbins MB, Moyle RG (2016) A genomic investigation of the putative contact zone between divergent Brown Creeper (*Certhia americana*) lineages: chromosomal patterns of genetic differentiation. Genome 59:115–125. https://doi.org/10.1139/gen-2015-0093

Marshall Graves JA (2015) Evolution of vertebrate sex chromosomes and dosage compensation. Nat Rev Genet 17:33–46. https://doi.org/10.1038/nrg.2015.2

Matsubara K, Tarui H, Toriba M, Yamada K, Nishida-Umehara C, Agata K, Matsuda Y (2006) Evidence for different origin of sex chromosomes in snakes, birds, and mammals and step-wise differentiation of snake sex chromosomes. Proc Natl Acad Sci U S A 103:18190–18195. https://doi.org/10.1073/pnas.0605274103

McCormack JE, Maley JM, Hird SM, Derryberry EP, Graves GR, Brumfield RT (2012) Next-generation sequencing reveals phylogeographic structure and a species tree for recent bird divergences. Mol Phylogenet Evol 62:397–406. https://doi.org/10.1016/j.ympev.2011.10.012

Merilä J, Hendry AP (2014) Climate change, adaptation, and phenotypic plasticity: the problem and the evidence. Evol Appl 7(1):1–14

Patwardhan A, Roy A, Ray S (2014) Molecular markers in phylogenetic studies – a review. J Phylogenet Evol Biol 2. https://doi.org/10.4172/2329-9002.1000131

Peterson BK, Weber JN, Kay EH, Fisher HS, Hoekstra HE, Orlando L (2012) Double digest RADseq: an inexpensive method for de novo SNP discovery and genotyping in model and non-model species. PLoS One 7(5):e37135

Pigliucci M (2001) Phenotypic plasticity: beyond nature and nurture. JHU Press, Baltimore

Scanes CG (2015) Sturkie's avian physiology. Elsevier

Schlötterer C (2004) The evolution of molecular markers – just a matter of fashion? Nat Rev Genet 5:63–69

Semagn K, Bjornstad A, Ndjiondjop MN (2006) An overview of molecular marker methods for plants. Afr J Biotechnol 5:2540–2568. https://doi.org/10.5897/AJB2006.000-5110

Silva CNS, McFarlane SE, Hagen IJ, Rönnegård L, Billing AM, Kvalnes T, Kemppainen P, Rønning B, Ringsby TH, Sæther B-E, Qvarnström A, Ellegren H, Jensen H, Husby A (2017) Insights into the genetic architecture of morphological traits in two passerine bird species. Heredity 119:197–205. https://doi.org/10.1038/hdy.2017.29

Toraño EG, Garciá MG, Fernández-Morera JL, Ninõ-Garciá P, Fernández AF (2016) The impact of external factors on the epigenome: in utero and over lifetime. Biomed Res Int. https://doi.org/10.1155/2016/2568635

Wang Z, Gerstein M, Snyder M (2009) RNA-Seq: a revolutionary tool for transcriptomics. Nat Rev Genet 10:57–63. https://doi.org/10.1038/nrg2484

Weissensteiner MH, Pang AWC, Bunikis I, Höijer I, Vinnere-Petterson O, Suh A, Wolf JBW (2017) Combination of short-read, long-read, and optical mapping assemblies reveals large-scale tandem repeat arrays with population genetic implications. Genome Res 27:697–708. https://doi.org/10.1101/gr.215095.116

West-Eberhard MJ (2003) Developmental plasticity and evolution. Oxford University Press, Oxford

Chapter 4
Morphological Variation in Birds: Plasticity, Adaptation, and Speciation

Till Töpfer

Abstract The huge diversity of phenotypes and associated geographic patterns has made birds prime examples for studies in speciation. For this purpose, morphological approaches were first choice to assess the degree of relatedness between species and their intraspecific variation for centuries, until molecular genetic studies seriously challenged traditional morphology-based conclusions. However, the current development of multivariate statistics and the ease to blend morphological, phylogenetic, and ecological insight has gradually led to a reconsideration of morphology as a valuable tool for ornithological research. This chapter reviews the most important aspects of morphological variation in birds, how its plasticity can be assessed and to which extent phenotypic variation can be incorporated into a broader evolutionary framework that explains modifications of the avian body in the light of speciation processes.

Keywords Avian morphology · Phenotypic plasticity · Trait evolution · Geographic variation · Eco-morphology

4.1 General Aspects of Phenotypic Variation in Birds

Studying speciation means to analyze the variation of characters, no matter if on molecular or organismic level. In birds, a great deal of such variation concerns phenotypic traits, making morphological assessments an important tool to collect evidence both for identification and classification. Identification of birds is intuitively associated with their external appearance. In contrast to many other animal groups, birds are commonly perceived as more or less readily distinguishable on grounds of size, shape, coloration, song, and other behavior. Being abundant and popular for centuries, birds remain to be subjects of both popular pastime and scientific

---

T. Töpfer (✉)
Zoological Research Museum Alexander Koenig, Centre for Taxonomy and Evolutionary Research, Section Ornithology, Bonn, Germany
e-mail: t.toepfer@leibniz-zfmk.de

D. T. Tietze (ed.), *Bird Species*, Fascinating Life Sciences,
https://doi.org/10.1007/978-3-319-91689-7_4

occupation (Birkhead et al. 2014). At the core of this lies the study of differentiation of bird populations using comparative approaches. Although the distinction of certain species might differ if different conceptual principles are applied, phenotypic characters persist as relevant criteria for the delimitation of bird species.

In contrast to the common and oftentimes inevitable use of identification keys, as in most invertebrates, there is no comprehensive identification key of birds to date. A major reason is that in no other vertebrate group the legacy of ancient pictorial accounts remains as strongly as in birds—the huge number of contemporary identification guides, regardless of being in printed or digital form, speaks for itself. It is a compelling thought that such guides effectively combine depictions and descriptions in a time-honored way that have their roots in the narrative accounts of the first popular books of natural history. Even though modern analytical methods, spearheaded by molecular genetics, have shaken our understanding of species boundaries in birds, we still strongly rely on phenotypic assessment in species delimitation and classification. So, why are the external characteristics of birds so decisive that they still form a touchstone of species identity?

First and foremost, it is the enormous range of morphological variation in birds despite their relatively similar body plan. Even on lower taxonomic level, with resembling appearances of closely related taxa as a rule, the diversity of structural and coloration patterns is striking. Since many of these differentiations can be interpreted as functional adaptations to a certain lifestyle and/or to environmental conditions, they tell a lot about ecological causes of phenotypic differentiation. This, in turn, fosters our understanding of the evolutionary constraints that may eventually lead to speciation. Considering the current development of multidisciplinary approaches in evolutionary research, it is worthwhile to have a detailed look at the different aspects of morphological variation revolving around the problem of speciation in birds.

4.2 The Historical Role of Morphological Criteria for Species Delimitation

Morphology in its broadest sense, i.e., comprising all manifestations of structure and form, has been most influential for the classification of biological beings over centuries. As a basic ordering principle, morphology was applied to organize a seemingly chaotic variety of life forms on higher ("macrotaxonomic") levels first (Mayr 1982). When the scientific interest focused on lower, species-level ("microtaxonomic") classification, the desire grew to also reflect phylogenetic relationships in taxonomy. In any case, morphological approaches remained unrivalled to assess the degree of relatedness between species and their intraspecific variation until the 1990s, when techniques of molecular genetics, particularly DNA sequencing, became broadly available. This held true for ornithology as well. Except for bioacoustics (Baptista and Kroodsma 2001), that had a substantial influence on testing species boundaries experimentally (at least in vocally active birds), there was hitherto no other noteworthy method to gather relevant evidence (Box 4.1).

The advent of molecular methods and the rapidly growing computational power enabled a previously nonexistent phylogenetic interpretation of extensive data sets and stimulated an extremely fruitful discourse about speciation in birds by challenging various traditional morphology-based classifications. During the subsequent phase of rapid upheavals in avian systematics that also had a massive impact on the huge ornithological community beyond academic circles, the importance of morphology for species identification and classification has been widely challenged. This was actually a renewed flaring-up of the old discussion to which extent morphology actually reflects phylogeny. However, having meanwhile entered the era of genomics that produces fascinating deeper insight into avian evolution, it turned out that interdisciplinary approaches are the means of choice to further elucidate the riddles of speciation in birds. The development of multivariate statistics and the ease to blend morphological, phylogenetic, and ecological parameters has gradually led to a reconsideration of morphology as a valuable tool for ornithological research.

Box 4.1 The Importance of Ornithological Collections

Bird collections, either in museums or in private hands, have been the cradle of comparative studies on character variation from the beginning of scientific ornithology. Providing a well-documented and geographically widespread material basis for analyses of differentiation patterns, they have fueled the debates about species delimitation and conceptual issues of speciation and classification (Bruce 2003). Having previously been considered sources for morphological and geographical information in the first place, the use of ornithological collections has experienced a major extension when bird specimens have been shown to represent invaluable resources for molecular genetic research as well. As a consequence, many bird collections store additional samples for molecular genetic analyses together with classic skin or skeleton specimens nowadays. Considering the growing interest in integrative research approaches (e.g., Alström et al. 2016), combining studies of variation on very different biological levels, ornithological collections will continuously contribute to scientific progress by further archiving and documenting vouchers of contemporary avian diversity (Webster 2017).

4.3 Phenotypic Variation and Plasticity of Characters

The fascinating diversity of their phenotypes and associated geographic patterns has made birds prime examples for studies in speciation. However, like in other groups of animals, there is not always an immediate connection between external appearance and species identity. In order to correctly distinguish between interspecific differences and intraspecific variation, it is inevitable to scrutinize the extent of geographic and individual variability within and between potential species (Mayr 1969). As the limits of such variation may shift temporarily as a result of adaptive processes (e.g., Grant and Grant 2014), character plasticity has to be considered likewise. Of course, this does not only concern morphological traits. A plethora of molecular studies has shown that the variation of genetic characters (i.e., DNA sequence polymorphisms) is a powerful tool to define groups of organisms even on a very fine scale. This is especially valuable when combining morphological and molecular evidence. In the following, however, only the variation of external characters will be in focus.

Geographic Variation Geographic variation refers to the differentiation between different populations (Mayr 1969). Obviously, geographic variation covers all aspects of phenotypic differentiation on various geographic scales (Cicero and Remsen 2007). There is no rule as to how large geographic distances have to be until morphological differentiation sets in, but evidently at least some (historical) amount of geographic isolation (vicariance) appears necessary (Price 2008). Consequently, geographic variation often is not only pronounced in birds with a wide distribution but also in those that inhabit ecologically fragmented ranges on smaller geographic scales. Traditionally, populations of these polytypic species are grouped as subspecies based on the particular composition of morphological traits at a given locality (Winker and Haig 2010).

Geographic differentiation can either be continuous (clinal) or discontinuous. While the latter is commonly found in patterns of insular distributions or wide geographic range separations (disjunct distribution), the former frequently occurs in more or less continuous continental ranges. Clinal variation, however, can well be incongruent between different phenotypic traits indeed. As a result of such incongruently directed character changes, each bird population at any intersection of different clines represents a certain morphological entity with a particular trait combination.

As with other forms of morphological variability, geographic variation is not restricted to particular traits and can basically affect any aspect of a bird's physique. What makes it even more interesting in evolutionary respect is the often-found correlation between geographic variation and ecogeographic rules (e.g., Sun et al. 2017), indicating the selective effect of environmental conditions on the avian body. Moreover, eco-morphological insight, such as the connection between wing structure and flight performance or migratory distance, may be likewise reflected in inter- and intraspecific geographic variation (Winkler and Leisler 2005).

Individual Variation By definition, individual variation comprises all phenotypic manifestations at the same locality (Mayr 1969). Such local variation, however, does not necessarily need to be less pronounced than geographic variation. There are different forms of individual variation and all of these forms may be represented by interindividually variable characters again. Moreover, the forms of individual variation are not exclusive as they might occur even in the very same individual at different ages or seasons. The most frequent forms of individual variation in birds are sexual, age-dependent, and seasonal variation, complemented by local eco-morphological differentiation within populations.

Sexual Variation Sexual variation includes all aspects of phenotypic differentiation between the two sexes. This is of course most obvious in those birds that exhibit a pronounced sexual dimorphism, be it in size, coloration (patterns), or in both. Sexual dimorphism does not need to be constant throughout the year, as more conspicuous breeding plumages might differ more strongly between the sexes than non-breeding plumages, indicating the overlap with seasonal variation. In addition, sexual dimorphism may not only be due to sexual selection alone but might also have an ecological component. In many raptors (Falconidae, *Accipiter*), there is a substantial size difference between the two sexes with the larger females apparently exploiting a different food spectrum than the smaller males. Within females and males, again, there is variation in the respective characters that defines the extent of sex-specific characteristics. Other special cases are those birds in which the external appearance of one sex is strongly correlated with specific aspects of the mate choice process and where strong sexual selection leads to extreme phenotypic diversity within one sex. The prime example is the Ruff *Philomachus pugnax*, whose males are individually distinct in plumage patterns and coloration. But there is not only the males' extreme plumage variability (that can be subdivided into main character groups) but also the existence of female-mimicking males that also form an intermediate size group in wing length between "proper" males and females (Jukema and Piersma 2006).

Age-dependent Variation Regarding age-dependent variation, a broad range of character differences have to be considered. As far as the feather coat is concerned, consecutive plumage generations of the same individual are involved. A precise knowledge of molt sequence is thus necessary to properly assess this kind of phenotypic variation (e.g., Jenni and Winkler 1994). It is the nature of things that during molt transitional plumage stages occur and that particularly in long-lived birds (e.g., eagles and vultures) persistent composite plumages exist for years that comprise feathers of several generations. As a rule of thumb, a distinction can be made between plumages worn before becoming reproductively active (juvenile and so-called immature plumages) and those worn by reproductively active birds (adult plumages). Nonreproductive plumages might differ in terms of dimensions, structure, and coloration and patterns in being shorter, differently shaped (particularly in the large flight feathers), appearing more fluffy and being less strikingly colored and patterned. In addition, it is commonly the case that nonreproductive plumages are very similar between sexes even if the adult plumage is clearly sexually dimorphic. Moreover, there is growing evidence that bird plumages may also be subject to senescence ("progressive greying," van Grouw 2013). As birds apparently remain reproductively active until their death, however, this stage has to be considered a representation of the adult plumage as well. But not only plumage traits are subject to age-dependent variation. Other physical features like iris, leg, or bill color as well as structural differences because of not yet full-grown characters (e.g., bill size and shape) might be concerned likewise.

Seasonal Variation Another often-found form of variation is seasonal: In the course of the year, many birds acquire two different plumages of which one usually is more conspicuous. Since in most cases such conspicuousness is explained by the demands of mate choice and breeding, the breeding plumage tends to be more prominent – at

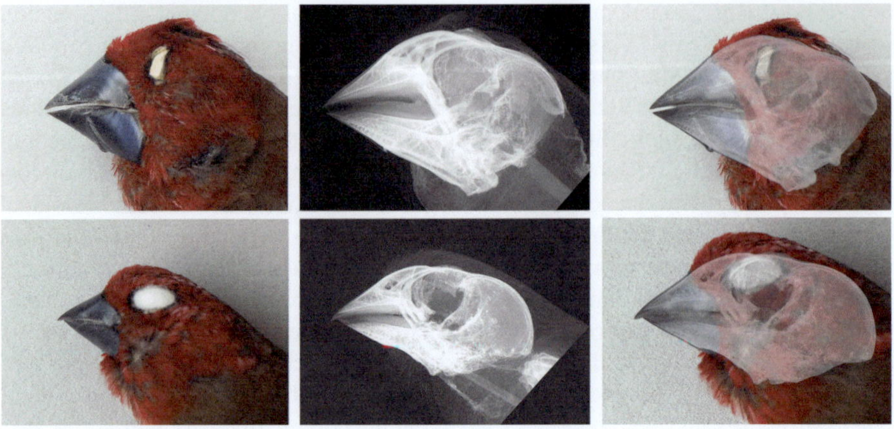

Fig. 4.1 Variation of bill dimensions in the *Pyrenestes* seedcrackers (specimens from the Zoological Research Museum A. Koenig). Top row, *P. ostrinus*; lower row, *P. minor*. Specimen profiles are shown in external view, in X-ray view, and as a superimposition of external and X-ray views, respectively. Such extreme intraspecific variation of an external character is also reflected in the internal anatomy as obvious from the X-ray pictures of the individual skulls

least in one sex (cf. sexual variation). In the more unobtrusively colored or patterned non-breeding plumages, possible phenotypic differences between males and females might become even less pronounced. In a number of birds, particularly among nonmigratory species of the temperate regions, seasonal variation in external characteristics is an adaptive way to cope with substantially changing environmental conditions. One of the most striking examples is the difference between the summer plumage and the winter plumage of some ptarmigan *Lagopus* species which permits a seasonally efficient camouflage throughout the year. Similar to the effect in age-dependent plumages, there are also transitional plumage stages consisting of feathers of two different, although recurringly molted, feather generations.

Resource Polymorphism As described above, patterns of phenotypic variation can be complex and causally overlapping. To this adds the fact that in some birds, apparently as an adaptation to quickly changing resources within habitats, there is a corresponding swift eco-morphological response between and within species, as most prominently described for Darwin's finches (Grant and Grant 2014). The so-called resource polymorphism has also been studied extensively in *Pyrenestes* seedcrackers (Fig. 4.1) in which the size and structure of their bills differ at the same time between individuals of a single population that have adapted to different food items (Smith 1990).

Phenotypic Plasticity The preceding paragraphs show that morphological variability concerns different manifestations of external differentiation on individual, population, or geographic level and that sometimes several forms of variation might even overlay. Collectively, they represent the phenomenon of phenotypic plasticity. Premising the idea that selection acts on the phenotype, it is important to keep in mind that character variation as such is not a result of evolutionary adaptation in the first instance rather than a mean to flexibly enhance the evolutionary success of a population by stochastically offering variants of morphological traits for selective processes. The phenotype resulting from natural selection of advantageous traits may then represent a particular fraction of previous generations' character variation and again might have its very own range of variation. As phenotypes do also reflect a response to local or temporal ecological conditions, which are not fixed over time themselves, a great deal of morphological plasticity in time can be likewise observed within or between populations and species (e.g., Grant and Grant 2014). Ultimately, shifts in character variation between populations can (but do not have to) lead to evolutionary distinct units that might carry the germ of new species (cf. Yousefi et al. 2017). Therefore, particularly if preserved specimens are part of studies of avian speciation, researchers have to carefully select the geographic and temporal origin of the specimens.

4.4 Assessing Morphological Variation

Just as diverse as morphological variation itself is the number of methods to assess it properly. As soon as ornithologists became aware of the evolutionary and ecological importance of understanding phenotypic variation that hitherto had been documented mainly descriptively with a focus on identification, the need for objective and reproducible methods grew. Through the common use of multivariate statistics, a great deal of stimulating functional, ecological, and evolutionary interpretations emerged. The technical progress of the last decades further supported the application of computer-based assessments of morphological traits. For the assessment of size and proportions, an array of two- (linear) and three-dimensional methods of measurement is currently available and gets constantly modified and extended. Combined with sophisticated statistical analyses, they allow complex descriptions of the avian body and its modifications. In a similar attempt to objectivize color assessments of plumage and bare parts, advanced methods of spectral analyses are applied, and the description of color patterns is currently being modernized by the application of pattern recognition software (Bostwick et al. 2017; Burns et al. 2017).

Linear Measurements As a classic method of assessing morphological variation, linear measurements are routinely taken in almost every scientific ornithological study. Explanations and recommendations of methods are well-documented in the literature (e.g., Baldwin et al. 1931; Svensson 1992; Eck et al. 2011). However, considering the fact that most biometric methods often rely on a set of more or less established measurements, students of morphological variation need to be aware that such standard measurements might capture only certain morphological aspects of the avian body (Box 4.2). Nonetheless, these measurements are very helpful to assess not only variation per se but will also generate data allowing the assessment of functionally relevant complexes such as the flight, walking, and feeding apparatus (Fig. 4.2).

Box 4.2 Avian Body Size

Although seemingly trivial, the comparative assessment of a bird's body size is not entirely straightforward. As the avian body construction is generally shaped to reduce body mass, there is a massive constructional pressure on the trunk, whereas limb and neck bones as well as the large flight feathers, though often extensive in their dimensions and functionally important, may contribute considerably less to the overall body mass in many birds.

Having arisen from traditional studies of preserved specimens, a relatively small number of mostly linear standard measurements are routinely taken in ornithological research, both for live and preserved birds (Eck et al. 2011). This is due to the fact that the techniques of bird taxidermy (essentially skinning the bird and coating an artificial replacement body with that skin, e.g., Winker 2000) allow only a limited number of reliable comparative measurements, once the study skin has been made (Fig. 4.2).

As a consequence, some biologically relevant measurements cannot be taken any more (Fig. 4.2). Therefore, irretrievable morphometric data (e.g., wingspan, total body length, body mass) need be documented during the preservation process of bird specimens in order to guarantee the comparability of data with those taken from live birds. Because of the scarcity of such (statistically meaningful) data in many groups, authors have used body mass data as a proxy for body size instead. Due to the computational power of contemporary statistical analyses, however, a combination of several morphological traits, including morphometric data and body mass, appears to be the most promising avenue to come to biologically sound conclusions.

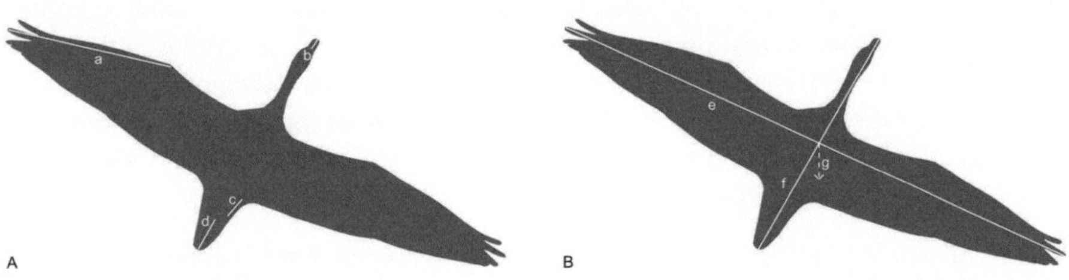

Fig. 4.2 Morphometric assessment of avian body size. (**A**) Classical partial measurements, usually taken from study skins; (**B**) biologically relevant total measurements, meaningful for flight performance analyses. While measurements under (**A**) can be re-collected from specimens virtually at any time, those under (**B**) are irretrievable from study skins and must be taken from the fresh dead or live bird. a, wing length; b, bill length; c, tarsus length; d, tail length; e, wing span; f, total body length; g, body mass

Three-dimensional Measurements Three-dimensional scanning techniques that complement classic linear measurements are not only desired because of the reproducibility and precision of digital data but also because they allow more complex reconstructions of morphological shapes and proportions that are hardly accessible by linear measurements. Although a great deal of work is still done manually by setting digital landmarks for final calculations, it is foreseeable that these processes will become more and more automated, thus further accelerating the speed of analyses. As three-dimensional visualizations principally allow minimally invasive analyses of the inner body as well (e.g., via CT scanning), a number of new findings will additionally enlighten the connection between phenotype and internal characteristics of birds (James 2017).

Color Assessment By contrast, the assessment of plumage and soft part colors traditionally is a descriptive one. Because the individual perception of colors is often incongruent between observers, in particular regarding more subtle color differences, color charts have been developed to objectivize the assessment. There is a number of such charts available, sometimes exclusively designed to meet the requirements of biologists (e.g., Ridgway 1912), but practical problems remain as different editions of the same charts might differ because of the respective printing processes. A fine-scaled comparability of results is only guaranteed if the charts from the very same edition are compared (and if the charts were stored under similar conditions). Meanwhile, spectrophotometric analyses have complemented descriptive methods (Burns et al. 2017). By studying the reflectance spectra of color patches, the presence of pigments and light-reflecting feather microstructures can be inferred based on wavelength-specific reflection patterns that also allow a quantification of involved pigments. Another important aspect of spectrophotometry is the detection of ultraviolet components of plumage color that are not visible by humans but apparently play a role in avian signaling as birds are able to see in the UV range. Spectral color data can be incorporated in similar statistical analyses like morphometric data, making virtual reconstructions of so-called color spaces possible. However, spectral analyses have not replaced classic descriptive approaches completely yet. Not only that for some research questions it would simply be exaggerated to expend high-end technological effort, its application is still relatively laborious considering data collection and analysis. It is thus likely that in the future a well-balanced combination of both traditional and spectral color analyses will be performed instead of a complete replacement of descriptive methods.

Pattern Analyses Nonetheless, color assessment does not only comprise the precise identification of a color but also its distribution on a bird's body. Even more than color identification, pattern analysis is mostly a descriptive procedure. Considering the huge complexity of many bird plumage (or egg shell) patterns, it is obviously difficult to apply a single simple method for comparative studies. Therefore, researchers have chosen different approaches to describe the shape and distribution of color patterns depending on the focus of their study and the group of birds considered (Bostwick et al. 2017; Burns et al. 2017). The more complex these patterns are, the more often partitioned approaches are applied, concentrating on analyzing parts of the avian body first before recomposing the whole pattern information. With the aid of pattern recognition software, however, a great leap forward was made to effectively analyze and compare external patterns in birds. To date, most of the studies still focus on the analysis of two-dimensional data, using digital photographs of objects. Nonetheless, the most useful way to apply digital techniques to pattern recognition in ornithology is the assessment of three-dimensional data in order to reflect the true distribution of patterns on the avian body properly. Considering the current speed of technical advance in the field of three-dimensional pattern recognition in general, such technology will likely become available for scientific bird research very soon.

4.5 Disentangling Phylogenetic and Adaptive Constraints

Probably the most arduous challenge in the study of phenotypic variation is to detect if it reflects phylogenetic or adaptive constraints. In evolutionary respect, morphological plasticity within and between bird species is a result of natural selection that might produce convergent phenotypes under similar extrinsic pressures for functional reasons. Consequently, externally resembling bird taxa do not necessarily need to be closely related (and vice versa), and morphological uniformity might disguise cryptic diversity on species level. Indeed, both polymorphism and cryptic diversification can obstruct our view on species distinction. While the former phenomenon nowadays poses a much smaller difficulty in terms of species assessment because of molecular genetic counterchecking, the latter still is less easy to detect and often only discovered if other methods have provided additional evidence. The task of disentangling phylogenetic and adaptive constraints when studying morphological variation is all the more compounded by the fact that phylogeny and adaptation are usually not independent of each other. During trait evolution, as a response to selective pressures, adaptive processes shape the morphological composition of a population which becomes subsequently fixed itself during phylogeny. It is therefore reasonable to assume that present morphological variation reflects a causally mixed result of trait evolution within an avian lineage. Apparently, there is no infinite plasticity of characters within avian lineages. Once a given set of characters is acquired within a lineage, a certain extent of modification can be observed, but rarely revolutionary reconstructions.

Regarding the question if a morphological character is a relevant feature of species distinctness or just an adaptation to similar ecological conditions, current multidimensional statistical analyses can identify potential phylogenetic effects in a morphological data set. However, credible phylogenetic hypotheses have to be available a priori which usually are based on molecular genetic data. Potentially, the emerging field of population genomics might provide other reliable data (Botero-Castro et al. 2017). However, without the support of additional information, the interpretation of morphological variability as species-delimiting criteria still lies in the eye of the beholder.

4.6 A Contemporary Perspective on Morphological Variation

We have seen that studies of morphological variation are important elements of multidisciplinary studies by combining classic and modern approaches. Having become modernized themselves by sophisticated statistical analytic tools, they do not only complement other cutting-edge methodologies but add considerable new input to our contemporary understanding of bird evolution and adaptation. In combination with growing insight into phylogenetics and ecology of birds, reconstructions of trait differentiation are not limited to the study of few avian lineages but extendable to eco-morphological analyses within and between communities at any geographic dimension ("integrated eco-morphology", Leisler and Schulze-Hagen 2011). It is to be expected that such analyses will become possible even back in time provided enough comparative data (from historical specimens or fossils) can be accessed. The rapidly evolving field of (population) genomics will surely do its share here. In the long run, it would be desirable to integrate data on phenotypic variation into a broader evolutionary framework that also explains modifications of the avian body in the light of speciation processes. By all means, the study of morphological variation will remain a crucial element of ornithological studies, not only when dealing with speciation. The current technical advances allow a broad range of investigations that will deliver important insight into the evolutionary differentiation of birds way beyond than just telling species apart.

References

Alström P, Rasmussen PC, Zhao C, Xu J, Dalvi S, Cai T, Guan Y, Zhang R, Kalyakin MV, Lei F, Olsson U (2016) Integrative taxonomy of the Plain-backed Thrush (*Zoothera mollissima*) complex (Aves, Turdidae) reveals cryptic species, including a new species. Avian Res 7:1

Baldwin SP, Oberholser HC, Worley LG (1931) Measurements of birds. Sci Publ Cleveland Mus Nat Hist 2:1–165

Baptista LF, Kroodsma DE (2001) Avian bioacoustics. In: del Hoyo J, Elliott A, Sargatal J (eds) Handbook of the birds of the world, vol 6. Lynx Edicions, Barcelona, pp 11–52

Birkhead TR, Wimpenny J, Montgomerie B (2014) Ten thousand birds. Ornithology since Darwin. Princeton University Press, Princeton

Bostwick KS, Harvey TA, Scholes E III (2017) Leveraging diverse specimen types to integrate behavior and morphology. In: Webster MS (ed) The extended specimen. Emerging frontiers in collection-based research. CRC Press, Boca Raton, pp 75–88

Botero-Castro F, Figuet E, Tilak MK, Nabholz B, Galtier N (2017) Avian genomes revisited: hidden genes uncovered and the rates versus traits paradox in birds. Mol Biol Evol 34:3123–3131

Bruce M (2003) A brief history of classifying birds. In: del Hoyo J, Elliott A, Sargatal J (eds) Handbook of the birds of the world, vol 8. Lynx Edicions, Barcelona, pp 11–43

Burns KJ, McGraw KJ, Shultz AJ, Stoddard MC, Thomas DB (2017) Advanced methods for studying pigments and coloration using avian specimens. In: Webster MS (ed) The extended specimen. Emerging frontiers in collection-based research. CRC Press, Boca Raton, pp 23–55

Cicero C, Remsen JV (2007) Festschrift for Ned K. Johnson: geographic variation and evolution in birds. Ornithol Monogr 63:1–114

Eck S, van den Elzen R, Fiebig J, Fiedler W, Heynen I, Nicolai B, Töpfer T, Winkler R, Woog F (2011) Measuring birds. Vögel vermessen. Wilhelmshaven, Deutsche Ornithologen-Gesellschaft

Grant PR, Grant BR (2014) 40 years of evolution. Darwin's finches on Daphne Major Island. Princeton University Press, Princeton and Oxford

James HF (2017) Getting under the skin. A call for specimen-based research on the internal anatomy of birds. In: Webster MS (ed) The extended specimen. Emerging frontiers in collection-based research. CRC Press, Boca Raton, pp 11–22

Jenni L, Winkler R (1994) Moult and ageing of European passerines. Academic, London

Jukema J, Piersma T (2006) Permanent female mimics in a lekking shorebird. Biol Lett 2:161–164

Leisler B, Schulze-Hagen K (2011) The reed warblers: diversity in a uniform bird family. KNNV Publishing, Zeist

Mayr E (1969) Principles of systematic zoology. McGraw-Hill, New York

Mayr E (1982) The growth of biological thought. Belknap Press, Cambridge, MA

Price T (2008) Speciation in birds. Roberts and Company, Greenwood Village

Ridgway R (1912) Color standards and color nomenclature, Washington, DC

Smith TB (1990) Natural selection on bill characters in the two bill morphs of the African finch *Pyrenestes ostrinus*. Evolution 44:832–842

Sun Y, Li M, Song G, Lei F, Li D, Wu Y (2017) The role of climate factors in geographic variation in body mass and wing length in a passerine bird. Avian Res 8:1

Svensson L (1992) Identification guide to European Passerines, 4th edn. Ugga, Stockholm

van Grouw H (2013) What colour is that bird? Br Birds 106:17–29

Webster MS (2017) The extended specimen. Emerging frontiers in collection-based research. CRC Press, Boca Raton

Winker K (2000) Obtaining, preserving, and preparing bird specimens. J Field Ornithol 71:250–297

Winker K, Haig SM (2010) Avian Subspecies. Ornithol Monogr 67:1–200

Winkler H, Leisler B (2005) To be a migrant: ecomorphological burdens and chances. In: Greenberg R, Marra PP (eds) Birds of two worlds: the ecology and evolution of migration. John Hopkins University Press, Baltimore, pp 79–86

Yousefi M, Kaboli M, Eagderi S, Mohammadi A, Rezai A, Nourani E (2017) Micro-spatial separation and associated morphological adaptations in the original case of avian character displacement. Ibis 159:883–891

Chapter 5
Song: The Learned Language of Three Major Bird Clades

Martin Päckert

Abstract Vocal learning has evolved several times independently in mammals and three major orders of birds. Of these only hummingbirds and passerine birds have complex songs, whereas the large vocal repertoires of parrots comprise various call types associated with different behavioral contexts. Generally, bird song has two major functions: territorial defense and mate attraction. In the latter context, particularly in songbirds (Oscines), the evolution of male song repertoires has strongly been driven by sexual selection: Song complexity and repertoire size have been shown to be indicators of male quality and are thus crucial traits for female choice. Today, the age of phylogenetics provides new methods for the study of the mode and tempo of organismic diversification and of trait evolution, e.g., of vocal learning. As a striking result, song learning seems to be associated with species richness across the avian tree of life. This provided recent evidence for the theory that song learning can act as a pacemaker of evolution. The spatial variation of song dialects is often correlated with genetic diversification. Extreme cases are small isolated populations, for example, on islands. In the field, the effect of song diversification as a barrier for gene flow can best be observed in zones of secondary contact between close relatives. Even in cases of hybridization, differences of song might affect female choice and thus lead to assortative mating and prevent gene flow in one or even in both directions. These are probably the most exciting case studies, where speciation in action (or in reverse) can be studied in the wild.

Keywords Dialects · Cultural evolution · Passeriformes · Sexual selection

M. Päckert (✉)
Senckenberg Natural History Collections, Dresden, Germany
e-mail: martin.paeckert@senckenberg.de

D. T. Tietze (ed.), *Bird Species*, Fascinating Life Sciences,
https://doi.org/10.1007/978-3-319-91689-7_5

5.1 Eager Birds: The Advanced Learners

The capacity of vocal learning is found only in three larger branches of the avian tree of life: hummingbirds (Trochilidae), parrots (Psittacidae), and perching birds (Passeriformes). These three advanced learners represent quite different groups of birds with respect to anatomy, behavior, and ecology. Indeed, parrots and passerines are closest relatives, but hummingbirds belong to a totally different branch of the bird phylogeny: They are sister to swifts (Apodidae) that dispose of rather simple-structured vocal repertoires. This basically means that in birds, vocal learning and the associated brain structures have evolved at least twice independently (Jarvis 2006; Suh et al. 2011; Nowicki and Searcy 2014). Even in perching birds (Passeriformes), not every species is capable of vocal learning: Only the true songbirds of the suborder Passeri are the real vocal learners: the Oscines. Apart from this species-rich group, two tropical suborders of passerines dispose of mainly innate vocalizations: the New Zealand wrens (Acanthisittidae) and the highly diverse tyrant flycatchers (Tyranni, also called Suboscines). Learning capacity has been shown to correspond with similarities in forebrain areas of hummingbirds and Oscines that were not found in their non-learning sister groups (swifts and Suboscines; Gahr 2000).

Of the three vocal learners, passerines have by far the most complex vocalizations with respect to the diversity and combinations of single sounds in complex compositions such as territorial songs. Let's say passerines are the greatest performers. For their immense diversity, passerine vocal repertoires probably show the greatest variation in space and time. Therefore, in the following, I mainly focus on these most eager learners among birds.

5.2 Passerine Song

Almost all passerine birds have a large repertoire of highly diverse vocalizations displayed in various behavioral contexts. We distinguish two main classes of vocalizations: calls and songs (Catchpole and Slater 1995). Calls are simple rather short vocalizations for communication among mates, rivals, parents and young, within a flock and even among different species like in case of alarm calls (Marler 2004b). The characteristic "rain calls" of chaffinches (Fig. 5.1) are highly variable among regions, even among different quarters of a city. Their function is not fully understood; however, for the resemblance of these regularly repeated buzzing sounds with those of other finch species, they have sometimes been considered an energy-saving rudimentary song type (Constantine 2006).

In contrast, songs are long and complex vocalizations that in many species are composed of many different notes and motifs. The sequential order of notes or song parts (syntax) is a characteristic for most passerine species. Generally, the single bird displays many songs in succession during song bouts that can last over several

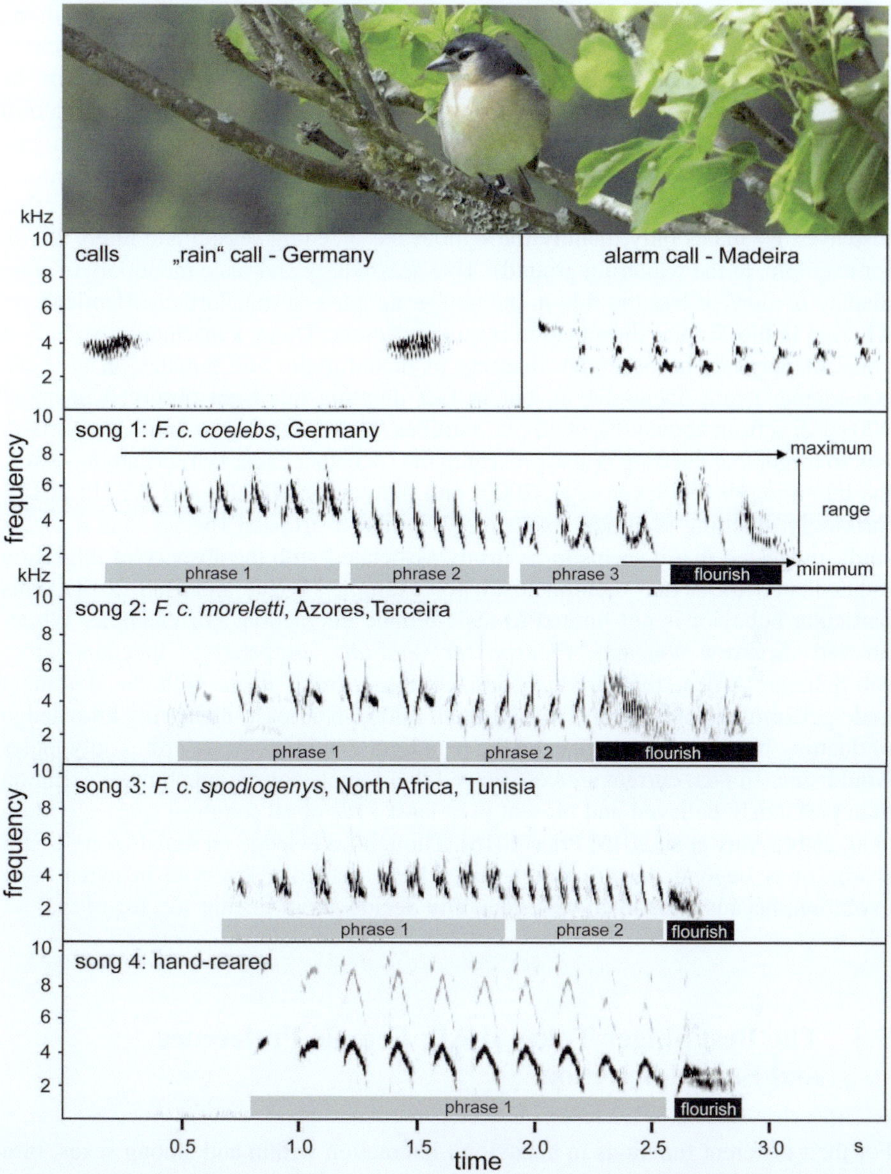

Fig. 5.1 Sonagraphic analysis of bird vocalizations. A sonagram is a spectrographic image of sound events, i.e., a plot of frequency (*y*-axis) versus time (*x*-axis). Different frequency and time parameters can be measured for scientific analysis (see song 1). Each sound event (note) is represented by a continuous black line. The plate shows different calls and dialect song types of the Common Chaffinch *Fringilla coelebs* (picture: *F. c. moreletti*, São Miguel Azores; M. Päckert 2017). Typical chaffinch songs are composed of different phrases (stereotyped repetitions of one element or motif) and a variable terminal flourish. All recordings of M. Päckert except song 3 (S. Eck) and song 4: a hand-reared bird recorded by Thorpe in the year 1954 (from Marler and Slabbekoorn 2004)

minutes. The successive songs can have more or less the same structure like in the Common Chaffinch *Fringilla coelebs* (Fig. 5.1), or they can show a great variation of many different song types like in the Common Blackbird *Turdus merula* or the Common Starling *Sturnus vulgaris*. The entirety of all different song types of a bird is termed a repertoire. Unlike calls, songs have two main functions and receivers: territorial defense in communication between males and mate attraction in communication between males and females. As a consequence, territorial song is often displayed by males only, mainly throughout the breeding season and rarely heard, for example, in the wintering grounds. That seasonality and male monopoly of song display is more or less the rule in the temperate parts of the Northern Hemisphere, whereas in the Tropics the situation is quite different. There, a fascinating variant of vocal display can be observed: duetting of mated males and females. It has been considered a rare phenomenon, but in fact duetting has been observed in about 400 species from about 40% of all bird families (Hall 2009; Logue and Krupp 2016). Not to forget that duetting is also present in the two other eager learners among birds, the hummingbirds (Ficken et al. 2000) and the parrots (Dahlin and Wright 2012). Strikingly, duetting is almost exclusively limited to tropical species, and a recent study suggested that it seems to be firmly associated with the absence of migration rather than with sexual dichromatism, for example (Logue and Hall 2014). This particular behavior is not limited to male-female interaction: For example, White-browed Sparrow-Weavers *Plocepasser mahali*, cooperative breeders from sub-Saharan Africa, regularly perform complex group duets with the dominant male performing solo displays (Voigt et al. 2006). Not least, increasing knowledge of duetting behavior has challenged the paradigm that in most passerines only males would sing. In fact, current studies showed that female song is much more common than previously believed and present in at least 71% of all songbird species (Odom et al. 2014; Amy et al. 2015; Price 2015). Though knowledge on female song is still scarce, it is beyond dispute that females have played a key role in avian song evolution, because it is them who generally decide about mating success of males.

5.3 The Best Singer Takes It All: Female Preference and Sexual Selection

For their different functions in behavioral interaction within and among sexes, bird songs are considered both armaments and sexually selected ornaments (Tobias et al. 2011). It was indeed Darwin, who first developed the theory of sexual selection and considered that vocalizations could play a major role for female choice. Originally, Darwin's view on female choice was merely an "aesthetic" conception of bird song and this has been immediately criticized by Wallace, who postulated that from the perspective of females, song would be an indicator of male vigor and viability (review in Prum 2012).

There is good knowledge that across a broad spectrum of vertebrate species, mate preference is learned by both males and females during a phase of sexual imprinting (Irwin and Price 2009; Verzijden et al. 2012). Also in birds, females discriminate between different variations of songs, and like male song female choice is learned from different tutors. Likewise, female birds were shown to react more positively and intensively toward song types of their own mate (O'Loghlen and Beecher 1997) or toward the local dialect from their own population (Danner et al. 2011). Furthermore, in many species like the Song Sparrow *Melospiza melodia*, female preference is firmly associated with repertoire size (Reid et al. 2004). Another famous example comes from reed warblers of the genus *Acrocephalus*: Their long warbling songs displayed from dense riverine vegetation can comprise a multitude of different notes. For example, Marsh Warblers *A. palustris* are great learners, because in their African wintering grounds, they can pick up vocalizations of many tropical bird species and remember them next spring in their Eurasian breeding territories (Dowsett-Lemaire 1979). Likewise, a reed warbler male in its first year can easily double its vocal repertoire by mimicking other species (Kroodsma 2004: Box 18). Repertoire size and composition can accordingly change between successive breeding seasons (Wegrzyn and Leniowski 2009). Female reed warblers seem to have a preference for long and complex songs, because apparently the males with the largest repertoires get mated first and have a greater breeding success (Catchpole 1980, 1986; Catchpole and Leisler 1996; Kroodsma 2004: Box 17; Darolová et al. 2012). These classical studies became a textbook example for sexual selection on repertoire size. However, reservations are standing, because mating and breeding success are influenced by a few more factors than female choice only. About 10 years later, Forstmeier and Leisler (2004) meticulously repeated those experiments by Catchpole and his team on Great Reed Warblers *A. arundinaceus* and came to an astonishing result: In the same Swedish population, male repertoire sizes were positively correlated with male pairing success and clutch size in the 1980s, but not in the 1990s. The authors suggested that recent population decline of Scandinavian Great Reed Warblers might have changed the role of territory quality and thus of repertoire sizes as a putative indicator. We will later see that a change of ecological factors (both biotic and abiotic) might lead to remarkable changes of local song types. However, despite the critical reevaluation of Catchpole's studies (see also Darolová et al. 2012), there is still good evidence that female birds can infer reliable information on male quality from their songs (Nowicki et al. 1998). The so-called good-genes hypothesis originally referred only to visual ornamental traits such as coloration. However, soon the assumption was made that this hypothesis would apply to bird vocalizations, too. For example, according to general belief, animals tend to avoid inbreeding depression, i.e., they avoid mating among members of close kinship or even with individuals from an inbred population (Pusey and Wolf 1996). Just recently, researchers found out that captive female Atlantic Canaries *Serinus canaria* indeed discriminate between inbred and outbred birds by their different songs. As a consequence, they mated less often with an inbred male, and, if they did, breeding success was lower than that of couples with an outbred male (deBoer et al. 2016). We will see that male quality does not relate only to inherited but also to acquired traits in a later paragraph; but let's first step back on a brief walk through history of bioacoustic research.

5.4 How It All Began: A Brief History of Bioacoustic Studies

In the second half of the twentieth century, experimental studies on song learning in birds were developed by behaviorist scholars like the neurobiologist Fernando Nottebohm or the zoologist William H. Thorpe, who pursued a comparative ethological approach. He was a pioneer of bioacoustic studies, and he was also the first scientist to use spectrography (Fig. 5.1)—a newly emerging technique in these days—for analysis of bird song (Marler 2004a). Generally, the core question of the early behaviorist studies was to disentangle learned and innate components of bird song; however, their experimental designs differed.

For example, classical neuroscientific experiments focused on test birds that were surgically deafened a few days after hatching (Nottebohm 1970) or whose innervation of the vocal apparatus (the syrinx) had been manipulated (Nottebohm 1970, 1971). Ethological experiments had a more simplified setting: They studied cage birds that were hand-reared in acoustic isolation. For the deprived situation of the test birds, these studies later became famous as Kaspar-Hauser experiments, named after the mysterious foundling from Nürnberg of the early nineteenth century. The simple design with completely isolated individuals was expected to unmistakably unveil the ancestral innate structure of a bird's song. However, in a second step, scientists tried to find out more about a birds learning capacity by exposing the test birds to different tutors (of the same and of different species), to tape recordings or even to human whistles that resulted in exact copies of a German folk tune by Eurasian Bullfinches *Pyrrhula pyrrhula* (Nicolai et al. 2014). The probably most popular model species of early ethologists was the Common Chaffinch (review in Riebel et al. 2015). Figure 5.1 shows the comparison of two fully developed songs from wild birds with a recording of a hand-reared bird by Thorpe (1958). Evidently, the Kaspar-Hauser song is more simplified in structure and note variation, and it lacks the characteristic partition into distinct phrases. However, the test bird's song still bares striking similarities with the song of wild birds: The frequency course is slightly descending toward the simplified terminal flourish at the end of the strophe.

Based on their experimental results, most scientists agreed that birds have a sensitive phase for the acquisition of fully developed (crystallized) song (Nottebohm 1968, 1969, 1970). The duration of this phase can vary extremely among species (Hultsch and Todt 2004) from only a few days in the nest with the father as tutor to the first breeding season with further tutorial input from the territory neighbor. These "age-limited learners" (or "close-ended learners") are opposed to the so-called "open-ended learners" among them the Common Starling (Eens et al. 1992).

However, lifelong learning is also known from the other two eager learners of birds: It has been shown for some hummingbird species (Araya-Salas and Wright 2013) and of course for parrots above all from the studies of Irene Pepperberg on Grey Parrots *Psittacus erithacus*. Her most famous parrot Alex proved to be capable even of referential learning of human speech (Pepperberg 2004, 2010). Generally, learning mode seems to be correlated with repertoire size in a way that open-ended learners have the larger repertoires (Creanza et al. 2016)—which is not too surprising, because they simply have more time to increase their "vocabulary" annually. The cage experiment was soon complemented by a great number of studies on wild bird populations in the field, and their basic conclusion can be summarized in brief: Bird song is subject to cultural evolution. Because several different cultural traits, such as song learning or female mating preference are involved and mutually affect each other, that complex evolutionary process was recently suggested to end up in the reconstruction of a cultural niche (Creanza et al. 2016).

5.5 Telltale Songs: Evolution and Phylogenetic Information of Vocalizations

As one result of cultural evolution, the song of a bird species bares ultimate diagnostic traits that allow for exact species determination by the scientist and for effective species recognition among the birds themselves. Nevertheless, species of the same genus, for example, may share common song features that clearly indicate close kinship. Kinglets are among the smallest passerines, and therefore most species have a rather high-pitched voice. All of the six *Regulus* species have clearly structured songs that are composed of two or three parts (Martens et al. 1998; Päckert et al. 2003). Each of these parts is defined by characteristic sound constraints: For example, all species share one song part (Fig. 5.2) that is steadily increasing within the same frequency range. Because this rising sequence contains similar whistling notes in all species, it can be considered an ancestral song part that all kinglet species have in common. Only three species complete their songs with a terminal flourish, and a single Goldcrest *R. regulus* male, for example, can have about 20 different individual variations of this end part. The molecular family tree of kinglets in Fig. 5.2 shows that the three flourish-singers are closest relatives and that among these only the Goldcrest and its sister species *R. goodfellowi* on Taiwan share a common song introduction. This basically means that some components of song give us a clue on relatedness, whereas others do not. In terms of scientists, we say that songs carry phylogenetic information (Rheindt et al. 2014). This has been demonstrated for treecreepers *Certhia* (Tietze et al. 2008) and leaf warblers, Phylloscopidae (Päckert et al. 2004; Mahler and Gil 2009; Tietze et al. 2015). These case studies also showed that highest phylogenetic signal was found for innate components of songs like, for example, call-like elements or the general syntax. For example, like in the previous Chaffinch example, hand-reared Goldcrests always displayed the typical bi- or tripartite

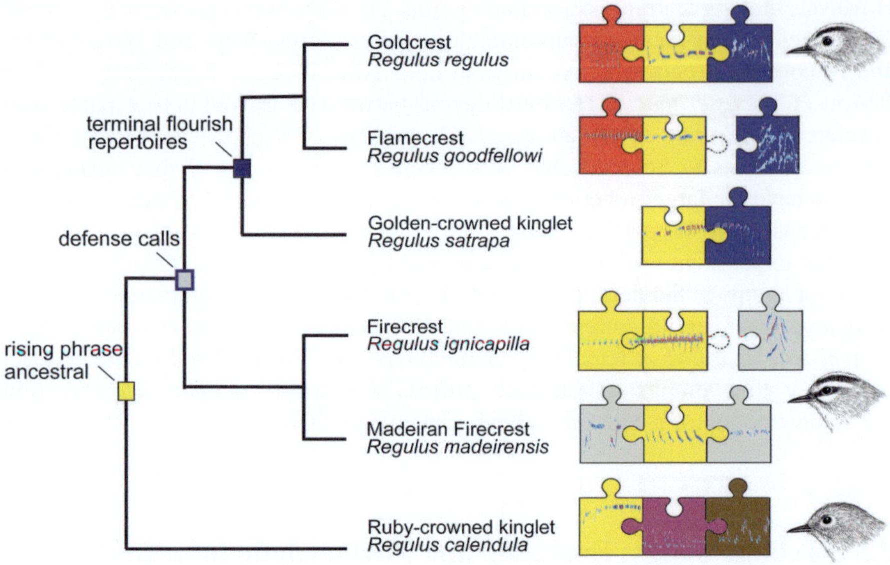

Fig. 5.2 Evolution of territorial song in kinglets *Regulus*; the schematized molecular tree is modified after Päckert et al. (2003, 2009). For each species, the characteristic and invariable bi- or tripartite song structure is shown as jigsaw puzzles. Homologous song parts are indicated by pieces of the puzzle with the same color with typical sonagrams included. Gray pieces indicate innate vocalizations such as calls integrated in the song of some species. Two species only occasionally attach a terminal part to their song (unconnected pieces): *R. ignicapilla* (invariable terminal call) and *R. goodfellowi* (variable terminal flourish). In contrast, songs of *R. regulus* and *R. satrapa* regularly include a variable terminal flourish (connected pieces). From the many diverse dialects of the Goldcrest, a recording from Japan (Hokkaido; *R. regulus japonensis*) was chosen for the strong similarities of the yellow phrase (repetition of whistle notes) with that of North American *R. satrapa* (see Martens et al. 1998)

partitioned song with a simple flourish motif at the end (Thaler 1990). On the other hand, Common Firecrests *R. ignicapilla* have rather simple songs that can contain call elements usually displayed in a different behavioral context, e.g., in close aggressive contact. Only very occasionally Common Firecrests end their songs with such calls that pretty much resemble single motifs in the terminal flourish repertoires of *R. regulus* and *R. satrapa*. In so far, crests and kinglets provide a stunning example of vocal evolution of complex song structures and repertoires derived from ancestral simple vocalizations such as calls (Fig. 5.2).

5.6 Vocal Learning as a Pacemaker of Evolution

Strikingly, passerines represent by far the greatest part of the global avian biodiversity: About 60% of all bird species belong to the order of Passeriformes. At the same time, their top position in the avian tree of life makes passerines the "youngest" bird order, and on average they have higher diversification rates than nonpasserines (Jetz et al. 2012). This basically means that the immense passerine species richness has evolved over a rather short period of time compared to other bird orders. Since long, ornithologists and evolutionary biologists have been trying to find an explanation for this overwhelming success of passerine birds. It was the famous evolutionary biologist Ernst Mayr, who came up first with the idea that evolutionary change can be effectively caused by newly acquired behavior (reviewed in Corning 2014). Later, it was Gerhard Thielcke (1970, 1972), another pioneer in the field of bioacoustics, who adopted Mayr's ideas and hypothesized that song learning can be a "pacemaker of evolution." Like behaviorists of the 1960s, Thielcke based his ideas on the results of Kaspar-Hauser experiments that he carried out with captive treecreepers of species *Certhia familiaris* and *C. brachydactyla*. Since then, the pacemaker theory has been lively and controversially discussed, but the original idea was kept alive: If songs of two birds differ to a certain degree, they will not consider each other as a member of their own species, i.e., song variation affects species recognition in both males and females. Therefore, avian vocalizations can be an effective premating barrier and cause reproductive isolation (to a certain degree) among closest relatives (Martens 1996, 1998).

These theories have been recently put to test with modern analytical methods. The basic question was: Are song-learners greater speciators than those birds that do not learn? This can be done in a totally theoretical approach using simulations under different evolutionary models. To put it simple, in theory, the learning model always produced greater mean differentiation of songs as a response of males to female preferences and sexual selection (Lachlan and Servedio 2004). Further evidence came from phylogenetic studies posing the basic question: Do song-learners generate more branches of the avian tree of life over time than non-learners? A comparison of two Neotropical passerine families yielded two striking results (Mason et al. 2017): Each burst of speciation (i.e., accelerated emergence of new branches) was associated with increased diversification of songs in both groups. However, the mean speciation rates were greater in the oscine study family with learned songs (tanagers and allies, Thraupidae) than in the non-learning suboscine study family (antbirds, Thamnophilidae).

5.7 Dialects: Spatial Variation

In analogy to human language, spatially separated song types of the same species are called dialects. Possibly the most comprehensive fine-scale monitoring of passerine song dialects is available for the Yellowhammer *Emberiza citrinella*: Across its European breeding range, a mosaic distribution of distinct song types and the corresponding dialect boundaries has been documented over the last decades (Petrusková et al. 2015). Just recently, an exciting newsflash announced that lost birdsong had been discovered in introduced Yellowhammer populations from New Zealand. There, far beyond the species' native breeding range on the other side of the globe, some dialect types were registered that apparently went lost in the British area of origin of the species (Pipek et al. 2017). In analogy to population genetic processes, such timely change of local song repertoires has been considered a merely stochastic effect termed cultural drift (Lynch 1996). That means just by chance, rare song types become either fixed in a local population, or they do completely vanish from the local repertoire. As stochastic processes have greatest effects on small samplings, cultural drift is believed to act most effectively on small isolated populations such as on remote islands (Potvin and Clegg 2015). However, under the assumption of neutral evolution, no general pattern or directionality of song evolution on islands can be hypothesized, and this has been confirmed by some empirical comparative studies (Baker et al. 2006). Taking the Macaronesian archipelagic chain as an example, we find both increase of acoustic variation and greater song repertoires on islands like in Goldcrests on the Azores (Päckert and Martens 2004) as well as impoverishment of song and loss of syntactical structure in Common Chaffinch island populations (Lachlan et al. 2013). Like on oceanic islands, cultural drift has been suggested to act on passerine song divergence among sky islands such as isolated mountain chains (Purushotham and Robin 2016).

An extreme case of stochastic effect on island song repertoires is so-called withdrawal of learning. This assumes a scenario where a small group of young birds without fully crystallized song colonizes an island and in succession develops a specific island song type composed of innate vocalizations (Thielcke 1970). In a way, this theory postulates a Kaspar-Hauser experiment under natural conditions. Madeira Firecrests *R. madeirensis* provide a striking example. In contrast to songs of European *R. ignicapilla*, songs on the island of Madeira are composed of innate vocalizations only (Fig. 5.2). The typical ascending kinglet phrase is accompanied by two different call types that are also displayed in different behavioral contexts: contact calls at the beginning and aggressive calls at the end of the song (Päckert et al. 2001; Constantine 2006).

All predictions on the effect of cultural drift are based on the a priori assumption of merely stochastic processes on selectively neutral variants of song. However, as we have already seen, sexual selection plays a key role on individual and spatial song variation and on the evolution of bird song in general. Moreover, there are further environmental selective pressures shaping song variation within and among species.

5.8 Competition for Acoustic Space: The Role of Ecology

A strong pressure of natural selection on bird vocalizations is ecology, i.e., habitat characteristics, community structure, intra- and interspecific concurrence, etc. (reviews in: Catchpole and Slater 1995: pp. 71–92; Slabbekoorn and Smith 2002; Slabbekoorn 2004). Optimal sound transmission strongly depends on habitat characteristics above all on vegetation density. This is actually one reason why many birds choose elevated song perches to bring their message across densely vegetated areas. In fact, in dense tropical forests, songs of closely related passerine species differ in predictable adaptive parameters according to perch height and vegetation cover (Nemeth et al. 2001; Barker and Menill 2009). Apparently, in a given local assemblage, different bird species compete for acoustic space (Luther 2009; Azar and Bell 2016). For the same reasons, song of one and the same species can strongly vary in frequency and time characteristics among open habitat and densely covered vegetation (Hunter and Krebs 1979). In high mountain systems such as the Andes and the Himalayas, strong ecological segregation of closely related species has also triggered divergence of their songs according to specific local habitat characteristics along both latitudinal and elevational gradients (Caro et al. 2013; Singh and Price 2015). Separation of gene pools can be a consequence.

We shortly return to the island example once more, because effects of any selective pressure on phenotype, behavior, and local gene pools are strongest and most visible in small isolated populations. In a comparative study, island songs strikingly did not differ in complexity from those of adjacent continental populations but included fewer features associated with aggressive behavior such as rattles, buzzes, and fast trills (Morinay et al. 2013). This stunning observation has been explained by reduced territoriality due to a lesser need of discrimination against closely related species in species-poor communities. Further insight on how competition might affect song variation came from a long-term study of Darwin's finches from the Galapagos Islands (Grant and Grant 2010). On Daphne Major, the finch community changed dramatically when in 1983 an alien species, Large Ground Finch *Geospiza magnirostris*, colonized the island and rapidly increased in numbers. In response to the invasion, the songs of the two native species (Common Cactus Finch *G. scandens* and Medium Ground Finch *G. fortis*) dramatically changed within a few years with respect to frequency bandwidth, trill rate, and song duration. These selective shifts of song characteristics went along with significant decrease of beak sizes due to competition with a larger species. This reciprocal effect is well-known, because in Darwin's finches, beak size correlates strongly with song characteristics (Podos 2001; Podos et al. 2004). A different story is told from the conquest of Galapagos by another invasive species: a parasitic fly, *Philornis downsi*, which in the 1960s was accidentally introduced to the archipelago. Finches infected by the parasites showed in many cases massive beak deformations, and mortality of nestlings and fledglings increased. Infected survivors had greater beaks; thus not only natural selection would act on feeding and foraging strategies, but also sexual selection would act on songs that would modify along with beak modifications (Kleindorfer and Dudaniec 2016). The reason is bird song is considered an honest signal of male quality (Gil and Gahr 2002; Spencer et al. 2003; Murphy et al. 2008). In our example, the infected males would be recognized by their aberrant songs, and females would thus avoid a risky parental investment and mate with a fitter male (Spencer et al. 2005; Bischoff et al. 2009). Such parasite-mediated sexual selection has been previously found for visual signals (e.g., phenotypical traits such as plumage coloration) and became better known as the Hamilton-Zuk hypothesis (Balenger and Zuk 2014).

5.9 Dialects as a Language Barrier and Isolating Mechanism

Across a species' breeding range, variation of songs can change gradually and continuously along a cline. For example, in many species, frequency and time parameters of songs can clinally change with latitude or longitude (Tietze et al. 2011, 2015; Singh and Price 2015; see also ring species example in Fig. 5.3). This holds also true for comparisons across different species (Weir and Wheatcroft 2011; Tietze et al. 2015). On the other hand, large distribution gaps are often correlated with strong character discontinuities of songs, the most extreme examples being found on islands (see above). Already before the era of molecular systematics, a correlation between bioacoustic and genetic divergence has been assumed based on the example of North American White-crowned Sparrows *Zonotrichia leucophrys* (Baker 1975). Meanwhile, this interrelationship has been demonstrated in numerous case studies that came to a concordant conclusion: Divergence of song dialects contributes to speciation processes, e.g., in White-crowned Sparrows (MacDougall-Shackleton and MacDougall-Shackleton 2001; Toews 2017; Lipshutz et al. 2017) or in Mediterranean Subalpine Warblers *Sylvia cantillans* and allies (Brambilla et al. 2008). Another extreme and controversial example was reported from a so-called ring species, the Greenish Warbler *Phylloscopus trochiloides*. Ring species have a ring-shaped distribution around a geographic barrier such as continental waters or big mountains, like in our bird example the Tibetan Plateau (Fig. 5.3; for the theoretical background, see Martens and Päckert 2007). On each side of the barrier, biometric and bioacoustic traits change gradually along a chain of Greenish Warbler populations that are connected by gene flow. However, in a Siberian contact zone north of the plateau, gene pools and songs of the two terminal populations (*viridanus* in the West and *plumbeitarsus* in the East) differ dramatically (Irwin 2000; Irwin et al. 2001). Character divergence is as high that apparently gene flow between eastern and western Greenish Warblers is strongly limited due to a strong reproductive barrier (Alcaide et al. 2014). Strikingly, the effect of song dialects is not as strong as predicted by Irwin (2000), because a comprehensive study across the Siberian contact zone provided clear evidence of clinal variation of several frequency and time parameters of songs along a transect of recording sites (Kovylov et al. 2012). So, why would we find restricted gene flow despite apparent admixture of song dialects? In the next paragraph, we gain further insight in these processes from another Siberian contact zone.

5.10 Sympathy in Sympatry: Bilingual Birds in a Hybrid Zone

Spatial patterns of song dialects give us a clue not only on intraspecific variation but also putative premating barriers in areas of overlap where populations "of different languages" get into contact. The middle Amur valley of Far Eastern Russia is where two tit species meet: the yellow-bellied western Great Tit *Parus major* and the white-

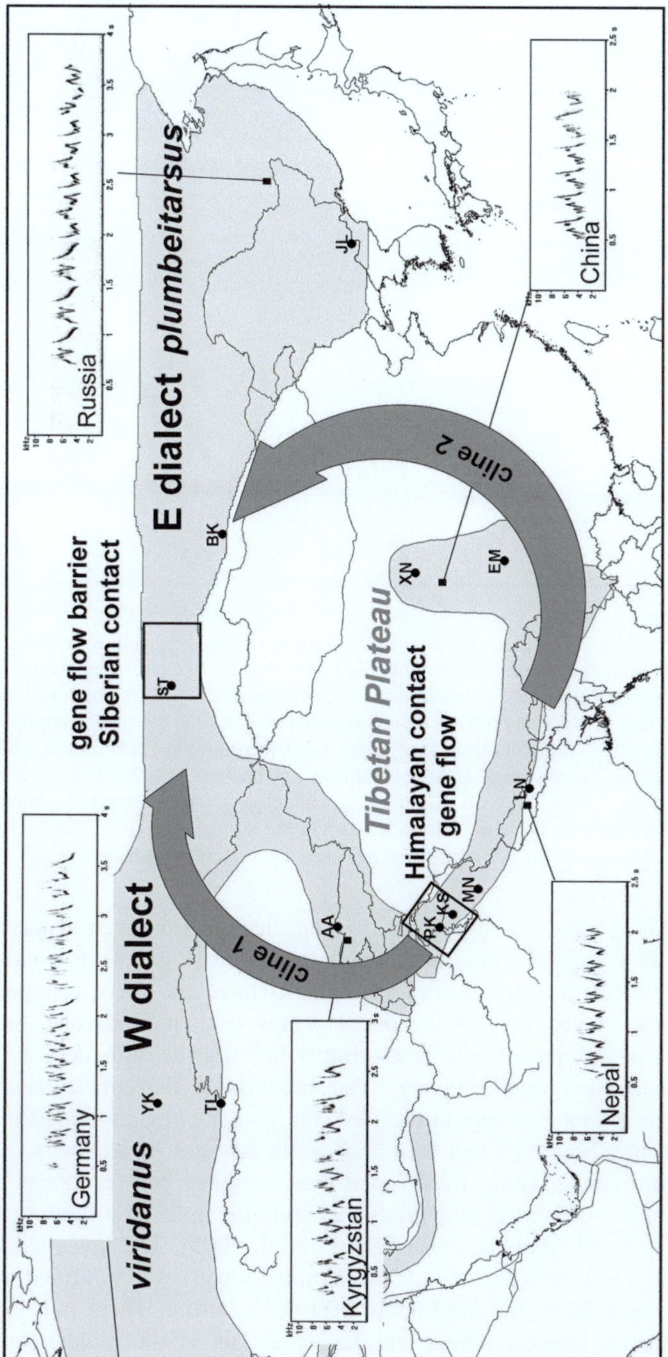

Fig. 5.3 Clinal song variation in the Greenish Warbler *Phylloscopus trochiloides* (modified from Martens and Päckert (2007), including information from Irwin (2000), Irwin et al. (2001); sampling locations indicated). Along two clines East and West of the Tibetan Plateau, vocalizations gradually change from short and simply structured songs in the South toward longer and more complex songs in the North. Where populations of the two clines meet in secondary contact, there is extensive gene flow in the Himalayas but almost completely restricted gene flow in central Siberia (Alcaide et al. 2014). sonagrams: all recordings by J. Martens, except Far East Russia by M. Päckert

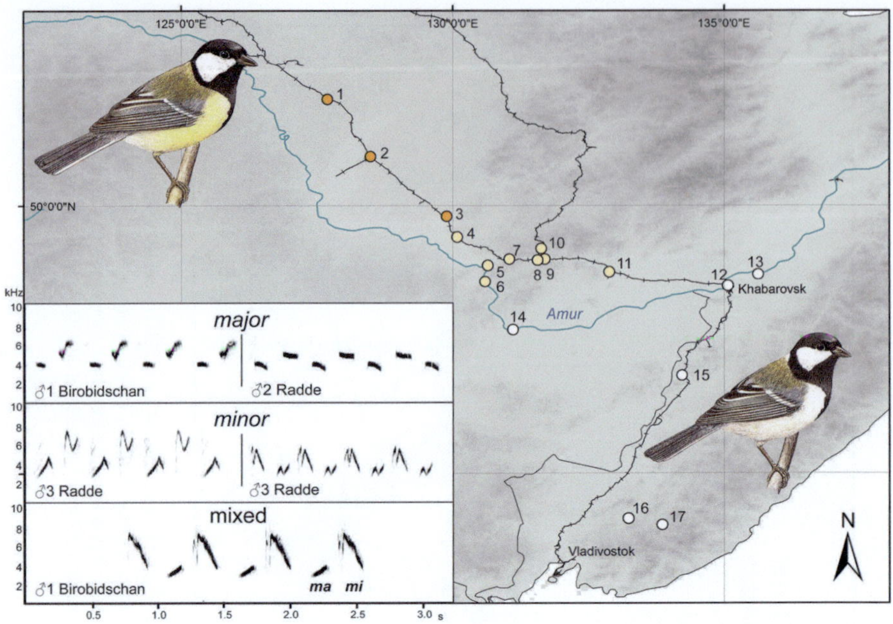

Fig. 5.4 Bilingual song repertoires in the Far East Russian hybrid zone of the yellow-bellied Great Tit *Parus major* (map, orange dots) and the white-bellied Japanese Tit *P. minor* (map, white dots; core zone of overlap, pale yellow dots). In the core zone of overlap, males can sing both *major* and *minor* song types, and they also display mixed songs composed of both note types; all recordings of M. Päckert, June 1996; population data from study sites 1–18 according to Päckert et al. (2005), Kvist and Rytkönen (2006), Fedorov et al. (2009); localities: 1 = Shimanovsk, 2 = Belogorsk, 3 = Novobureysk, 4 = Akhara, 5 = Pashkovo, 6 = Radde, 7 = Obluchje, 8 = Izvestkovyi, 9 = Birakan, 10 = Kuldur, 11 = Birobidzhan (spelled "Birobidschan" in the sonograms), 12 = Khabarovsk, 13 = Maleshevo, 14 = Amurzet, 15 = Bikin, 16 = Arsenev, 17 = Buluga-Fadeyevo

bellied Japanese Tit *P. minor*. Songs of the two species are similar in syntax but differ fundamentally in the tonal quality of alternating notes (Fig. 5.4). Beyond the Amur valley, the different note repertoires seem to have an effect on species recognition, because playback of Asian *minor* songs does not provoke strong territorial reactions in European Great Tits—simply because the birds do not consider the Asian song their own language. The situation in the contact zone is different: There, from their first day in the nest, the young birds are reared in an acoustic environment where they can hear both European and Asian songs. As a consequence, in the middle Amur, male repertoires comprise both European and Asian song types and even mixed song types that combine notes of different tonal quality (Fig. 5.4; see also Martens 1996; Päckert et al. 2005). The juvenile birds effectively received a bilingual education! These mixed repertoires confirmed what could already be observed in the field about 100 years before: There is a lot of hybridization going on between Great and Japanese Tits in the middle Amur, because local populations include a high percentage of individuals with intermediate plumage coloration—a yellow-tinge of all tones.

Analysis of local song repertoires revealed another striking thing: All bilingual males had yellow plumage to a certain degree, whereas almost all purely white *minor* birds sang pure Asian *minor* dialect (Päckert et al. 2005). So for some reason, the yellow *major* birds and their hybrids are better learners. This apparently has an effect on female choice, because like in other bird species, female Great Tits seem to prefer males with larger and more variable repertoires (McGregor and Krebs 1982; Baker et al. 1986). It was thus not surprising that most solitary males in the middle Amur populations were white (Nazarenko et al. 1999): The lazy learners were apparently less attractive to females and had to go solo. Olofsson and Servedio (2008) postulated that female preference for mixed repertoires could act as a positive feedback mechanism and as a selective pressure on repertoire composition. In conclusion, they considered sympatry an effective driver of cultural evolution.

Such bias in female mating preferences is also called assortative mating, and this is rather the rule than the exception in avian hybrid zones. At a closer look, the situation of the two Russian tit species is perfectly paralleled in other hybrid zones, such as in the eastern European contact area of the European Pied Flycatcher *Ficedula hypoleuca* and the Collared Flycatcher *F. albicollis*. There, the majority of male Pied Flycatchers (65%) include various parts of Collared Flycatcher song in their song repertoire, but not vice versa (Qvarnström et al. 2006). From individual mixed repertoires, researchers could make exact predictions on hybridization: female Collared Flycatchers only paired with Pied Flycatchers when they were mixed singers (Haavie et al. 2004).

Limited or asymmetric gene flow due to song-biased assortative mating has also been shown in two contact zones between two species of chiffchaffs in the Pyrenees (*Phylloscopus collybita*, *P. ibericus*: Helbig et al. 2001) and between two subspecies of *P. collybita* in western Siberia (Shipilina et al. 2017). All these examples impressively demonstrate a strong effect of the variation of learned vocal signals on species recognition, mate choice, and thus gene flow among populations of what one observer would consider the same species whereas others would speak of two species. In fact, these borderline cases are the most fascinating examples to study speciation in action in the field.

Acknowledgments I would like to thank Jochen Martens for his helpful comments on the manuscript.

References

Alcaide M, Scordato ESC, Price TD, Irwin DE (2014) Genomic divergence in a ring species complex. Nature 511(7507):83–85

Amy M, Salvin P, Naguib M, Leboucher G (2015) Female signalling to male song in the domestic canary, *Serinus canaria*. R Soc Open Sci 2(1):140196

Araya-Salas M, Wright T (2013) Open-ended song learning in a hummingbird. Biol Lett 9 (5):20130625

Azar JF, Bell BD (2016) Acoustic features within a forest bird community of native and introduced species in New Zealand. Emu 116(1):22–31

Baker MC (1975) Song dialects and genetic differences in white-crowned sparrows (*Zonotrichia leucophrys*). Evolution 29(2):226–241

Baker MC, Bjerke TK, Lampe H, Espmark YO (1986) Sexual response of female great tits to variation in size of males' song repertoires. Am Nat 128(4):491–498

Baker MC, Baker MSA, Tilghman LM (2006) Differing effects of isolation on evolution of bird songs: examples from an island-mainland comparison of three species. Biol J Linn Soc Lond 89 (2):331–342

Balenger SL, Zuk M (2014) Testing the Hamilton-Zuk hypothesis: past, present and future. Inegr Comp Biol 54(4):601–613

Barker NKS, Menill DJ (2009) Song perch height in Rufous-and-white Wrens: does behaviour enhance effective communication in a tropical forest? Ethology 115(9):897–904

Bischoff LL, Tschirren B, Richner H (2009) Long-term effects of early parasite exposure on song duration and singing strategy in great tits. Behav Ecol 20(2):265–270

Brambilla M, Janni O, Guidali F, Soraces A (2008) Song perception among incipient species as a mechanism for reproductive isolation. J Evol Biol 21(3):651–657

Caro LM, Caycedo-Rosales PC, Bowie RC, Slabbekoorn H, Cadena CD (2013) Ecological speciation along an elevational gradient in a tropical passerine bird? J Evol Biol 26(2):357–374

Catchpole CK (1980) Sexual selection and the evolution of complex songs among European warblers of genus *Acrocephalus*. Behaviour 74(1):149–166

Catchpole CK (1986) Song repertoires and reproductive success in the great reed warbler *Acrocephalus arundinaceus*. Behav Ecol Sociobiol 19(6):439–445

Catchpole CK, Leisler B (1996) Female aquatic warblers (*Acrocephalus paludicola*) are attracted by playback of longer and more complicated songs. Behaviour 133(15):1153–1164

Catchpole CK, Slater PJB (1995) Bird song – biological themes and variations. Cambridge University Press, Cambride

Constantine & The Sound Approach (2006) The sound approach to birding – a guide to understanding bird sound. The Sound Approach, Dorset

Corning PA (2014) Evolution 'on purpose': how behaviour has shaped the evolutionary process. Biol J Linn Soc 112(2):242–260

Creanza N, Fogarty L, Feldman MW (2016) Cultural niche construction of repertoire size and learning strategies in songbirds. Evol Ecol 30(2):285–305

Dahlin CR, Wright TF (2012) Duet function in the yellow-naped amazon, *Amazona auropalliata*: evidence from playbacks of duets and solos. Ethology 118(1):95–105

Danner JE, Danner RM, Bonier F, Martin PR, Small TW, Moore IT (2011) Female, but not male, tropical sparrows respond more strongly to the local song dialect: implications for population divergence. Am Nat 178(1):53–63

Darolová A, Krištofík J, Hoi H, Wink M (2012) Song complexity in male marsh warblers: does it reflect male quality? J Ornithol 153(2):431–439

deBoer RA, Eens M, Müller W (2016) 'Out of tune': consequences of inbreeding on bird song. Proc R Soc Lond B 283(1835):20161142

Dowsett-Lemaire F (1979) The imitative range of the song of the marsh warbler, *Acrocephalus palustris*, with special reference to imitations of African birds. Ibis 121(4):453–468

Eens M, Pinxten R, Verheyen RF (1992) Song learning in captive European starlings, *Sturnus vulgaris*. Anim Behav 44(6):1131–1134

Fedorov VV, Surin VL, Valchuk OP, Kapitonova LP, Kerimov AB, Formozov NA (2009) Maintaining morphological specificity and genetic introgression in populations of the great tit *Parus major* and the Japanese tit *P. minor* in the Middle Amur region. Russ J Genet 45 (7):771–780

Ficken MS, Rusch KM, Taylor AJ, Powers DR (2000) Blue-throated hummingbird song: a pinnacle of nonoscine vocalizations. Auk 117(1):120–128

Forstmeier W, Leisler B (2004) Repertoire size, sexual selection, and offspring viability in the great reed warbler: changing patterns in space and time. Behav Ecol 15(4):555–563

Gahr M (2000) Neural song control system of hummingbirds: comparison to swifts, vocal learning (Songbirds) and nonlearning (Suboscines) passerines, and vocal learning (Budgerigars) and nonlearning (Dove, owl, gull, quail, chicken) nonpasserines. Comp Neurol 426(2):182–196

Gil D, Gahr M (2002) The honesty of bird song: multiple constraints for multiple traits. Trends Ecol Evol 17(3):133–141

Grant BR, Grant PR (2010) Songs of Darwin's finches diverge when a new species enters the community. PNAS 107(47):20156–20163

Haavie J, Borge T, Bureš S, Garamszegi LZ, Lampe HM, Moreno J, Qvarnström A, Török J, Sætre GP (2004) Flycatcher song in allopatry and sympatry – convergence, divergence and reinforcement. J Evol Biol 17(2):227–237

Hall ML (2009) A review of vocal duetting in birds. Adv Study Behav 40:67–21

Helbig AJ, Salomon M, Bensch S, Seibold I (2001) Male-biased gene-flow across an avian hybrid zone: evidence from mitochondrial and microsatellite DNA. J Evol Biol 14(2):277–287

Hultsch H, Todt D (2004) Learning to sing. In: Marler P, Slabbekoorn H (eds) Nature's music – the science of birdsong. Elsevier Academic, London, pp 80–107

Hunter ML, Krebs JR (1979) Geographical variation in the song of the great tit (*Parus major*) in relation to ecological factors. J Anim Ecol 48(3):759–785

Irwin DE (2000) Song variation in an avian ring species. Evolution 54(3):998–1010

Irwin DE, Price T (2009) Sexual imprinting learning and speciation. Heredity 82(4):347–354

Irwin DE, Bensch S, Price TD (2001) Speciation in a ring. Nature 409(6818):333–337

Jarvis ED (2006) Evolution of brain structures for vocal learning in birds: a synopsis. Act Zool Sin 52(suppl):85–89

Jetz W, Thomas GH, Joy JB, Hartmann K, Mooers AO (2012) The global diversity of birds in space and time. Nature 491(7424):444–448

Kleindorfer S, Dudaniec RY (2016) Host-parasite ecology, behavior and genetics: a review of the introduced fly parasite *Philornis downsi* and its Darwin's finch hosts. BMC Zool 1:1

Kovylov NS, Marova IM, Ivanitsky VV (2012) Variation of song and plumage of the western (*Phylloscopus trochiloides viridanus*) and eastern (*Phylloscopus trochiloides plumbeitarsus*) forms of greenish warbler in sympatry zone: whether the hypothesis of ring speciation is true? Biol Bull 39(9):729–740

Kroodsma D (2004) The diversity and plasticity of birdsong. In: Marler P, Slabbekoorn H (eds) Nature's music – the science of birdsong. Elsevier Academic, London, pp 108–131

Kvist L, Rytkönen S (2006) Characterization of a secondary contact zone of the Great tit *Parus major* and the Japanese tit *P. minor* (Aves: Passeriformes) in Far Eastern Siberia with DNA markers. Zootaxa 1325:55–73

Lachlan RF, Servedio MR (2004) Song learning accelerates allopatric speciation. Evolution 58 (9):2049–2063

Lachlan RF, Verzijden MN, Bernhard CS, Jonker P-P, Koese B, Jaarsma S, Spoor W, Slater PJB, ten Cate C (2013) The progressive loss of syntactical structure in bird song along an island colonization chain. Curr Biol 23(19):1896–1901

Lipshutz SE, Overcast IA, Hickerson MJ, Brumfield RT, Derryberry EP (2017) Behavioural response to song and genetic divergence in two subspecies of white-crowned sparrows (*Zonotrichia leucophrys*). Mol Ecol 26(11):3011–3027

Logue DM, Hall ML (2014) Migration and the evolution of duetting in songbirds. Proc R Soc Lond B 281(1782):20140103

Logue DM, Krupp DB (2016) Duetting as a collective behavior. Front Ecol Evol 4:7

Luther D (2009) The influence of the acoustic community on songs of birds in a neotropical rain forest. Behav Ecol 20(4):864–871

Lynch A (1996) The population memetics of birdsong. In: Kroodsma DE, Miller EH (eds) Ecology and evolution of acoustic communication in birds. Cornell University Press, Ithaca, NY, pp 221–240

MacDougall-Shackleton EA, MacDougall-Shackleton SA (2001) Cultural and genetic evolution in mountain white-crowned sparrows: song dialects are associated with population structure. Evolution 55(12):2568–2575

Mahler B, Gil D (2009) The evolution of song in the *Phylloscopus* leaf warblers (Aves: Sylviidae): a tale of sexual selection, habitat adaptation, and morphological constraints. Adv Study Behav 40:35–66

Marler P (2004a) Science and birdsong: the good old days. In: Marler P, Slabbekoorn H (eds) Nature's music – the science of birdsong. Elsevier Academic, London, pp 1–38

Marler P (2004b) Bird calls: a cornucopia for communication. In: Marler P, Slabbekoorn H (eds) Nature's music – the science of birdsong. Elsevier Academic, London, pp 132–177

Marler P, Slabbekoorn H (eds) (2004) Nature's music – the science of birdsong. Elsevier Academic, London

Martens J (1996) Vocalization and speciation of Palearctic birds. In: Kroodsma DE, Miller EH (eds) Ecology and evolution of acoustic communication in birds. Cornell University Press, Ithaca, NY, pp 221–240

Martens J (1998) Geographische Variabilität der Lautäußerungen von Sperlingsvögeln – Auswirkungen auf Artbildung und das Artkonzept (Aves: Passeriformes: Oscines). In: Eck (ed) 100 Jahre Artkonzepte in der Biologie. Zool Abh Mus Tierk Dresden 50(suppl):35–50

Martens J, Päckert M (2007) Ring species – do they exist in birds? Zool Anz 246(4):315–324

Martens J, Päckert M, Nazarenko A, Valchuk O, Kawaji N (1998) Comparative bioacoustics of territorial song in the Goldcrest (*Regulus regulus*) and its implications for the intrageneric phylogeny of the genus *Regulus*. Zool Abh Mus Tierkd Dresden 50(7):99–128

Mason NA, Burns KJ, Tobias JA, Claramunt S, Seddon N, Derryberry EP (2017) Song evolution, speciation and vocal learning in passerine birds. Evolution 71(3):786–796

McGregor PK, Krebs JR (1982) Mating and song types in the great tit. Nature 297:60–61

Morinay J, Cardoso GC, Doutrelant C, Covas R (2013) The evolution of birdsong on islands. Ecol Evol 3(16):5127–5140

Murphy MT, Sexton K, Dolan ACD, Redmont LJ (2008) Dawn song of the eastern kingbird: an honest signal of male quality? Anim Behav 75(3):1075–1084

Nazarenko AA, Valchuk O, Martens J (1999) Secondary contact and overlap of *Parus major* and *Parus minor* populations in the middle Amur River basin. Zool Zhurn 78(3):371–381

Nemeth E, Winkler H, Dabelsteen T (2001) Differential degradation of antbird songs in a neotropical rainforest: adaptation to perch height? J Acoust Soc Am 110(6):3263–3274

Nicolai J, Gundacker C, Teeselink K, Güttinger HR (2014) Human melody singing by bullfinches (*Pyrrhula pyrrhula*) gives hints about a cognitive note sequence processing. Anim Cogn 17 (1):143–155

Nottebohm F (1968) Auditory experience and song development in the chaffinch *Fringilla coelebs*. Ibis 110(4):549–568

Nottebohm F (1969) The "critical period" for song learning in birds. Ibis 111(3):386–387

Nottebohm F (1970) Ontogeny of bird song. Science 167(3920):950–956

Nottebohm F (1971) Neural lateralization of vocal control in a passerine bird song. J Exp Zool 177 (2):229–262

Nowicki S, Searcy WA (2014) The evolution of vocal learning. Curr Opin Neurobiol 28:48–53

Nowicki S, Peters S, Podos J (1998) Song learning, early nutrition and sexual selection in songbirds. Am Zool 38(1):179–190

O'Loghlen AL, Beecher MD (1997) Sexual preferences for mate song types in female song sparrows. Anim Behav 53(4):835–841

Odom KJ, Hall ML, Riebel K, Omland KE, Langmore NE (2014) Female song is widespread and ancestral in songbirds. Nat Commun 5:3379

Olofsson H, Servedio MR (2008) Sympatry affects the evolution of genetic versus cultural determination of song. Behav Ecol 19(3):596–604
Päckert M, Martens J (2004) Song dialects on the Atlantic islands: goldcrests of the Azores (*Regulus regulus azoricus*, *R. r. sanctae-mariae*, *R. r. inermis*). J Ornithol 145(1):23–30
Päckert M, Martens J, Hofmeister T (2001) Lautäußerungen der Sommergoldhähnchen von den Inseln Madeira und Mallorca (*Regulus ignicapillus madeirensis*, *R. i. balearicus*). J Ornithol 142(1):16–29
Päckert M, Martens J, Nazarenko AA, Kosuch J, Veith M (2003) Phylogenetic signal in song of crests and kinglets (Aves: *Regulus*). Evolution 57(3):616–629
Päckert M, Martens J, Sun YH, Veith M (2004) The radiation of the *Seicercus burkii* complex and its congeners – molecular genetics and bioacoustics. Org Divers Evol 4(4):341–364
Päckert M, Martens J, Nazarenko AA, Valchuk O, Petri B, Eck S, Veith M (2005) The great tit, *Parus major*, a misclassified ring species. Biol J Linn Soc 86(2):153–174
Päckert M, Martens J, Severinghaus LL (2009) The Taiwan Firecrest (*Regulus goodfellowi*) belongs to the Goldcrest assemblage (*Regulus regulus* s. l.) – evidence from mitochondrial DNA and territorial song of the Regulidae. J Ornithol 150(1):205–220
Pepperberg IM (2004) Grey parrots: learning and using speech. In: Marler P, Slabbekoorn H (eds) Nature's music – the science of birdsong. Elsevier Academic, London, pp 363–373
Pepperberg IM (2010) Vocal learning in grey parrots: a brief review of perception, production, and cross-species comparisons. Brain Lang 115(1):81–91
Petrusková T, Diblíková L, Pipek P, Frauendorf E, Procházka P, Petrusek A (2015) A review of the distribution of Yellowhammer (*Emberiza citrinella*) dialects in Europe reveals the lack of a clear macrogeographic pattern. J Ornithol 156(1):263–273
Pipek P, Petrusková T, Petrusek A, Diblíková L, Eaton MA, Pyšek P (2017) Dialects of an invasive songbird are preserved in its invaded but not native source range. Ecography 40:001–009
Podos J (2001) Correlated evolution of morphology and vocal signal structure in Darwin's finches. Nature 409(6817):185–188
Podos J, Southall JA, Rossi-Santos MR (2004) Vocal mechanics in Darwin's finches: correlation of beak gape and song frequency. J Exp Biol 207(4):607–619
Potvin DA, Clegg SM (2015) The relative roles of cultural drift and acoustic adaptation in shaping syllable repertoires of island bird populations change with time since colonization. Evolution 69 (2):368–380
Price JJ (2015) Rethinking our assumptions about the evolution of bird song and other sexually dimorphic signals. Front Ecol Evol 3:Article 40
Prum RO (2012) Aesthetic evolution by mate choice: Darwin's really dangerous idea. Philos Trans R Soc B 367(1600):2253–2265
Purushotham CB, Robin VV (2016) Sky island bird populations isolated by ancient genetic barriers are characterized by different song traits than those isolated by recent deforestation. Ecol Evol 6 (20):7334–7343
Pusey A, Wolf M (1996) Inbreeding avoidance in animals. Trend Ecol Evol 11(5):201–206
Qvarnström A, Haavie J, Sæther SA, Eriksson D, Pärt T (2006) Song similarity predicts hybridization in flycatchers. J Evol Biol 19(4):1202–1209
Reid JM, Arcese P, Cassidy ALE, Hieberts SM, Smith JNM, Stoddard PK, Marr AB, Keller LF (2004) Song repertoire size predicts initial mating success in male song sparrows, *Melospiza melodia*. Anim Behav 68(5):1055–1063
Rheindt FE, Grafe TU, Abouheif E (2014) Rapidly evolving traits and the comparative method: how important is testing for phylogenetic signal? Evol Ecol Res 6(3):377–396
Riebel K, Lachlan RF, Slater PJB (2015) Learning and cultural transmission in chaffinch song. In: Naguib M, Brockmann HJ, Mitani JC, Simmons LW, Barrett L, Healy S, PJB S (eds) Advances in the study of behavior, vol 47. Elsevier Science, San Diego, pp 181–227
Shipilina D, Serbyn M, Ivanitskii V, Marova I, Backström N (2017) Patterns of genetic, phenotypic, and acoustic variation across a chiffchaff (*Phylloscopus collybita abietinus/tristis*) hybrid zone. Ecol Evol 7(7):2169–2180

Singh P, Price TD (2015) Causes of the latitudinal gradient in birdsong complexity assessed from geographical variation within two Himalayan warbler species. Ibis 157(3):511–527
Slabbekoorn H (2004) Singing in the wild: the ecology of birdsong. In: Marler P, Slabbekoorn H (eds) Nature's music – the science of birdsong. Elsevier Academic, London, pp 178–206
Slabbekoorn H, Smith TB (2002) Bird song, ecology and speciation. Philos Trans R Soc Lond B 357(1420):493–503
Spencer KA, Buchanan KL, Goldsmith AR, Catchpole CK (2003) Song as an honest signal of developmental stress in the zebra finch (*Taeniopygia guttata*). Horm Behav 44(2):132–139
Spencer KA, Buchanan KL, Leitner S, Goldsmith AR, Catchpole CK (2005) Parasites affect song complexity and neural development in a songbird. R Proc Soc Lond B 272(1576):2037–2043
Suh A, Paus M, Kiefmann M, Churakov G, Franke FA, Brosius J, Kriegs JO, Schmitz J (2011) Mesozoic retroposons reveal parrots as the closest living relatives of passerine birds. Nat Commun 2:443
Thaler E (1990) Die Goldhähnchen. Neue Brehm Bücherei, Lutherstadt Wittenberg
Thielcke G (1970) Lernen von Gesang als möglicher Schrittmacher der Evolution. Z Zool Syst Evolutionsforsch 8(1):309–320
Thielcke G (1972) On the origin of divergence of learned signals (songs) in isolated populations. Ibis 115(4):511–516
Thorpe WH (1958) The learning of song patterns by birds, with especial reference to the song of the chaffinch *Fringilla coelebs*. Ibis 100(4):535–570
Tietze DT, Martens J, Sun Y-H, Päckert M (2008) Evolutionary history of treecreeper vocalisations (Aves: *Certhia*). Org Divers Evol 8(4):305–324
Tietze DT, Martens J, Sun YH, Severinghaus LL, Päckert M (2011) Song evolution in the Coal Tit. J Avian Biol 42(3):214–230
Tietze DT, Martens J, Fischer BS, Sun Y-H, Klussmann-Kolb A, Päckert M (2015) Evolution of leaf-warbler songs (Aves: Phylloscopidae). Ecol Evol 5(3):781–798
Tobias JA, Gamarra-Toldeo V, Carcía-Olachea D, Pulgarín PC, Seddon N (2011) Year-round resource defence and the evolution of male and female song in suboscine birds: social armaments are mutual ornaments. J Evol Biol 24(10):2118–2138
Toews DPL (2017) From song dialects to speciation in white-crowned sparrows. Mol Ecol 26 (11):2842–2844
Verzijden MN, ten Cate C, Servedio MR, Kozak GM, Boughman JW, Svensson EI (2012) The impact of learning on sexual selection and speciation. Trends Ecol Evol 27(9):511–519
Voigt C, Leitner S, Gahr M (2006) Repertoire and structure of duet and solo songs in cooperatively breeding white-browed sparrow weavers. Behaviour 143(2):159–182
Wegrzyn E, Leniowski K (2009) Syllable sharing and changes in syllable repertoire size and composition within and between years in the great reed warbler. J Ornithol 151(2):255–267
Weir JT, Wheatcroft D (2011) A latitudinal gradient in rates of evolution of avian syllable diversity and song length. Proc R Soc Lond B 278(2712):1713–1720

Chapter 6
Timing Matters: Allochronic Contributions to Population Divergence

Barbara Helm and Robyn Womack

Abstract Globally, most birds reproduce to some extent seasonally, and the timing of their annual breeding events involves regulation by biological clocks. Biological clocks also regulate diel activities, including song and other courtship behaviors which occur at certain times of day. Differences between individuals in the timing of display and breeding (i.e., allochrony) can drive genetic divergence, contribute to isolation between populations, and ultimately lead to speciation. Isolation by breeding time is thought to be common in areas where reproductive seasons differ greatly over short distances, for example in tropical regions. Here we introduce the ways biological clocks of birds function, and review evidence for variation between individuals and between populations. The potential importance of allochrony for the speciation of birds is underpinned by the periodic growth and regression of their reproductive organs, by their rigid migration programs, and possibly also by their learnt, well-timed mating behaviors (in particular song). We exemplify isolation by time in tropical songbirds and in migratory species. Potentially, further contexts that could promote allochronic isolation in birds include the differentiation between urban and rural populations, as urbanization is commonly associated with modified timing of breeding and song.

Keywords Isolation by time · Biological rhythm · Circannual · Circadian · Seasonality · Chronotype · Assortative mating · Song · Tropical · Migration

6.1 Timing Is Everything!

A number of environmental rhythms repeat faithfully and are highly predictable, because they arise from astronomical cycles. These include annual and daily rhythms caused by the orbit of the Earth around the Sun and the rotation of the

---

B. Helm (✉) · R. Womack
Institute of Biodiversity, Animal Health and Comparative Medicine, University of Glasgow, Glasgow, UK
e-mail: barbara.helm@glasgow.ac.uk

D. T. Tietze (ed.), *Bird Species*, Fascinating Life Sciences,
https://doi.org/10.1007/978-3-319-91689-7_6

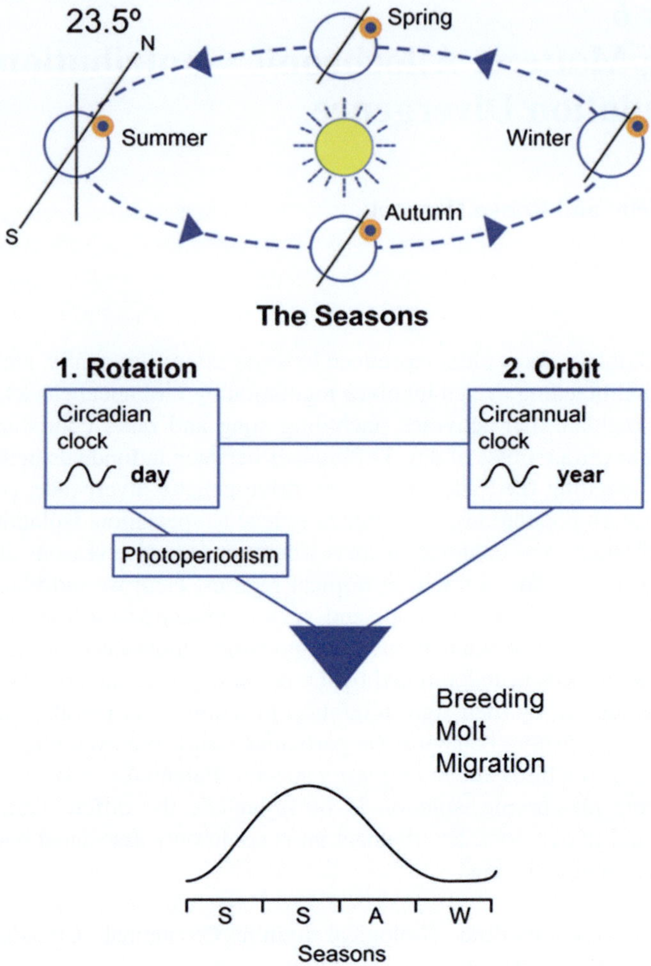

Fig. 6.1 Diel timing every 24 hours aligns with the rotation of the Earth (line S/N indicates the rotational axis, and the filled circle represents a specific time point of reference) and annual timing aligns with the Earth's orbit around the Sun every 12 months (shown by the eclipse). These astronomical cycles have favored the evolution of internal circadian (approximately daily) and circannual (approximately annual) clocks. Circadian rhythms regulate the timing of many daily activities, including song and courtship. For annual timing, the ability to respond to changes in day length (photoperiodism) depends on the circadian clock system. Birds utilize both types of internal, innate timing mechanism to regulate reproduction, molt, migration, and other processes [adapted from Helm and Lincoln (2017)]

Earth around its axis (Fig. 6.1). Likewise, the orbit of the Moon around the Earth shapes lunar and tidal rhythms. Many humans underestimate the challenges of living in a deeply rhythmic world, because they can extend their day into the night using artificial lights and are sheltered from cold, rain, and storms by housing, heating, and protective gear. In contrast, for a bird, a temperate-zone forest is a very different habitat at night than at daytime and during the summer compared to winter (Foster and Kreitzman 2005; Helm and Lincoln 2017). Changes in its environment include those in abiotic factors like light and temperature, in biotic opportunities and risks like food availability and exposure to predators and parasites, and in social factors like its forest bird community and the behavior of its conspecifics (Helm et al. 2006, 2017). Likewise, for a bird in a tropical forest or savannah, alternation between dry and rainy seasons can also fundamentally modify its habitat (Moore et al. 2005; Goymann and Helm 2014). It is therefore unsurprising that birds have structured their lives by these rhythms. Because the many, sometimes competing, demands on timing require careful balancing, species, populations, and individuals within populations may differ in the precise solution of timing challenges.

In this chapter, we argue that biological timekeeping, in particular on an annual and daily time scale, can promote isolation between differently timed individuals and can be a driver of the diversification of birds (Hendry and Day 2005; Taylor and Friesen 2017). We will first introduce basic background on biological rhythms and then demonstrate the diversity of avian timing, highlighting putative isolation mechanisms.

6.2 Clockworks

Because astronomically based environmental rhythms are so predictable (Fig. 6.1), organisms have evolved a range of preparatory strategies for coping with them. A big step early in the evolution of life (Hut and Beersma 2011) was anticipation of environmental changes, when cyanobacteria progressed from simply reacting to periodical changes to predicting them in advance. Since then, organisms have evolved sophisticated behaviors that rely on precise prediction of the periodic changes in the environment. For example, many migratory bird species leave their breeding grounds before they become inhospitable, and initiate their return movements in anticipation of the upcoming spring (Numata and Helm 2014). Likewise, birds are usually not simply woken by morning twilight but will have anticipated the upcoming day well in advance (Foster and Kreitzman 2005).

The remarkable achievement of anticipation is possible, because birds, like other organisms, have internalized timekeeping. They have evolved body clocks (referred to as biological rhythms) that are innate and thereby part of their inheritance from generation to generation. These clocks tick on even if the birds are sheltered from environmental changes, for example in a continuously lit room or under the continuous light of polar summer days (Foster and Kreitzman 2005; Ashley et al. 2014; Helm and Lincoln 2017). Birds will alternate each day between activity and rest, their body temperature will rise and fall, and they may rhythmically show behaviors, for example crowing, even under constant conditions (Shimmura and Yoshimura 2013; Shimmura et al. 2015). However, these biological clocks run at their own, internal speed, which often makes them drift a bit. If a bird cannot experience environmental cycles, its biological rhythm is usually somewhat faster, or slower, than the corresponding environmental cycles (hence, it is called a “circa” rhythm).

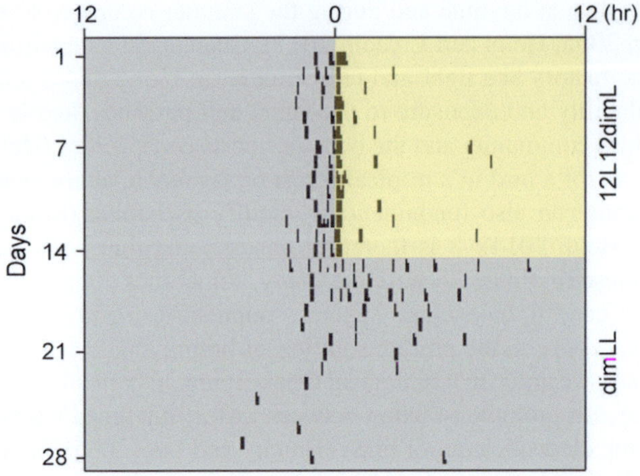

Fig. 6.2 Circadian regulation of rooster crowing. The figure shows crowing of a rooster (black marks) over 4 weeks. First, the bird was recorded under cycles of 12 hours of light followed by 12 hours of dimmed light at night, indicated by yellow and gray backgrounds, respectively. Then, the bird was recorded under constant conditions of dim light (gray background only). The onset of its crowing became progressively earlier under constant conditions. Time "0" on the *x*-axis shows start of light; time points "12" mark 12 hours before and after the start of light. Each line in the graph (*y*-axis) shows a day of recording [adapted from Shimmura and Yoshimura (2013), Shimmura et al. (2015)]

The best-known biological rhythm is that of the circadian clock which drives timing across the 24-hour day. This can be seen in Fig. 6.2 which shows the time of crowing in a rooster. While the rooster experiences changes between light (indicated by the yellow box) and darkness, it always starts crowing at a similar time of day, somewhat before dawn. But once the light is kept constant, the rooster crows a bit earlier each morning, and its rhythm drifts. This drift can be measured as the "period length" of a rooster, i.e., the time taken from one cycle (start of crowing) to the next. In Fig. 6.2, the rooster's clock is faster than 24 hours, and hence, it has a shorter period length. In their natural environment, such fast-clocked individuals are usually particularly early risers, whereas slow-clocked individuals tend to be late risers (Dominoni et al. 2013b).

For the circadian clock, the mechanisms that drive this rhythm have been intensively studied, highlighted by the recent award of the 2017 Nobel Prize in Physiology or Medicine. We now know that rhythms are generated within cells by a loop of so-called clock genes, which switch each other on and off to measure out the period of circa 24 hours. Many other genes are involved, for example by linking clocks to metabolism. Then, the millions of cellular clocks in a bird, fly or human need to be coordinated to produce useful body time. This is achieved, for example by nerve-cell coupling in brain centers and by hormones, such as melatonin which peaks at night (Foster and Kreitzman 2005). Through various links and feedback loops, organisms thus achieve coordinated rhythms within their bodies. Because clocks are so important to the life of organisms, they are very sensitive to important cues from the environment, in particular to changes of light and darkness. Birds perceive these changes and adjust ("synchronize") their biological rhythms so that they match the 24-hour daily cycle of the environment (Fig. 6.2).

Although only circadian rhythms are understood in detail, the same principles hold for cycles on other time scales (Numata and Helm 2014). Recent, exciting breakthroughs have shed light on circatidal clocks, which help coastal organisms anticipate the rising and falling of marine water levels (Zhang et al. 2013; Kaiser et al. 2016), and circalunar clocks, by which organisms anticipate the waxing and waning of the Moon (Zantke et al. 2013). Finally, the rhythms that relate most directly to avian reproduction are circannual rhythms (cycles repeating with a period length of circa one year; Gwinner 1986, 1996; Helm and Lincoln 2017). Circannual rhythms can regulate many processes, including preparations for breeding, molt, and migration. Most avian species breed at least to some extent seasonally (Goymann and Helm 2014). Likewise, birds generally molt at least once per year to replace their feathers, and many species carry out regular migrations which are often precisely timed (Gwinner 1996; Battley 2006; Helm et al. 2006; Newton 2008). Associated with their aerial lifestyle, birds undergo extreme annual changes: for example, they greatly reduce their reproductive organs (testes of males, ovaries and oviducts of females; Williams 2012) outside the breeding seasons. In small songbirds, testes typically grow to diameters of 0.5 cm during breeding but shrink in winter to below 10% of this size and contain no sperm (Helm 2009; Williams 2012). This makes reproduction in birds generally highly seasonal, preventing many species from spontaneously breeding, for example during warm weather in winter or after migrating to tropical areas. An example for innate circannual rhythms in reproductive condition is shown in Fig. 6.3 for Eurasian Blackcaps *Sylvia atricapilla*.

As described above for circadian rhythms, circannual cycles also typically continue if a bird is prevented from witnessing environmental changes (Gwinner 1996; Goymann and Helm 2014). For example, in stonechats (songbirds within the genus *Saxicola*), the growth and reduction of reproductive organs is driven by a circannual clock. When pairs were kept in aviaries under constant day length of 12 hours of light alternating with 12 hours of night, males and females reproduced if they came into reproductive condition at similar times (Gwinner 1996; Goymann and Helm 2014). Their sons and daughters never experienced rhythms in their environment, but nonetheless showed annual cycles of breeding condition and molt, driven by their circannual clocks. Under these conditions, breeding condition of the stonechats drifted from the annual cycle of 365 days, much like the crowing behavior of the rooster drifted from the 24-hour cycle shown in Fig. 6.2. Under natural conditions, circannual rhythms like circadian rhythms synchronize to environmental cycles (Gwinner 1986; Helm and Lincoln 2017). The most important synchronizing cue on an annual time scale is photoperiod (day length), although other cues can also play a role (Helm et al. 2006; Goymann and Helm 2014).

The precise timing of behavior and activities of birds under natural conditions is influenced by many additional factors that modify the outcome of biological timekeeping. For example, the time of song and breeding can be influenced by age,

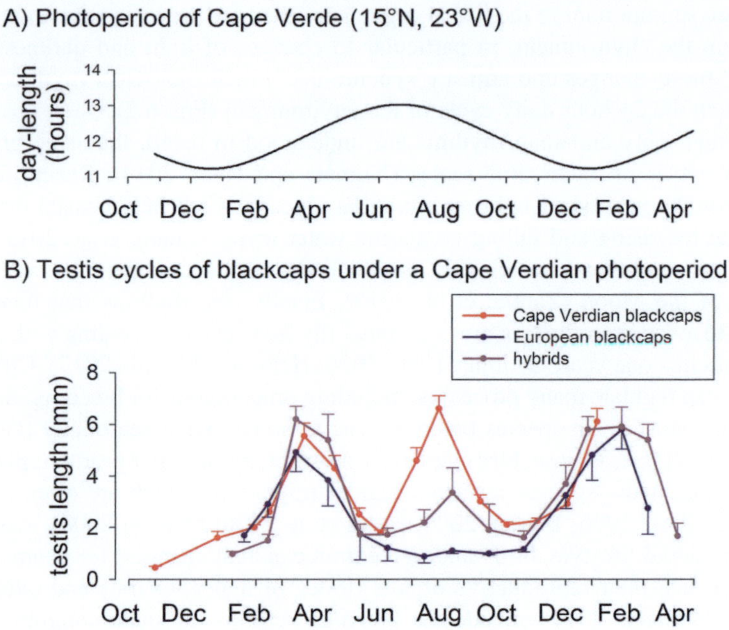

Fig. 6.3 Innate, internal cycles of testis growth and regression in male Eurasian Blackcaps *Sylvia atricapilla*. Blackcaps on the Cape Verde islands go through two reproductive cycles per year, whereas European Blackcaps go through one annual cycle. In captivity, the birds show the time patterns of their native populations, and hybrids are intermediate. Male Blackcaps were kept under the photoperiodic condition of the Cape Verde islands. (**a**) Number of daylight hours. (**b**) Testicular cycles of Blackcaps from Cape Verde (red), southern Germany (blue), and hybrids (purple) of the two populations maintained under identical conditions [data from Berthold and Querner (1993), figure adapted from Goymann and Helm (2014)]

experience, dominance, body condition, and various local environmental factors (Poesel et al. 2006; Helm et al. 2006; Shimmura et al. 2015). But importantly for the subject of avian speciation, innate biological rhythms provide a powerful substrate for evolutionary change (Fig. 6.3; Helm and Visser 2010; Taylor and Friesen 2017).

6.3 Allochrony: Differences in Timing Between Individuals, Populations, and Species

Birds that differ from each other in reproductive timing are said to be “allochronic” (i.e., they diverge in timing; Taylor and Friesen 2017). In the above description of clockworks, we explained that individuals may display consistently early or late timing behavior, captured by the term “chronotype” (Hau et al. 2017; Helm et al. 2017). Such variation is found between individuals, between populations, and

Fig. 6.4 Chronotype of individual songbirds schematically depicted against the backdrop of a Scottish forest woodland. The image shows three individuals during early morning hours which differ slightly from each other in their biological clock times (depicted by a clock imposed on each bird's head). For two of these individuals, subjectively, the time for starting dawn song (set to just after 03:00 hours in May in Scotland) is yet to come. The third individual, in contrast, has a slightly earlier biological rhythm. For this bird, the time for initiating song has already started (depicted by its biological clock showing 03:05 hours), and the bird is accordingly singing. Original artwork: Robyn Womack

between species. Within populations, some individuals are typically the first to become active in the morning, while others may start later but extend their activities into the evening. This is illustrated in Fig. 6.4, which schematically shows three conspecifics at the same time in the same forest patch. The three birds differ slightly in their biological rhythm, with their chronotype highlighted as clock time. Similarly, on an annual time scale, some individuals come into reproductive condition or return from migration early in the year, whereas others might be timed later. On both diel and annual time scales, there is evidence that individual differences are heritable (e.g., Nussey et al. 2005; Helm and Visser 2010; Caprioli et al. 2012).

Timing differences between individuals can have important reproductive consequences. Courtship and reproductive behaviors are often fitted to short phases of conducive environmental conditions. Therefore, reproductive success can differ between individuals performing these behaviors early or late in the day or year (Hau et al. 2017). For example, in Fig. 6.4, the bird whose clock is the most advanced is already singing, while the others are still preparing for the day. Several studies have shown that a male that starts dawn song particularly early in the day has high chances of enjoying increased reproductive success (Poesel et al. 2006; Hau et al. 2017). A recent review article (Hau et al. 2017) discusses to which extent the timing of behaviors may therefore be under sexual selection. As detailed by Hau and colleagues, for the timing of behaviors such as song to be a candidate for sexual selection, it would need to increase matings but also impose a cost on the singing bird (e.g., shortened sleep duration). For sexual selection to take effect, there also needs to be a genetic contribution to the specific timing of an individual, which can then be modified by evolution (Helm and Visser 2010). Clock genes, and potentially their epigenetic modification, have been shown to play a role in such modifications in a number of taxa (Caprioli et al. 2012; Hau et al. 2017; Taylor and Friesen 2017; Saino et al. 2017).

However, sexual selection is not the only factor that could drive and maintain genetic variation in the timing of individuals. It is very likely that there are also other pressures, such as natural selection for greater fecundity or survival (Hau et al. 2017). For example, for both males and females, waking early could enable improved feeding opportunities, and an early return from migration could provide access to superior breeding territories (Bearhop et al. 2005; Hau et al. 2017). Thereby, individuals which due to a mutation rise or return earlier could enjoy greater fecundity or survival. Maintaining diversity in timing however requires that such benefits are not universal. For example, early risers may sometimes suffer energetic costs on brisk mornings or predation by nocturnal predators, counterbalancing natural selection for earlier onset of activity. Likewise, early-returning migrants can get caught in severe weather in late winter, from which late-returning migrants are spared (Newton 2008). A role of biological clocks in the contexts of sexual and natural selection has been demonstrated in a range of organisms, especially in insects and fish (Kaiser et al. 2016; Taylor and Friesen 2017).

For variation between individuals to drive divergence within a population, there must be isolating mechanisms. For example, because reproductive behaviors often occur during short phases only, males and females that are similarly timed might preferentially mate with each other (Fig. 6.4). Such "assortative mating" could then accelerate genetic changes in the clockworks, further driving divergence within a population (Bearhop et al. 2005; Dominoni et al. 2013a, b). An example is once again Blackcaps (Bearhop et al. 2005). In this species, some individuals began making a genetically determined shorter migration journey and returned to the breeding grounds earlier in the spring (Newton 2008). Males and females carrying out this shorter migration preferentially mated with each other soon after arrival on the breeding grounds, before conspecifics taking a longer route arrived (Bearhop et al. 2005). This difference in mating times could have contributed to the observed, fast spread of the novel migration behavior.

Variation in chronotype within populations connects seamlessly to differences between populations, if time-structuring factors isolate parts of a population from others. Migration, as described above, can be one such factor. Others can be important ecological factors that differ in timing. Here, the most convincing examples come from insects, for example, from fruit flies that have specialized on host plants that differ in timing of fruition (Taylor and Friesen 2017). In birds, most examples of allochronic divergence involve some separation in space at the time of breeding. Spatial distances do not need to be large because habitats can differ in optimal timing even over short distances, for example in tropical environments.

Thus, breeding timing differs dramatically between populations of Rufous-collared Sparrows *Zonotrichia capensis* that breed on the western and eastern slopes of the Ecuadorian Andes, which in turn differ in the timing of rainy seasons (Moore et al. 2005). These populations do not overlap in the timing of breeding and have already started to diverge (see below).

On larger spatial scales, populations may differ systematically in reproductive timing, for example across latitudinal gradients, where more northerly breeding populations typically reproduce later in the year (Helm 2009). One extreme example is stonechat populations from different geographic origins that differ so starkly in reproductive timing (Helm 2009) that across the distribution range, some local populations will be in breeding condition at any time of the year. Another example is the blackcaps shown in Fig. 6.3. Blackcaps on the Cape Verde islands go through two full annual cycles of gonadal growth and regression and breed during two distinct breeding seasons, unlike Eurasian Blackcaps that have a single gonadal cycle and breeding season per year. Figure 6.3 shows that under identical captivity conditions, these differences were retained: Each population showed its characteristic reproductive cycle. Moreover, the differences were heritable, as indicated by the intermediate patterns in hybrids (Berthold and Querner 1993; Goymann and Helm 2014).

With greater taxonomic distance, such timing differences generally become even more evident, and links to underlying biological clocks are starting to be revealed (Taylor and Friesen 2017). For example, differences in breeding time have been associated with latitudinal clines in clock gene variants, with the most convincing examples relating to fish and insects (Hut et al. 2013; Taylor and Friesen 2017).

6.4 Isolation by the Clock

If differences in timing are underwritten by differences in the birds' biological rhythms, as in the stonechats and Blackcaps, they can become powerful isolation mechanisms and ultimately lead to speciation. For example, stonechats from differently timed populations mentioned above were crossbred with each other under identical captive conditions (Helm 2009). Like the Blackcaps shown in Fig. 6.3, the captive stonechats displayed the typical timing of their populations. Only when the breeding phases of the two populations overlapped were pairs from different populations successful in producing viable offspring (Helm 2009). Thus in the wild, such allochronic populations might already be limited in possible gene flow if they were to meet, for example through dispersing individuals.

In contrast to the well-established concept of "isolation by space," such "isolation by time" (Hendry and Day 2005) has only more recently been established as a formal idea. In wild birds, some convincing examples for a role of allochrony for speciation come from seabirds that nest colonially on islands (Taylor and Friesen 2017), for example Band-rumped Storm Petrel *Oceanodroma castro* (Friesen et al. 2007). On some islands, different parts of the Band-rumped Storm Petrel population breed sympatrically, but at completely different times of the year, and thereafter leave the colonies for the rest of the year (Friesen et al. 2007). As a consequence, these allochronically breeding petrels never meet, and the island population shows substantial genetic substructure by reproductive timing. In this example, breeders on one and the same island are in effect allopatric during reproductive seasons and have greatly diverged from each other.

Initial isolation of allochronically breeding populations has also been shown for wild birds that reproduce in neighboring locations. The Andean Rufous-collared Sparrows described above have diverged genetically, but furthermore, they show also signs of divergence in sexual signals. When song from the allochronic populations of the opposite Andean slope was played back to the local population, the receiving birds responded more hesitantly than when they heard songs from synchronic populations (Danner et al. 2011). This might suggest that avian song, which is only performed during the breeding season, may diverge quickly between populations that never hear each other sing, and thereby further accelerate allochronic isolation (Wilkins et al. 2013). Allochronic divergence of birds over short distances might be relatively common in tropical regions, where seasonality is often very local. A recent study involving 57 New World bird species related speciation to allochrony in response to different seasonal timing of food availability and ultimately precipitation (Quintero et al. 2014). Findings of comparative analyses were seen to support a so-called asynchrony of seasons hypothesis (Martin et al. 2009), which proposes that allochrony is an important contributor to the high speciation rates in the tropics.

6.5 Conclusions

Most studies on divergence between populations and on speciation have focused on spatial segregation, but recent evidence emphasizes also a role for temporal segregation. The evolutionary mechanisms are probably somewhat different because unlike spatial behavior, temporal behavior is based on innate, internal biological mechanisms. Because biological rhythms differ somewhat between individuals, they offer rich substrate for evolutionary change and can contribute to isolation and speciation, for example by mating time. The potential importance of allochrony for avian speciation might be increased by seasonally displayed mating behaviors, in particular by song, which may function as an additional isolation factor (Wilkins et al. 2013). Examples for allochronic divergence of bird populations that breed in relative proximity, so far, stem mainly from migratory species and from tropical areas. However, allochronic isolation might be more widespread and could occur, so to speak, at our door step. For example, birds breeding in urban habitats often sing much earlier in the day and reproduce far earlier in the year than conspecifics in nearby rural areas. Such behavioral differences on both time scales have been linked to the birds' biological rhythms (Dominoni et al. 2013a, b) and could contribute to decreasing gene exchange between urban and rural populations (Gill and Brumm 2013).

Further Reading

Hau M, Dominoni D, Casagrande S, Buck CL, Wagner G, Hazlerigg D et al (2017) Timing as a sexually selected trait: the right mate at the right moment. Philos Trans R Soc B Biol Sci 372

Moore IT, Bonier F, Wingfield JC (2005) Reproductive asynchrony and population divergence between two tropical bird populations. Behav Ecol 16:755–762

Numata H, Helm B (2014) Annual, lunar and tidal clocks: patterns and mechanisms of nature's enigmatic rhythms. Springer Japan, Tokyo

Taylor RS, Friesen VL (2017) The role of allochrony in speciation. Mol Ecol 26:3330–3342

References

Ashley NT, Ubuka T, Schwabl I, Goymann W, Salli BM, Bentley GE et al (2014) Revealing a circadian clock in captive arctic-breeding songbirds, lapland longspurs (*Calcarius lapponicus*), under constant illumination. J Biol Rhythm 29:456–469

Battley PF (2006) Consistent annual schedules in a migratory shorebird. Biol Lett 2:517–520

Bearhop S, Fiedler W, Furness RW, Votier SC, Waldron S, Newton J et al (2005) Assortative mating as a mechanism for rapid evolution of a migratory divide. Science 310:502–504

Berthold P, Querner U (1993) Genetic and photoperiodic control of an avian reproductive cycle. Cell Mol Life Sci 49(4):342

Caprioli M, Ambrosini R, Boncoraglio G, Gatti E, Romano A, Romano M et al (2012) Clock gene variation is associated with breeding phenology and maybe under directional selection in the migratory Barn Swallow. PLoS One 7:e35140

Danner JE, Danner RM, Bonier F, Martin PR, Small TW, Moore IT (2011) Female, but not male, tropical sparrows respond more strongly to the local song dialect: implications for population divergence. Am Nat 178:53–63

Dominoni D, Quetting M, Partecke J (2013a) Artificial light at night advances avian reproductive physiology. Proc Biol Sci 280:20123017

Dominoni DM, Helm B, Lehmann M, Dowse HB, Partecke J (2013b) Clocks for the city: circadian differences between forest and city songbirds. Proc Biol Sci 280:20130593

Foster RG, Kreitzman L (2005) In rhythms of life: the biological clocks that control the daily lives of every living thing. Yale University Press, New Haven, CT

Friesen VL, Smith AL, Gómez-Díaz E, Bolton M, Furness RW, González-Solís J et al (2007) Sympatric speciation by allochrony in a seabird. Proc Natl Acad Sci U S A 104:18589–18594

Gill D, Brumm H (2013) Avian urban ecology: behavioural and physiological adaptations. Oxford University Press, Oxford

Goymann W, Helm B (2014) Seasonality of life-cycles in tropical birds – circannual rhythms and Zeitgeber. In: Numata H, Helm B (eds) Annual, Lunar and tidal clocks: patterns and mechanisms of nature's enigmatic rhythms. Springer Japan, Tokyo, pp 237–275

Gwinner E (1986) Circannual clocks. Springer, Berlin

Gwinner E (1996) Circannual clocks in avian reproduction and migration. Ibis 138:47–63

Hau M, Dominoni D, Casagrande S, Buck CL, Wagner G, Hazlerigg D et al (2017) Timing as a sexually selected trait: the right mate at the right moment. Philos Trans R Soc B Biol Sci 372:20160249

Helm B (2009) Geographically distinct reproductive schedules in a changing world: costly implications in captive Stonechats. Integr Comp Biol 49:563–579
Helm B, Lincoln GA (2017) Circannual rhythms anticipate the earth's annual periodicity. In: Kumar V (ed) Biological timekeeping: clocks, rhythms and behaviour. Springer India, New Delhi, pp 545–569
Helm B, Visser ME (2010) Heritable circadian period length in a wild bird population. Proc B Biol Sci 277:3335–3342
Helm B, Piersma T, Van der Jeugd H (2006) Sociable schedules: interplay between avian seasonal and social behaviour. Anim Behav 72:245–262; 1215
Helm B, Visser ME, Schwartz W, Kronfeld-Schor N, Gerkema M, Piersma T et al (2017) Two sides of a coin: ecological and chronobiological perspectives of timing in the wild. Philos Trans R Soc B Biol Sci 372:20160246
Hendry AP, Day T (2005) Population structure attributable to reproductive time: isolation by time and adaptation by time. Mol Ecol 14:901–916
Hut RA, Beersma DGM (2011) Evolution of time-keeping mechanisms: early emergence and adaptation to photoperiod. Philos Trans R Soc B Biol Sci 366:2141–2154
Hut RA, Paolucci S, Dor R, Kyriacou CP, Daan S (2013) Latitudinal clines: an evolutionary view on biological rhythms. Proc R Soc Lond B Biol Sci 280:20130433
Kaiser TS, Poehn B, Szkiba D, Preussner M, Sedlazeck FJ, Zrim A et al (2016) The genomic basis of circadian and circalunar timing adaptations in a midge. Nature 540:69–73
Martin PR, Bonier F, Moore IT, Tewksbury JJ (2009) Latitudinal variation in the asynchrony of seasons: implications for higher rates of population differentiation and speciation in the tropics. Ideas Ecol Evol 2
Moore IT, Bonier F, Wingfield JC (2005) Reproductive asynchrony and population divergence between two tropical bird populations. Behav Ecol 16:755–762
Newton I (2008) The migration ecology of birds. Academic, London
Numata H, Helm B (2014) Annual, lunar and tidal clocks: patterns and mechanisms of nature's enigmatic rhythms. Springer Japan, Tokyo
Nussey DH, Postma E, Gienapp P, Visser ME (2005) Selection on heritable phenotypic plasticity in a wild bird population. Science 310:304–306
Poesel A, Kunc HP, Foerster K, Johnsen A, Kempenaers B (2006) Early birds are sexy: male age, dawn song and extrapair paternity in blue tits, *Cyanistes* (formerly *Parus*) *caeruleus*. Anim Behav 72:531–538
Quintero I, González-Caro S, Zalamea P-C, Cadena CD (2014) Asynchrony of seasons: genetic differentiation associated with geographic variation in climatic seasonality and reproductive phenology. Am Nat 184:352–363
Saino N, Ambrosini R, Albetti B, Caprioli M, De Giorgio B, Gatti E et al (2017) Migration phenology and breeding success are predicted by methylation of a photoperiodic gene in the barn swallow. Sci Rep 7:45412
Shimmura T, Yoshimura T (2013) Circadian clock determines the timing of rooster crowing. Curr Biol 23:R231–R2R3
Shimmura T, Ohashi S, Yoshimura T (2015) The highest-ranking rooster has priority to announce the break of dawn. Sci Rep 5:11683
Taylor RS, Friesen VL (2017) The role of allochrony in speciation. Mol Ecol 26:3330–3342
Wilkins MR, Seddon N, Safran RJ (2013) Evolutionary divergence in acoustic signals: causes and consequences. Trends Ecol Evol 28:156–166. https://doi.org/10.1016/j.tree.2012.10.002
Williams TD (2012) Physiological adaptations for breeding in birds. Princeton University Press, Princeton

Zantke J, Ishikawa-Fujiwara T, Arboleda E, Lohs C, Schipany K, Hallay N et al (2013) Circadian and circalunar clock interactions in a marine annelid. Cell Rep 5:99–113

Zhang L, Hastings MH, Green EW, Tauber E, Sladek M, Webster SG et al (2013) Dissociation of circadian and circatidal timekeeping in the marine crustacean Eurydice pulchra. Curr Biol 23:1863–1873

Chapter 7
(Micro)evolutionary Changes and the Evolutionary Potential of Bird Migration

Miriam Liedvogel and Kira Delmore

Abstract Seasonal migration is the yearly long-distance movement of individuals between their breeding and wintering grounds. Individuals from nearly every animal group exhibit this behavior, but probably the most iconic migration is carried out by birds, from the classic V-shape formation of geese on migration to the amazing nonstop long-distance flights undertaken by Arctic Terns *Sterna paradisaea*. In this chapter, we discuss how seasonal migration has shaped the field of evolution. First, this behavior is known to turn on and off quite rapidly, but controversy remains concerning where this behavior first evolved geographically and whether the ancestral state was sedentary or migratory (Fig. 7.1d, e). We review recent work using new analytical techniques to provide insight into this topic. Second, it is widely accepted that there is a large genetic basis to this trait, especially in groups like songbirds that migrate alone and at night precluding any opportunity for learning. Key hypotheses on this topic include shared genetic variation used by different populations to migrate and only few genes being involved in its control. We summarize recent work using new techniques for both phenotype and genotype characterization to evaluate and challenge these hypotheses. Finally, one topic that has received less attention is the role these differences in migratory phenotype could play in the process of speciation. Specifically, many populations breed next to one another but take drastically different routes on migration (Fig. 7.2). This difference could play an important role in reducing gene flow between populations, but our inability to track most birds on migration has so far precluded evaluations of this hypothesis. The advent of new tracking techniques means we can track many more birds with increasing accuracy on migration, and this work has provided important insight into migration's role in speciation that we will review here.

Keywords Migratory traits · Orientation · Evolution · Selection · Candidate gene · High-throughput sequencing · Behavioral genomics

---

Both the authors contributed equally to this chapter.

M. Liedvogel (✉) · K. Delmore
MPRG Behavioral Genomics, Max Planck Institute for Evolutionary Biology, Plön, Germany
e-mail: liedvogel@evolbio.mpg.de

D. T. Tietze (ed.), *Bird Species*, Fascinating Life Sciences,
https://doi.org/10.1007/978-3-319-91689-7_7

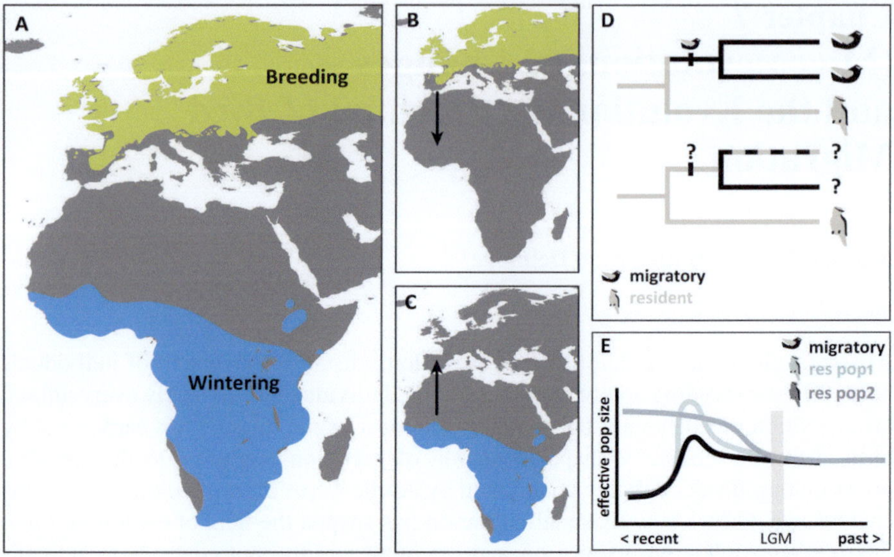

Fig. 7.1 Evolution of migratory and geographic ancestry. (**a**) Breeding and wintering ranges of a European migratory species. These geographical ranges may have evolved by either (**b**) a shift of the breeding range from higher latitudes to regions further south (northern-home hypothesis) or (**c**) a shift of breeding range to grounds further north in order to benefit from richer resources during the breeding season (southern-home hypothesis). Many European species (e.g., songbirds in genera *Sylvia*, *Saxicola*, *Phylloscopus*) are known to come from a speciation core placed around the Himalayan area, favoring the northern-home hypothesis—at least for Eurasian species. (**d**) Possible scenarios reconstructing the migratory phenotype: inherited from one common ancestor (upper panel) or through (one or multiple) gains and losses of migratory phenotype since a last shared common ancestor (lower panel). The latter scenario makes reconstruction of the migratory phenotype and estimates of phenotype transitions more difficult (redrawn after Zink 2011). (**e**) High-throughput genome sequencing data allow inferring demographic history like historical population sizes and divergence times (effective population size by time). This hypothetical demographic scenario summarizes patterns for effective population sizes of different populations; populations start to diverge after last glacial maximum (LGM, 20,000 years ago, indicated by range in gray). In this hypothetical scenario, the migratory population (migratory, in black) is ancestral, and two independent transitions to residence occur (res pop1 and res pop2 in gray), forming independently derived lineages. All three diverging populations show an increase in effective population size after divergence with population sizes of the migratory and resident population 1 dropping subsequently. This sketch for this hypothetical scenario of a migratory bird system leans on the monarch butterfly as one example illustrating how the availability of high-throughput sequencing data can rewrite our understanding of demographic events (see Zhan et al. 2014)

Migration is a fascinating phenomenon that takes place on a global scale and involves individuals from nearly all continents and throughout the animal kingdom. It is long accepted that, within birds, migration evolved in an early ancestor and now the machinery is present in all groups (e.g., Rappole 1995; Berthold 1999). This behavioral phenotype is particularly prominent and equally diverse in birds, ranging

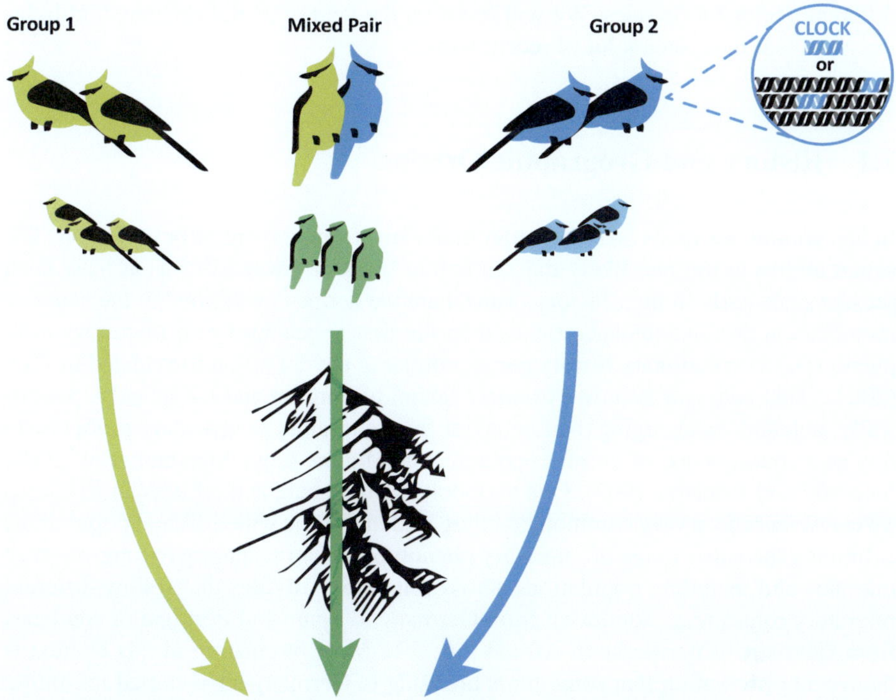

Fig. 7.2 Migratory phenotypes across divides and underlying genetic architecture of orientation. Migratory divides are natural scenarios where birds from two neighboring groups (these groups can be of different species, subspecies, or populations) with different migratory orientation meet and probably hybridize. Offspring of the same parental phenotype (phenotype reflected by color) follow the same migratory orientation phenotype as their parents (i.e., matching colors), usually optimized to lead them around ecological barriers. Offspring of mixed pairs have been shown to use an intermediate phenotype (mirrored as intermediate-color birds with intermediate heading) resulting in migratory routes that may lead them over significant barriers such as mountain ranges, oceans, or wide stretches of desert that a bird has to cross in nonstop flight. Migratory routes are depicted as arrows, coloration coding for differences in migratory phenotype. Attempts to characterize the underlying genomic architecture modulating this phenotypic difference have looked at different scales (here depicted for group 2 as circular inlay): Marker-based approaches focus on preselected candidate genes (exemplified for Clock here) and look for correlation between genetic variation at the marker region and individual phenotype. The advent of high-throughput sequencing data is currently allowing us to expand from marker-based approaches to the entire genome and no longer limits our analyses to restricted regions (depicted for group 2 in blue within genome-wide DNA in black) of the genome

from starlings flocking in impressive formations through the iconic pole-to-pole journeys of Arctic Terns *Sterna paradisaea* to the almost invisible movement of tiny passerine birds migrating at night completely on their own. These impressive journeys amaze laymen and professional scientists alike, and researchers across numerous disciplines have been studying various aspects of this phenomenon from different angles for decades. We will focus on the field of evolution here, highlighting areas that have seen a lot of recent work.

7.1 History and Geographic Origins

In this section, we focus on where migration came from, its origins or evolution. The actual ability to migrate likely existed before birds originated or has at least been present since early in their history. Thus, here we define "evolution" in the sense of modification and maintenance rather than the first appearance of a migratory trait/phenotype in evolutionary history per se, similar to the definition formulated in Zink (2002). This behavior is now extremely labile, turning on and off quickly. Specifically, populations changing their behavior from migratory or sedentary probably do this as a consequence of current ecological conditions (e.g., Alerstam et al. 2003; Salewski and Bruderer 2007). One textbook example here is the Eurasian Blackcap *Sylvia atricapilla*, a very common breeding bird in Europe with different populations exhibiting the entire range of migratory phenotypes, from sedentary to long-distance migrants and including populations across **migratory divides** that follow different migratory routes (e.g., southwest from Germany to Spain and Portugal or southeast from Germany to northeastern Africa; Fig. 7.2). Most fascinating in this context is the recent observation that some birds breeding in Germany have started migrating northwest to winter in the UK. This observation comes from the 1960s (Berthold et al. 1992; Rolshausen et al. 2009) and may be in response to an increasing number of bird feeders in the UK providing a great resource for overwintering Blackcaps (Plummer et al. 2015). In addition to within-species variation in migratory phenotype in the Blackcap, other species within the family Sylviidae show a complex phylogenetic pattern in the presence and absence of migratory behavior as well as migratory distance (Helbig 2003). Closest relatives of Blackcaps are African Hill Babbler *Pseudoalcippe abyssinica* (resident, r) and Garden Warblers *Sylvia borin* (migratory, m), but also more distantly related species show variation in migratory phenotype (such as Sardinian Warbler *S. melanocephala* (r), Common Whitethroat *S. communis* (m), Western Orphean Warbler *S. hortensis* (m), and Arabian Warbler *S. leucomelaena* (r)). This family-level variation in migratory behavior has allowed researchers to study phenotype-genotype associations between closely related species on different phylogenetic scales (Fig. 7.1d), e.g., de novo identified candidate loci for migratory traits can be tested for the presence or absence of a general pattern across migratory phenotypes on various phylogenetic scales (Helbig 2003; Pulido 2007; Voelker and Light 2011).

What remains controversial is the history (which state was ancestral? see Fig. 7.1d, e) and geographic origins (the tropics or the temperate region?) of these transitions. In the present day, most species winter in the tropics and migrate to temperate regions (Berthold 2001), but the remaining question concerns geographic ancestry: Did this pattern arise from shifts of breeding ranges in the tropical regions north or wintering grounds in the temperate region south (sometimes referred to as southern- vs. northern-home hypotheses; Fig. 7.1a–c) (Salewski and Bruderer 2007)? Recognizing this difference is relevant, because these alternate scenarios have implications for understanding evolutionary and demographic histories of many species. This topic is also interesting in terms of comprehending selection pressures behind transitions between migratory and sedentary states.

Overall, the southern-home hypothesis (i.e., assuming present-day nonbreeding sites as ancestral) has received the most support for migration patterns as we know them today (e.g., Rappole 1995; Berthold 2001): This scenario derives from the observation that seasonal abundances of resources like insects are high in the north and day length is longer. Accordingly, birds may be able to raise more offspring than their counterparts in the tropics. While the tropics would seem to provide these resources year round, there is a considerable amount of competition here that may have favored the expansion north for some populations, while others remained in the tropics. In support of this hypothesis, the center of diversity for many families—especially in North America—is in the tropics (see Brown 2014 for overview). Temperate regions have likely been affected by glacial cycles for millions of years (e.g., Hewitt 1999, 2000), and with every glacial retreat, we see evolution of migration. During these cycles, regions at higher latitudes get covered by ice sheets, which produced range contractions of birds around southern areas. As glaciers retreat, birds expand their ranges north again. A recent study consequently suggests glaciation as adaptive switch between migratory and sedentary behaviors over multiple such oscillations (Zink and Gardner 2017).

Results from several research angles provide support for these intuitive ideas (e.g., Cox 1968; Rappole and Jones 2002; Helbig 2003), including the observation above that the center of diversity for many groups is in the tropics. Many migrants have evolved from forms that originated in the tropics. What remains controversial is whether southern-home birds were resident or already migratory. The traditional view is that they were often sedentary (explicitly or implicitly) or partially migratory (e.g., Berthold 1999). But it may also be that environmental gradients associated with elevation may actually have allowed or even favored the persistence (or existence—if support for this idea accumulates) of migratory populations in southern latitudes. Consequently, it is important to distinguish between the origin of lineages and their current distribution (shaped by glaciation and ecological divergence) and the origin of the migratory behavior that may change within lineages.

Despite the intuitive potential of the southern-home hypothesis, many of the arguments are ambiguous or can be interpreted differently. For example, the center of diversity does not necessarily reflect the center of origin (e.g., Zink 2002), and the phylogenetic origin of a species does not necessarily have to coincide with the evolution of the behavior under study (Zink 2002; Salewski and Bruderer 2007; Bell 2005). In accordance with these arguments, several examples of recent work have provided strong support for the northern-home hypothesis. Winger et al. (2014) used a phylogenetic modeling approach, i.e., ranges of extant species and their evolutionary relationships, to infer geographic and demographic ancestry in the New World emberizoids, a very familiar group found in the Americas, comprising around 823 species, 120 of which are migratory. Their results suggest that migrants were more than twice as likely to evolve out of North America than the tropics, supporting the northern-home hypothesis—at least for this group of birds. Winger et al. (2014) further argue the transition from residency was facilitated by migration and species that are now resident in the tropics (in fact most of that focal family) have lost the ability to migrate. The authors conclude this represented a transition between residency in both north and south, i.e., birds may have started out moving south for the winter, but eventually all of them stayed. The northern-home scenario finds further support by a study in the Parulidae (Winger et al. 2012) as well as using molecular phylogenies across the complete bird tree of life (Rolland et al. 2014).

However, there is still controversy on this topic. Zink and Gardner (2017) criticize the model (in Winger et al. 2014) for not accounting for unequal living conditions at different latitudinal regions. In their view, it is crucial to take this covariate into account, as in the context of glacial oscillations, it seems very unlikely that living conditions were evenly distributed across latitudes. The authors advise to focus on using ranges at glacial maxima rather than phylogenetic reconstructions to understand the evolutionary history of migration, because range shifts occur more frequently than speciation events. Zink and Gardner (2017) conclude that migration promoted colonization of North America from the tropics. Continuously refined modeling approaches as well as advances in sequencing and tracking technologies will hopefully allow us to address these questions with much higher resolution (such as sketched in Fig. 7.1e). How these new tools can provide new and previously inaccessible insight to understand these evolutionary processes has been exemplified in a recent study on monarch butterflies *Danaus plexippus* with their iconic and impressive trans-generational migratory cycle (Zhan et al. 2014). Here the authors made use of high-throughput whole-genome re-sequencing data to characterize both the demography and geographical origin of migratory behavior in the monarch butterfly. Results from this study showed that ancestral populations of monarchs did not initially disperse out of Central or South but North America. Similar datasets for birds between and within different lineages will add a new level of understanding on the evolution of migration in birds.

7.2 Regulation

But how is this complex behavior regulated, integrated, and coordinated? Here it is important to realize that different bird species may use different migratory strategies. Most obviously, birds that migrate during the day and in flocks will face very different challenges when compared with songbirds that migrate at night and on their own. Juvenile day migrating birds may just follow their parents or other conspecifics on their first migratory journey and learn en route. This has been suggested for White Storks *Ciconia ciconia* (e.g., Schüz 1950; Chernetsov et al. 2004) and various species of North American waterfowl (Hochbaum 1955). This will not be possible for juvenile night-migrating songbirds like the Eurasian Blackcap we described above and many common songbirds in the temperate region like American Goldfinches *Spinus tristis* or European Robins *Erithacus rubecula*. Advances in tracking technology, including new GPS tags that transmit signals to the International Space Station and small archival tags that record light, are enabling researchers to track these birds on migration, providing fine-scale details about their migration and beginning to allow us to directly test hypotheses concerning the regulation of migratory behavior.

7.2.1 Variation in Migratory Strategy

Among migratory species, differences in how they orient on migration are especially prominent and related to the availability of reference systems. For example, flocking day migrants (particularly juvenile birds) likely follow the leader and use guiding as orientation strategy (Schüz 1943, 1951). With respect to compass references, the only ubiquitous reference system birds can use for orientation is the Earth's magnetic field (Wiltschko and Merkel 1966; Wiltschko and Wiltschko 1972). Additional celestial cues that can be used for orientation depend on time of day of the migratory journey and respective presence of the celestial cue. Day migrants can use the sun, while night-migrating songbirds may rely more on the stars and connected sunset cues such as the polarization pattern during nights (e.g., Wiltschko et al. 1987; Phillips and Moore 1992; Cochran et al. 2004). Experienced day migrants likely also make use of large-scale landmarks such as coastlines on a larger scale, but also more locally learned landmarks that can be used for fine-tuned navigation during the homing phase close to their final goal (i.e., breeding and wintering location, respectively), allowing for an additional layer of guidance and increased accuracy. This is not possible for first-time night-migrating songbirds that have to rely exclusively on inherited information. This arguably makes songbird migration more intriguing, especially given the fact that most birds return to the same territory year after year and based only on their genetic makeup.

7.2.2 Migratory Traits Are Inherited

Work with the Eurasian Blackcap beginning in the 1980s provided us with a considerable amount of information on the inheritance of migratory traits. For example, selection and cross-breeding experiments in this species have shown that migratory traits like distance (Berthold and Querner 1981) and direction (Helbig 1991, 1996; Berthold et al. 1992) are genetically encoded and (under strong artificial selection) have the potential to drastically change within a few generations. This is impressively illustrated by a selective breeding experiment on Blackcaps from a partially migratory population, i.e., a population where birds that migrate and resident birds co-occur. Here appropriate selection pressure resulted in almost complete resident/migratory individuals within three to six generations (Berthold et al. 1990), highlighting the important role of genetic factors in determining the migratory state. Displacement experiments provide further support that migratory traits are largely genetically determined in night migrants like songbirds (e.g., Perdeck 1958; Chernetsov et al. 2008). The classical displacement experiment by Perdeck beautifully illustrates different migratory strategies for adults (specifically: birds that have successfully completed the migratory journey to their wintering goals and back at least once and thus could have learned landmarks on the way) and juvenile birds on their first migratory journey heading to an area they have never been to before. The study is based on ringing recoveries of more than 10,000 ringed starlings (1/3 adults, 2/3 juveniles) and reveals that only experienced adults were able to compensate for the displacement and could actively navigate to their original wintering area. In contrast, inexperienced juveniles followed an innate clock and compass strategy (also called vector navigation), namely, sticking to an inherited program that told them when to leave the breeding grounds and how long to fly along an inherited migratory direction (Perdeck 1958). Some natural scenarios in the wild provide additional support, e.g., the Common Cuckoo *Cuculus canorus*, whose offspring never interacts with its parents, and still they know how to migrate to Sub-Saharan Africa.

Heritability estimates for migratory traits based on quantitative genetics assessment (disentangling the relative contribution of genetic variation and environment in explaining the variation of phenotypic traits) are moderately high; they vary depending on species and traits in focus (Pulido and Berthold 2003; Pulido 2007). As an example, heritability values in Blackcaps for timing of (autumn) migration range between 0.34 and 0.45 (Berthold and Pulido 1994; Pulido and Berthold 2010). Little is known on natural trait variation in wild populations, but given advances in tracking technology, these estimates may soon become much more accessible; in general, heritability estimates from quantified behavior of captive migrants seem to be comparable to heritabilities in the wild (Pulido and Berthold 2003).

7.2.3 Underlying Genetic Architecture: Simple and Common?

In addition to estimates of heritability for focal traits like orientation or timing of migration, cross-breeding experiments of Blackcaps also provide insight into the mode of inheritance. Specifically, studies focusing on the orientation preference of crossbred offspring from parental groups of Blackcaps from either side of the Central European migratory divide (i.e., the German breeders mentioned above that migrate either southeast or southwest) suggest an intermediate or additive mode of inheritance in this species. Orientation phenotype here was assessed via **orientation cage experiments**, where birds are placed in a circular cage that is lined with scratch-sensitive paper and tested for their migratory orientation. Scratches are quantified, and results suggest that directional preference of these birds is in very well accordance with the actual migratory direction of conspecifics in the wild that is quantified. Crossbred offspring (reared isolated from their parents to prevent any learning component from influencing the phenotype) in the first generation (F_1) exhibit a migratory direction intermediate to the parental phenotypes when tested in orientation cage experiments. Importantly, the orientation phenotype further segregates in the second generation (F_2). Specifically, the overall variance increases, but orientation preferences still cluster around both the parental and intermediate directions. This phenotype pattern suggests a simple genetic basis, i.e., only few genes with large effects influencing variability of this trait (Helbig 1991, 1996).

However, the number of examples investigating the genetics of migratory orientation is limited, and it is important to realize that different species may use different strategies and could exploit different processing cascades. The scenario in flycatchers (European Pied *Ficedula hypoleuca* and Collared *F. albicollis*), for example, suggests a different mode of inheritance. There are two species of flycatchers that breed next to one another in Central Europe and are believed to take different routes on migration, one going southwest to western Africa and the other southeast to eastern Africa (further south than the Blackcaps described above). Instead of using orientation experiments like in the Blackcap study, the orientation of these flycatchers is based on **stable isotope signatures** (Marra et al. 1998; Hobson 1999), which are used as a proxy for wintering location. Specifically this "signature" is coded in their feathers, reflecting the chemical composition of stable isotopes that is unique to the area where the bird resides during the time of feather growths, which in this species was on the wintering grounds. Feathers are particularly useful for stable isotope analyses as for most species they grow during a clearly defined and relatively short period of molt and then (in contrast to other tissue like claw or organs that have a higher turnover rate) become metabolically inert. In consequence, by plucking a feather of a flycatcher returning to its breeding grounds, the stable isotope signature it carries can be used as proxy to trace back its whereabouts during winter.

Interestingly and in contrast to the Blackcap with an intermediate phenotype, data from these flycatchers suggests genes controlling their migratory orientation are dominant, with hybrids between the two species wintering in the same location as European Pied Flycatcher (Veen et al. 2007). Further support for dominance of one migratory direction over the other comes from hybrids between Great Reed Warbler *Acrocephalus arundinaceus* and Clamorous Reed Warbler *A. stentoreus*, where again hybrids appear to follow the migratory orientation of Great Reed Warblers (Yohannes et al. 2011). These conflicting results on the phenotypic level call for caution in generalizing on the hypothesis of one common genetic basis for migratory orientation across species.

7.2.4 *Marker-Based Approaches: Candidate Genes for Migration*

Building on results from quantitative genetic analyses, orientation cages, and stable isotopes and benefitting from increasingly advanced molecular tools for non-model organisms, more recent work has started to focus on identifying actual genes that may control variability of migratory traits. One popular approach is the use of **candidate genes** to screen for signatures of selection for a focal trait. Candidate genes are often identified in model organisms (e.g., fruit flies *Drosophila* or *Arabidopsis*, weedy flowering plants from the mustard family) that may have a function related to the trait of interest, in this case migration. The use of candidate genes rests on the assumption that genes that have been shown to influence the expression of a certain phenotype might potentially influence a similar phenotype in other animals (Long 2001). Once a candidate has been identified, researchers go out into natural populations and screen them for variation at the gene that may in turn explain variation in the trait of interest. Again, here we focus on migration.

One series of studies focused on variation (allele frequencies) at a variable genomic region within the Clock gene, a key player entraining the daily rhythm (night/dark cycle) in many organisms. Variation at this focal genomic region was quantified and correlated with various photoperiod-entrained phenotypes, specifically latitudinal variation, timing of breeding, molt, and migration (i.e., correlated timing traits throughout the annual cycle). In brief, several of these studies suggest a regulatory role of Clock on timing and migratory phenotypes (specifically association with timing of migration in fish, O'Malley and Banks 2008; latitudinal clines in Eurasian Blue Tits, Johnsen et al. 2007; breeding phenology in Eurasian Blue Tits, Liedvogel et al. 2009, and Barn Swallows, Caprioli et al. 2012; timing of molt in Barn Swallows, Saino et al. 2013; migratory phenotypes both across and within populations, Peterson et al. 2013; Saino et al. 2015; timing of migration tracked by advanced technology for phenotype assessment, i.e., **light-level geolocators** in Barn Swallows, Bazzi et al. 2015), but results both within and across populations of different taxa are inconsistent and prohibit from drawing general conclusions (e.g., no correlation found for latitudinal clines in Bluethroats *Luscinia svecica*, Johnsen et al. 2007; Barn Swallows, Dor et al. 2011; *Tachycineta* swallows, Dor et al. 2012; breeding phenology, Liedvogel and Sheldon 2010, and Barn Swallows, Dor et al. 2011; migratory phenotypes in Blackcaps, Müller et al. 2011). Another candidate gene studied in the context of migration genetics is ADCYAP1, specifically length variation in one microsatellite (Müller et al. 2011). Studies in Blackcaps suggest a correlation of longer alleles associated with higher migratory activity in caged migrants as well as longer distances traveled by wild birds (Müller et al. 2011; Mettler et al. 2015). But similarly to studies focusing on Clock, results from studies on ADCYAP1 lack consistency across species (e.g., Peterson et al. 2013 in the genus *Junco*) in order to confirm its putative function regulating migratory behavior.

In conclusion, though candidate-gene analyses may provide important insight into the regulation of behavioral traits, this approach comes with a number of limitations and is—especially in the context of migratory behavior—clearly restricted by our lack of knowledge on the genetic architecture of this fascinating phenotype and consequently the number of candidate genes for assumed and predefined candidate traits (Liedvogel et al. 2011; Delmore and Liedvogel 2016). This skepticism is further emphasized by a recent study screening whole-genome data of migratory and non-migratory bird species for the presence or lack of consistency in genetic variation across a set of 25 hypothesized candidate genes for migratory traits (Lugo Ramos et al. 2017). This study fails to detect a signal for migratory phenotype using different approaches and in conclusion questions the applicability of candidate-gene approaches to understand the genetic architecture of migratory behavior (Lugo Ramos et al. 2017).

7.2.5 Enhancing Scale and Resolution: Genome-Wide Approaches

Recent advances in the field of genomics are allowing us to expand beyond candidate genes. Of primary importance was the development of **high-throughput sequencing techniques** where nearly all the DNA from a single individual can be sequenced quickly and relatively cheaply, allowing researchers who work with non-model organisms (including most of the migratory animals on Earth) to obtain genome-wide data (i.e., not simply focus on a target region such as a single candidate gene but the entire genome; Fig. 7.2) for their species. In essence, these technologies are allowing researchers to overcome the limitation of possibly not being able to pick the right gene candidates, as has been exemplified by identifying the underlying genetics contributing to beak shape among Darwin's finches (Lamichhaney et al. 2015). This is especially true for birds, where only few drops of blood are needed to extract DNA with high enough quantity and quality to generate thousands of sequence stretches allowing for de novo assembling its entire genome. Having a well-characterized reference genome of your focal evolutionary model at hand enables us to go ahead and de novo identify candidate genomic regions that may be involved in shaping migratory phenotypes.

The importance of applying high-throughput sequencing to questions of migration and its genetic basis has been exemplified in the Swainson's Thrush *Catharus ustulatus* (Delmore et al. 2016). This species includes two subspecies groups that hybridize in western North America and migrate along different routes. One subspecies hugs the coast of western North American to winter in Central America, and the other flies through the central United States and over the Gulf of Mexico to winter in South America. Previous work in the system using archival tags that record light intensity (geolocators) showed that hybrids take intermediate routes compared to their parents, leading them over ecological barriers like mountain chains that pure forms avoid (Delmore and Irwin 2014). Using high-throughput sequencing techniques, this variation in phenotype was linked to variation on the genome level. This approach enables the de novo characterization of genomic differentiation between migratory phenotypes and in this study allowed researchers to narrow in on a relatively small region of the genome (located on chromosome 4, covering ~60 genes including the previously discussed candidate gene Clock) that may govern this behavior (Delmore et al. 2016).

A comparable study investigating the genetic architecture of migratory orientation (phenotype assessment here was based on stable isotope signatures as a proxy for wintering location rather than geolocator tracks) was conducted in a songbird species with a Eurasian breeding range, the Willow Warbler *Phylloscopus trochilus*. Two subspecies of Willow Warbler form a migratory divide in central Scandinavia, and whole-genome re-sequencing data and SNP-chip data comparing these groups confirmed overall extremely high genetic similarity with significant differences restricted to two regions in the genome (on chromosomes 1 and 5) (Lundberg et al. 2017). The low level of genetic differentiation in Willow Warblers is in contrast to much more elevated differentiation levels in the Swainson's Thrush (Ruegg et al. 2014; Delmore et al. 2015a, 2016) indicative for early (low) or later stages of divergence (high levels of differentiation). Relatively few additional studies on this line of investigation are available, although work with the Barn Swallow (von Rönn et al. 2016; Safran et al. 2016; Scordato et al. 2017) has been conducted. In contrast to Swainson's Thrushes and Willow Warblers, Barn Swallows migrate during the day and in flocks. Accordingly, and as mentioned above, they may use different cues for migration, and there may be less of a genetic component to this behavior.

One surprising result from the limited work that has been conducted to date is that the de novo identified regions showing highest levels of differentiation with migratory orientation in the thrush and warbler are located in different regions of the genome. In addition, while the Clock gene may be involved in the orientation of thrushes, it was only one among many other genes that showed an association with migration in this system, and no such link was found in the Willow Warbler, suggesting there is no consistent pattern linking migratory traits with candidate genes or de novo identified regions across species. In summary, the lack of a consistent pattern across species (not only of marker-based but also using whole-genome approaches) calls for skepticism about the aforementioned hypothesized common and simple genetic basis of migratory traits. Consequently, we want to close this section with the interim conclusion that the genomic architecture of migratory traits may well be simple—but not necessarily common, i.e., shared across groups or species.

7.3 Population Differentiation and Speciation

Thus far we have discussed recent work on the origins of migratory behavior and how this is regulated. In this last section, we will focus on the role differences in migratory phenotype could play in **speciation**, the process by which one species evolves into two. This topic relates mostly to songbirds, the most diverse order of birds comprising more than 6000 species (i.e., half of all bird species). Among songbirds, 40% are migratory, many of which form migratory divides. Intuitively one might expect that migration would actually prevent speciation from occurring, as at the most basic level, this behavior increases movement both in the form of natal dispersal (where juvenile songbirds breed in locations distant from where they were born) and the annual movement between breeding and wintering grounds. There is some support for this idea. For example, Belliure et al. (2000) showed a negative relationship between dispersal and subspecies richness. Nevertheless, accumulating evidence suggests that migration may play a more important role in the formation of species.

There are two main ways differences in the migratory routes taken by songbirds can help maintain species boundaries: First, the use of different routes could cause populations to arrive on their breeding grounds at different times. If they mate assortatively based on arrival time, this difference will prevent mixed pairs from forming (i.e., pairs between individuals that migrate along different routes and following different timing schedules) and help maintain species boundaries (Irwin and Irwin 2005). Consider, for example, if one species arrives before the other and pairs immediately, there will only be birds following the same timing regime around to mate with. This scenario, where individuals are prevented from breeding with one another, is considered a pre-mating barrier to blending between species.

Bearhop et al. (2005) conducted one of the most elegant studies of this hypothesis in European Blackcaps. This work was conducted using the Blackcap with populations that migrate northwest to the UK or southwest to Spain and Portugal. Similar to work discussed above for flycatchers, Bearhop et al. used stable isotopes (of claw tissue in this case as Blackcaps do not have a complete molt on their wintering grounds, which would allow to use feathers) to identify the wintering grounds of Blackcaps breeding in Central Europe. Bearhop et al. noted that individuals from the northwestern population arrived earlier on the breeding grounds and mated assortatively based on the location of their wintering grounds. There are a few other studies supporting the role differences in arrival time that could play in maintaining species boundaries (e.g., Friesen et al. 2007 in a study focused on Band-rumped Storm Petrels *Oceanodroma castro* that breed in the Atlantic and Pacific Oceans). Nevertheless, it should be noted that there is normally some degree of overlap in arrival time between populations and only a few non-assortative pairs are needed to break down this barrier.

A more convincing way migration could contribute to speciation concerns the fitness of offspring produced by interbreeding between populations or (sub)species (i.e., hybrids) at these divides. As we have already discussed, migratory routes are largely genetically determined in songbirds. These routes are often optimized to take birds around large unsuitable areas that form ecological barriers (such as long stretches of deserts or water bodies that require nonstop flights and mountains). Accordingly, it has been predicted that (depending on the mode of inheritance) (1) hybrids at these divides use intermediate migratory routes, because they will have genes from both groups, and (2) that these routes will be inferior to those optimized routes of pure, parental forms as they will bring hybrids over the unsuitable areas that pure forms avoid (Helbig 1991; Irwin and Irwin 2005). Under this scenario, hybrids will likely survive at lower rates preventing gene flow between populations from breaking down species boundaries. Under this scenario, migration is considered a post-mating barrier as mating between populations has already occurred to produce hybrids.

Early work with orientation cages supported this hypothesis (e.g., Helbig 1991, 1996) but was evaluated more recently using archival tags (light-level geolocators) that record light intensity (Delmore and Irwin 2014). These devices were described above and provide researchers with year-round information on the location of birds, their breeding, stopover and wintering sites that are all captured along with movement days. This work was mentioned above, focusing on the Swainson's Thrush that forms a migratory divide in western North America. Hybrids in this system were fitted with geolocators and shown to take intermediate routes that brought them over several unsuitable areas, including deserts in the southwestern United States, the Coast, Cascade and Rocky Mountains (Delmore and Irwin 2014), confirming orientation cage experiments with the Blackcap. There is complimentary data from this system showing that hybrids are less fit than parental forms (e.g., population densities are lower at the center of the hybrid zone compared to the range of parental forms), supporting the idea that these intermediate routes reduce the fitness of hybrids and help maintain boundaries in this system.

All of the work we have discussed in the context of population differentiation and speciation thus far has been conducted in single systems. A few studies have begun to examine the role migration plays in speciation on a broader scale. For example, Møller et al. (2011) used data on ringing recoveries to identify migratory divides between passerines in Europe. Over one quarter of all species, they identified formed divides, and many showed evidence of local adaptation. In other words, they appear to have adapted to their local environment, and this fact likely helps maintain species boundaries. Rohwer and Irwin (2011) took this one step further, identifying 12 well-studied hybrid zones between birds in North America and showing that among these 12, 8 form migratory divides with substantial barriers (e.g., lower densities within hybrid zones). These studies established the plausibility that migratory divides could promote reproductive isolation on a broad scale but did not include a rigorous statistical analysis or a quantitative measure of reproductive isolation.

One recent attempt to fill this gap was conducted by Delmore et al. (2015b). In this study, morphology, color, and song for species pairs of North American songbirds that do or do not form divides were characterized using museum specimens and song databases. They used differences in these traits as a measure of species boundaries or isolation, assuming that the more isolated a pair is, the greater the differences in these traits. The underlying rationale was that, if migration does play a role in speciation, those pairs that form divides should exhibit greater differences in morphology, color, and/or song. But in contrary to this prediction, the authors found differences in morphology, color, and song which were greater in pairs that do not form migratory divides. This finding likely relates to a process called differentiation fusion, where species pairs that do not differ in migratory orientation were simply lost (blended together) and the only pairs that do not form divides that still exist are those that differed in another trait (morphology, color, and/or song). In other words, if a pair did not differ in migration, they had to differ in another trait to avoid blending into a single species. Under this scenario, differences in migration are still important in songbird speciation, though through a different process.

The comparative analyses discussed above suggest migration may play a primary role in maintaining species boundaries between songbirds. As a final note, and to place this idea in an even broader context, it is important to realize that recent work in the speciation literature has more generally advocated the importance of barriers deriving from ecological differences (versus those that related to intrinsic differences between species including things like genetic incompatibilities, Schluter 2001; Rundle and Nosil 2005; Nosil 2012; Butlin et al. 2017). Among birds, there has been even more emphasis placed on post-mating barriers (those that act after mating) as hybrids are very common in this group suggesting pre-mating barriers are not complete (Grant and Grant 1992; Price and Bouvier 2002; Rabosky and Matute 2013). Migration fills this role very well, as differences in migration are ecological in nature, and while they can act before mating (e.g., through differences in arrival time and subsequent assortative mating), they are likely more powerful after mating, reducing the fitness of hybrids that are forced to face huge barriers on migration.

Acknowledgments We thank Javier Pérez-Tris and Juan Carlos Illera for thoughts and discussion of the manuscript and Derek Caetano-Anolles for figure illustrations. Funding was provided by the Max Planck Society through a Max Planck Research Group grant.

References

Alerstam T, Hedenström A, Akesson S (2003) Long-distance migration: evolution and determinants. Oikos 103:247–260

Bazzi G et al (2015) Clock gene polymorphism and scheduling of migration: a geolocator study of the barn swallow *Hirundo rustica*. Sci Rep 5:12443. https://doi.org/10.1038/srep12443

Bearhop S et al (2005) Assortative mating as a mechanism for rapid evolution of a migratory divide. Science 310:502–504. https://doi.org/10.1126/science.1115661

Bell C (2005) The origin and development of bird migration: comments on Rappole and Jones, and an alternative evolutionary model. Ardea 93:115–123

Belliure J, Sorci G, Møller AP, Clobert J (2000) Dispersal distances predict subspecies richness in birds. J Evol Biol 13:480–487

Berthold P (1999) A comprehensive theory for the evolution, control and adaptability of avian migration. In: Adams NJ, Slotow RH (eds) Proceedings of the 22nd international ornithological congress, Durban. Ostrich 70(1):1–11

Berthold P (2001) Bird migration: a general survey, 2nd edn. Oxford University Press, Oxford

Berthold P, Pulido F (1994) Heritability of migratory activity in a natural bird population. Proc R Soc B Biol Sci 257:311–315. https://doi.org/10.1098/rspb.1994.0131

Berthold P, Querner U (1981) Genetic basis of migratory behavior in European warblers. Science 212:77–79. https://doi.org/10.1126/science.212.4490.77

Berthold P, Mohr G, Querner U (1990) Steuerung und potentielle Evolutionsgeschwindigkeit des obligaten Teilzieherverhaltens: Ergebnisse eines Zweiweg-Selektionsexperiments mit der Mönchsgrasmücke (*Sylvia atricapilla*). J Ornithol 131:33–45. https://doi.org/10.1007/BF01644896

Berthold P et al (1992) Rapid microevolution of migratory behaviour in a wild bird species. Nature 360:668–670. https://doi.org/10.1038/360668a0

Brown JH (2014) Why are there so many species in the tropics? J Biogeogr 41:8–22. https://doi.org/10.1111/jbi.12228

Butlin R et al, The Marie Curie Speciation Network (2017) What do we need to know about speciation? TREE 27:27–39. https://doi.org/10.1016/j.tree.2011.09.002

Caprioli M et al (2012) Clock gene variation is associated with breeding phenology and maybe under directional selection in the migratory barn swallow. PLoS One 7(4):e35140. https://doi.org/10.1371/journal.pone.0035140

Chernetsov N, Berthold P, Querner U (2004) Migratory orientation of first-year white storks (*Ciconia ciconia*): inherited information and social interactions. J Exp Biol 207:937–943

Chernetsov N, Kishkinev D, Mouritsen H (2008) A long-distance avian migrant compensates for longitudinal displacement during spring migration. Curr Biol 18:188–190. https://doi.org/10.1016/j.cub.2008.01.018

Cochran WW, Mouritsen H, Wikelski M (2004) Migrating songbirds recalibrate their magnetic compass daily from twilight cues. Science 304:405–408

Cox GW (1968) The role of competition in the evolution of migration. Evolution 22:180–192. https://doi.org/10.1111/j.1558-5646.1968.tb03461.x

Delmore KE, Irwin DE (2014) Hybrid songbirds employ intermediate routes in a migratory divide. Ecol Lett 17:1211–1218. https://doi.org/10.1111/ele.12326

Delmore KE, Liedvogel M (2016) Investigating factors that generate and maintain variation in migratory orientation: a primer for recent and future work. Front Behav Neurosci 10:3. https://doi.org/10.3389/fnbeh.2016.00003

Delmore KE et al (2015a) Genomic analysis of a migratory divide reveals candidate genes for migration and implicates selective sweeps in generating islands of differentiation. Mol Ecol 24:1873

Delmore KE, Kenyon HL, Germain RR, Irwin DE (2015b) Phenotypic divergence during speciation is inversely associated with differences in seasonal migration. Proc R Soc B Biol Sci 282:20151921

Delmore KE et al (2016) The genetics of seasonal migration and plumage color. Curr Biol 26: R1155–R1157

Dor R et al (2011) Low variation in the polymorphic Clock gene poly-Q region despite population genetic structure across barn swallow (*Hirundo rustica*) populations. PLoS One 6(12):e28843. https://doi.org/10.1371/journal.pone.0028843

Dor R et al (2012) Clock gene variation in *Tachycineta* swallows. Ecol Evol 2:95–105. https://doi.org/10.1002/ece3.73

Friesen VL, Smith AL, Gómez-Díaz E, Bolton M, Furness RW, González-Solís J, Monteiro LR (2007) Sympatric speciation by allochrony in a seabird. PNAS 104:18589–18594. https://doi.org/10.1073/pnas.0700446104

Grant PR, Grant BR (1992) Hybridization of bird species. Science 256:193–197. https://doi.org/10.1126/science.256.5054.193

Helbig AJ (1991) Inheritance of migratory direction in a bird species: a cross-breeding experiment with SE- and SW-migrating Blackcaps (*Sylvia atricapilla*). Behav Ecol Sociobiol 28:9–12. https://doi.org/10.1007/BF00172133

Helbig A (1996) Genetic basis, mode of inheritance and evolutionary changes of migratory directions in palaearctic warblers (Aves: Sylviidae). J Exp Biol 199:49–55

Helbig A (2003) Evolution of bird migration: a phylogenetic and biogeographic perspective. In: Berthold P, Gwinner E, Sonnenschein E (eds) Avian migration. Springer, Berlin, pp 3–20

Hewitt GM (1999) Post-glacial re-colonization of European biota. Biol J 68:87–112. https://doi.org/10.1111/j.1095-8312.1999.tb01160.x

Hewitt GM (2000) The genetic legacy of the Quaternatry ice ages. Nature 405:907–913. https://doi.org/10.1038/35016000

Hobson KA (1999) Tracing origins and migration of wildlife using stable isotopes: a review. Oecologia 120:314–326

Hochbaum HA (1955) Travels and traditions of waterfowl. University of Minnesota Press, Minneapolis, MN

Irwin DE, Irwin JH (2005) Siberian migratory divides: the role of seasonal migration in speciation. In: Greenberg R, Marra PP (eds) Birds of two worlds: the ecology and evolution of migration. Johns Hopkins University Press, Baltimore, MD, pp 27–40

Johnsen A et al (2007) Avian Clock gene polymorphism: evidence for a latitudinal cline in allele frequencies. Mol Ecol 16:4867–4880

Lamichhaney S et al (2015) Evolution of Darwin's finches and their beaks revealed by genome sequencing. Nature 518:371–375. https://doi.org/10.1038/nature14181

Liedvogel M, Sheldon BC (2010) Low variability and absence of phenotypic correlates of Clock gene variation in the great tit. J Avian Biol 41:543–550

Liedvogel M, Szulkin M, Knowles SCL, Wood M, Sheldon BC (2009) Phenotypic correlates of variation at the Clock gene in a wild blue tit population: evidence for a role in seasonal timing of reproduction. Mol Ecol 18:2444–2245

Liedvogel M, Åkesson S, Bensch S (2011) The genetics of migration on the move. TREE 26:561–569

Long A (2001) Candidate gene. Encycl Gen:263–264

Lugo Ramos JS, Delmore KE, Liedvogel M (2017) Candidate genes for migration do not distinguish migratory and non-migratory birds. J Comp Physiol A 203:383–397. https://doi.org/10.1007/s00359-017-1184-6

Lundberg M et al (2017) Genetic differences between willow warbler migratory phenotypes are few and cluster in large haplotype blocks. Evol Lett 1:155–168. https://doi.org/10.1002/evl3.15

Marra PP, Hobson KA, Holmes RT (1998) Linking winter and summer events in a migratory bird by using stable-carbon isotopes. Science 282:1884–1886. https://doi.org/10.1126/science.282.5395.1884

Mettler R, Segelbacher G, Schaefer HM (2015) Interactions between a candidate gene for migration (ADCYAP1), morphology and sex predict spring arrival in Blackcap populations. PLoS One 10 (12):e0144587. https://doi.org/10.1371/journal.pone.0144587

Møller AP, Garamszegi LZ, Peralta-Sánchez JM, Soler JJ (2011) Migratory divides and their consequences for dispersal, population size and parasite-host interactions. J Evol Biol 24:1744–1755. https://doi.org/10.1111/j.1420-9101.2011.02302.x

Müller JC, Pulido F, Kempenaers B (2011) Identification of a gene associated with avian migratory behaviour. Proc R Soc B Biol Sci 278:2848–2856. https://doi.org/10.1098/rspb.2010.2567

Nosil P (2012) Ecological speciation. Oxford University Press, Oxford

O'Malley KG, Banks MA (2008) A latitudinal cline in the Chinook salmon *Oncorhynchus tshawytscha* Clock gene: evidence for selection on PolyQ length variants. Proc R Soc B Biol Sci 275:2813–2821

Perdeck AC (1958) Two types of orientation in migrating starlings, *Sturnus vulgaris* L., and chaffinches, *Fringilla coelebs* L., as revealed by displacement experiments. Ardea 46:1–37

Peterson MP et al (2013) Variation in candidate genes CLOCK and ADCYAP1 does not consistently predict differences in migratory behavior in the songbird genus *Junco*. F1000Research:1–18

Phillips JB, Moore FR (1992) Calibration of the sun compass by sunset polarized light patterns in a migratory bird. Behav Ecol Sociobiol 31:189–193

Plummer KE, Siriwardena GM, Conway GJ et al (2015) Is supplementary feeding in gardens a driver of evolutionary change in a migratory bird species? Glob Chang Biol 21:4353–4363. https://doi.org/10.1111/gcb.13070

Price TD, Bouvier MM (2002) The evolution of F1 postzygotic incompatibilities in birds. Evolution 56:2083–2089. https://doi.org/10.1111/j.0014-3820.2002.tb00133.x

Pulido F (2007) The genetics and evolution of avian migration. Bioscience 57:165–174

Pulido F, Berthold P (2003) Quantitative genetic analysis of migratory behavior. In: Berthold P, Gwinner E, Sonnenschein E (eds) Avian migration. Springer, Berlin, pp 53–77

Pulido F, Berthold P (2010) Current selection for lower migratory activity will drive the evolution of residency in a migratory bird population. Proc Natl Acad Sci U S A 107:7341–7346

Rabosky DL, Matute DR (2013) Macroevolutionary speciation rates are decoupled from the evolution of intrinsic reproductive isolation in *Drosophila* and birds. Proc Natl Acad Sci U S A 110:15 354–15 359. https://doi.org/10.1073/pnas.1305529110

Rappole J (1995) The ecology of migrant birds: a neotropical perspective. Smithsonian Institution Press, Washington, DC

Rappole JH, Jones P (2002) Evolution of old and new world migration systems. Ardea 90:525–537

Rohwer S, Irwin DE (2011) Molt, orientation, and avian speciation. Auk 128:419–425. https://doi.org/10.1525/auk.2011.10176

Rolland J, Condamine FL, Jiguet F, Morlon H (2014) Faster speciation and reduced extinction in the tropics contribute to the mammalian latitudinal diversity gradient. PLoS Biol 12(1): e1001775. https://doi.org/10.1371/journal.pbio.1001775

Rolshausen G, Segelbacher G, Hobson KA, Schaefer HM (2009) Contemporary evolution of reproductive isolation and phenotypic divergence in sympatry along a migratory divide. Curr Biol 19:2097–2101. https://doi.org/10.1016/j.cub.2009.10.061

Ruegg K et al (2014) A role for migration-linked genes and genomic islands in divergence of a songbird. Mol Ecol 23:4757–4769

Rundle H, Nosil P (2005) Ecological speciation. Ecol Lett 8:336–352

Safran RJ et al (2016) Genome-wide differentiation in closely related populations: the roles of selection and geographic isolation. Mol Ecol 25:3865–3883

Saino N et al (2013) Timing of molt of barn swallows is delayed in a rare Clock genotype. PeerJ 1:e17

Saino N et al (2015) Polymorphism at the Clock gene predicts phenology of long-distance migration in birds. Mol Ecol 24:1758–1773. https://doi.org/10.1111/mec.13159

Salewski V, Bruderer B (2007) The evolution of bird migration: a synthesis. Naturwissenschaften 94:268–279. https://doi.org/10.1007/s00114-006-0186-y

Schluter D (2001) Ecology and the origin of species. TREE 16:372–380

Schüz E (1943) Versuche über die Bindung des Storchs an seinen Aufzuchtsort. Vogelzug 14:137–141

Schüz E (1950) Die Früh-Auflassung ostpreussischer Jungstörche in West-Deutschland durch die Vogelwarte Rossitten 1933. Vogelwarte 15:63–78

Schüz E (1951) Überblick über die Orientierungsversuche der Vogelwarte Rossitten (jetzt: Vogelwarte Radolfzell). Proceedings of the Xth international ornithological congress, Uppsala, Sweden, pp 249–268

Scordato ESC et al (2017) Genomic variation across two barn swallow hybrid zones reveals traits associated with divergence in sympatry and allopatry. Mol Ecol 26:5676–5691. https://doi.org/10.1111/mec.14276

Veen T et al (2007) Does migration of hybrids contribute to post-zygotic isolation in flycatchers? Proc R Soc B 274:707–712. https://doi.org/10.1098/rspb.2006.0058

Voelker G, Light JE (2011) Palaeoclimatic events, dispersal and migratory losses along the Afro-European axis as drivers of biogeographic distribution in *Sylvia* warblers. BMC Evol Biol 11:163. https://doi.org/10.1186/1471-2148-11-163

von Rönn JA, Shafer AB, Wolf JB (2016) Disruptive selection without genome-wide evolution across a migratory divide. Mol Ecol 25:2529–2541

Wiltschko W, Merkel FW (1966) Orientierung zugunruhiger Rotkehlchen im statischen Magnetfeld. Verh Deutsche Zool Ges 59:362–367

Wiltschko W, Wiltschko R (1972) Magnetic compass of European robins. Science 176:62–64

Wiltschko W, Daum P, Fergenbauer-Kimmel A, Wiltschko R (1987) The development of the star compass in Garden Warblers, *Sylvia borin*. Ethology 74:285–292

Winger BM, Lovette IJ, Winkler DW (2012) Ancestry and evolution of seasonal migration in the Parulidae. Proc R Soc B 279:610–618. https://doi.org/10.1098/rspb.2011.1045

Winger BM, Barker K, Ree RH (2014) Temperate origins of long-distance seasonal migration in New World songbirds. PNAS 111:12115–12120. https://doi.org/10.1073/pnas.1405000111

Yohannes E, Lee RW, Jochimsen MC, Hansson B (2011) Stable isotope ratios in winter-grown feathers of Great Reed Warblers *Acrocephalus arundinaceus*, Clamorous Reed Warblers *A. stentoreus* and their hybrids in a sympatric breeding population in Kazakhstan. Ibis 153:502–508

Zhan S et al (2014) The genetics of monarch butterfly migration and warning colouration. Nature 514:317–321. https://doi.org/10.1038/nature13812

Zink R (2002) Towards a framework for understanding the evolution of avian migration. J Avian Biol 33:433–436. https://doi.org/10.1034/j.1600-048X.2002.03081.x

Zink RM, Gardner AS (2017) Glaciation as a migratory switch. Sci Adv 3:e1603133

Chapter 8
Avian Diversity and Distributions and Their Evolution Through Space and Time

Manuel Schweizer and Yang Liu

Abstract Most modern orders of birds evolved and diversified during the last 65 million years following the demise of the nonavian dinosaurs and pterosaurs at the Cretaceous-Paleogene boundary. Diversification rates in birds increased from c. 50 million years ago onward driven by significant rate increases in different clades scattered throughout the entire phylogeny. No slowdown in the overall diversification rate has been identified, and equilibrium diversity might not have been reached. Birds breed on all of the continents on Earth and have adapted to almost every habitat. Substantial variation in distribution patterns occurs among the different species, ranging from narrow-range endemics restricted to a single oceanic island or to a particular habitat within a small geographic area to species with a near-cosmopolitan distribution, breeding on almost all continents. As in most groups, diversity of bird species is greatest in tropical regions near the equator and decreases toward the poles. This pattern, termed the latitudinal diversity gradient, cannot be causally linked to a single mechanism and might be influenced by both evolutionary and ecological processes. Species richness within a given area is basically the result of speciation, extinction, and dispersal. Speciation commences with the accumulation of genetically based divergence between populations and is completed by the development of reproductive isolation among them. This usually involves a phase of geographic separation of populations without contact, a process termed allopatric speciation. Speciation with ongoing gene flow between populations, i.e., parapatric speciation, and especially the evolution of reproductive isolation without geographic separation, i.e., sympatric speciation, appear rare in birds. Distribution patterns of different bird groups particularly across the Southern Hemisphere have for a long time been interpreted as being the result of vicariance evolution. Vicariance is considered to be the split of a geographical range of a species via a barrier caused

M. Schweizer (✉)
Naturhistorisches Museum der Burgergemeinde Bern, Bern, Switzerland
e-mail: manuel.schweizer@nmbe.ch

Y. Liu
State Key Laboratory of Biocontrol, Department of Ecology, School of Life Sciences, Sun Yat-sen University, Guangzhou, P. R. China

D. T. Tietze (ed.), *Bird Species*, Fascinating Life Sciences,
https://doi.org/10.1007/978-3-319-91689-7_8

by a historical event like montane uplift or the formation of oceans through tectonic rifting. The formation of such barriers should promote episodes of allopatric speciation in multiple clades, generating congruent biogeographic patterns among them. Using dated phylogenetic hypotheses, however, several studies have recently revealed discordance between sequences of geological events and phylogenetic patterns. Consequently, only past dispersal events, often over long distances and across oceans, can explain the current distribution patterns of several avian groups. In general, landscape changes might not result in congruent temporal diversification patterns among different bird groups. It can be assumed that the older an avian lineage is, the more time it has to colonize an area across a barrier. This increases the likelihood of dispersal across the barrier and subsequent diversification on either side of the divide. In addition, bird groups with lower dispersal abilities are expected to accumulate genetic differences among populations at a higher rate than lineages with higher dispersal capability. Consequently, diversification patterns are the result of interactions between ecological properties of different avian lineages and their environment as well as the age of a given lineage. Geographic ranges of birds are generally limited by a suite of biotic and abiotic factors. Range expansion is not only an important first step in speciation but also influences the number of coexisting species and thereby shapes the turnover of biodiversity in space and time.

Keywords Avian diversification · Avian diversity · Avian distribution · Allopatric speciation · Parapatric speciation · Sympatric speciation · Vicariance · Dispersal · Colonization · Range evolution

8.1 Spatiotemporal Diversification of Modern Birds

Birds possess little resemblance to a *Tyrannosaurus* or a *Stegosaurus*; nonetheless modern birds, Neornithes, are the only surviving lineage of dinosaurs, forming a well-supported monophyletic lineage (Mayr 2017). The initial divergence event within Neornithes comprised the split of Palaeognathae, the ratites and tinamous, from the remaining lineages called Neognathae (Fig. 8.1). Within the latter, there is a separation between the Galloanseres, containing all land- and waterfowl, and Neoaves, a clade encompassing all other extant birds. These basal phylogenetic relationships of modern birds are robustly supported by both morphological and molecular data (e.g., Cracraft 2001; Hackett et al. 2008; Jarvis et al. 2014; Livezey and Zusi 2007; Mayr 2011; Prum et al. 2015).

The clade of Neoaves comprises the majority of extant bird species and such ecologically and morphologically diverse groups as gulls, herons, hummingbirds, owls, parrots, penguins, pelicans, raptors, and songbirds. The phylogenetic relationships of the different basal lineages within Neoaves are still controversial and have not unambiguously been resolved, even when applying genome-wide molecular data (Jarvis et al. 2014; Prum et al. 2015). The difficulty in resolving the early branching pattern in the evolution of the different Neoaves lineages might be caused by their rapid and apparently near-simultaneous radiation following the Cretaceous-Paleogene (K-Pg) mass extinction of 65 million years ago (Ma) (Suh 2016).

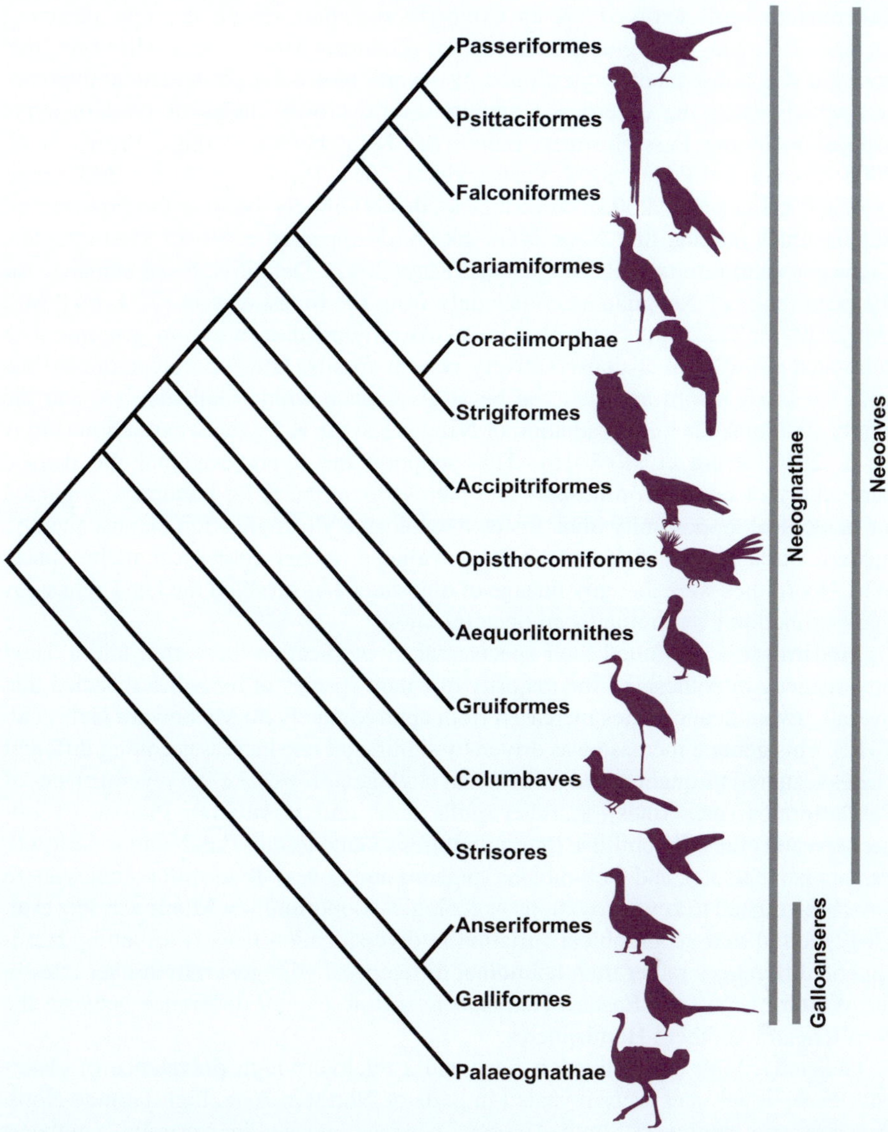

Fig. 8.1 Cladogram illustrating the phylogenetic relationships of the major clades of modern birds (Neornithes) redrawn and modified after Prum et al. (2015)

The mass extinction at the end of the Cretaceous was one of the three largest of the Phanerozoic (the last 541 million years) (Alroy 2008) and was triggered by the impact of a massive asteroid at Chicxulub, Mexico. It led to the disappearance of all dinosaur groups except modern birds and, among others, of pterosaurs, ammonites, and rudists (Schulte et al. 2010). A study of the fossil record led to the hypothesis that modern birds evolved via an explosive radiation among the few surviving lineages following this global perturbation (Feduccia 1995, 2003). However, this scenario was subsequently questioned by several molecular phylogenetic hypotheses, which dated the divergence among several crown clades of modern avian groups, including Passeriformes, before the K-Pg boundary (e.g., Brown et al. 2008; Cooper and Penny 1997; Pacheco et al. 2011; Pereira and Baker 2006; Pratt et al. 2009; Slack et al. 2006). Nevertheless, this is unlikely, because the fossil record demonstrates that the first Neornithine-like birds appeared c. 80–85 Ma in the late Cretaceous and resemble Palaeognathae (Mayr 2017). Definitive fossil evidence for the occurrence of Neognathae comes only from the Maastrichtian (72.1–66.0 Ma) (Mayr 2017). The results of recent molecular phylogenies based on genomic data calibrated using several conservatively chosen fossils, however, are more in line with the fossil record and date the basal divergence within Galloanseres, and the nearly simultaneous initial radiation of Neoaves, at the K-Pg mass extinction (Jarvis et al. 2014; Prum et al. 2015). This supports the hypothesis that the demise particularly of nonavian dinosaurs and pterosaurs at the K-Pg boundary generated an ecological opportunity that fostered explosive diversification. Consequently, modern birds rapidly filled the newly available vacant ecological niche space. Why Neornithes were the only lineage of dinosaurs that survived the late Cretaceous mass extinction event remains an open question.

Modern birds continued their spectacular diversification thereafter, and a dated phylogenetic hypothesis of the majority of extant species of birds has revealed that overall, diversification rates increased from approximately 50 Ma onward (Jetz et al. 2012). This general increase was driven by significant rate increases among different clades scattered throughout the entire avian phylogenetic tree, i.e., in several clades of Passeriformes (passerines), Laridae (gulls terns and skimmers), Picidae (woodpeckers) and allies, Trochilidae (hummingbirds), Galloanseres (land- and waterfowl), Psittacidae (parrots), and Columbidae (pigeons and doves). Rate shifts often seem to have been related to key innovations or ecological opportunities. Moreover, Jetz et al. (2012) found that geographical variation in diversification rates is driven by hemispheric differences rather than latitudinal differences, with generally higher rates in the Western versus the Eastern Hemisphere and no overall difference between the Northern and Southern Hemispheres.

Generally, high diversification rates and a relatively high prevalence of clades with exceptional rates were revealed in parts of Northern Asia, high-latitude North America, and southwest South America, whereas lineages in Australia, Southeast Asia, Africa, and Madagascar display overall comparatively low diversification rates. Given an Australian or African origin for several ancient radiations, this may be the result of density-dependence effects on diversification. It is generally assumed that diversity (species richness) within a clade does not increase continuously with age, but might reach an equilibrium diversity regulated by ecological limits in a given area, resulting in diversity dependence of speciation and extinction rates and to a decoupling of clade age and species richness (Rabosky 2009). Overall, however, Jetz et al. (2012) revealed that opportunities for diversification in birds since the late Eocene (c. 50 Ma) appear not to be exhausted and that equilibrium diversity may not yet have been reached. Such a lack of any slowdown in diversification might be a general feature of (adaptive) radiations among single clades of birds at larger geographic scales, such as in dynamic continental ecosystems with continuous ecological opportunity (Bryson et al. 2014; Derryberry et al. 2011; Liu et al. 2016; Schweizer et al. 2014). However, a slowdown in diversification reflecting ecological saturation through competition for niche space might still be prevalent in clades at geographically confined scales (Jonsson et al. 2012; Phillimore and Price 2008; Price et al. 2014; Rabosky and Lovette 2008; Reddy et al. 2012; Weir 2006).

8.2 Global Distribution and Diversity Patterns

Birds are ubiquitous on Earth today, breed on all continents, and have adapted to almost all types of habitats. However, no single bird species can be considered to be truly cosmopolitan, i.e., occurring on all continents. Nonetheless, species such as the Peregrine Falcon *Falco peregrinus*, which breeds on all continents except Antarctica (del Hoyo and Collar 2014), or the Osprey *Pandion haliaetus*, which breeds on five continents and visits a sixth, South America, as a nonbreeding visitor, are virtually cosmopolitan. Additionally, some long-distance migrants and pelagic birds visit larger parts of the world during their annual migration cycles. On the other hand, species with only a very small distribution are called narrow endemics and are restricted to a small oceanic island or to a particular habitat within a small geographic area. As just one example, the Sombre Rock Chat *Oenanthe dubia* is today probably restricted to a small area in the Upper Awash Valley in Ethiopia where it lives only on rocky slopes with some grass and bushes (Collar and Sharpe 2017). Famous examples of distinctive oceanic island endemic passerine species with very restricted ranges and very tiny populations include Hawaiian honeycreepers and white-eyes on islands of the Pacific Ocean.

The early naturalists already realized that species are not randomly distributed, and they tried to divide the world into different regions characterized by distinct faunal (or floral) assemblages. Such a biogeographic regionalization can be considered one of the central aims of biogeography (Kreft and Jetz 2010). In 1858, the British ornithologist P.L. Sclater published the first global regionalization based on distributions of birds. This scheme was later adapted by A.R. Wallace (1867) for the global mammal fauna. Surprisingly, this original zoogeographic regionalization is still widely used, although several modifications have been published, implementing not only up-to-date species distribution information but also phylogenetic data in some cases and using different criteria for the delimitation of regions (e.g., Ficetola et al. 2017; Holt et al. 2013; Kreft and Jetz 2010; Vilhena and Antonelli 2015) (Fig. 8.2). The world is thereby divided into six principal biogeographic regions, the Nearctic, Neotropical, Palearctic, Afrotropical, Oriental, and Australasian regions. In terms of the number of endemic families, the Neotropical region is most unique (42 endemic families), followed by the Australasian (38), Afrotropical (27), Oriental (10), Palearctic (5), and Nearctic (1) regions (Winkler et al. 2015). Of course,

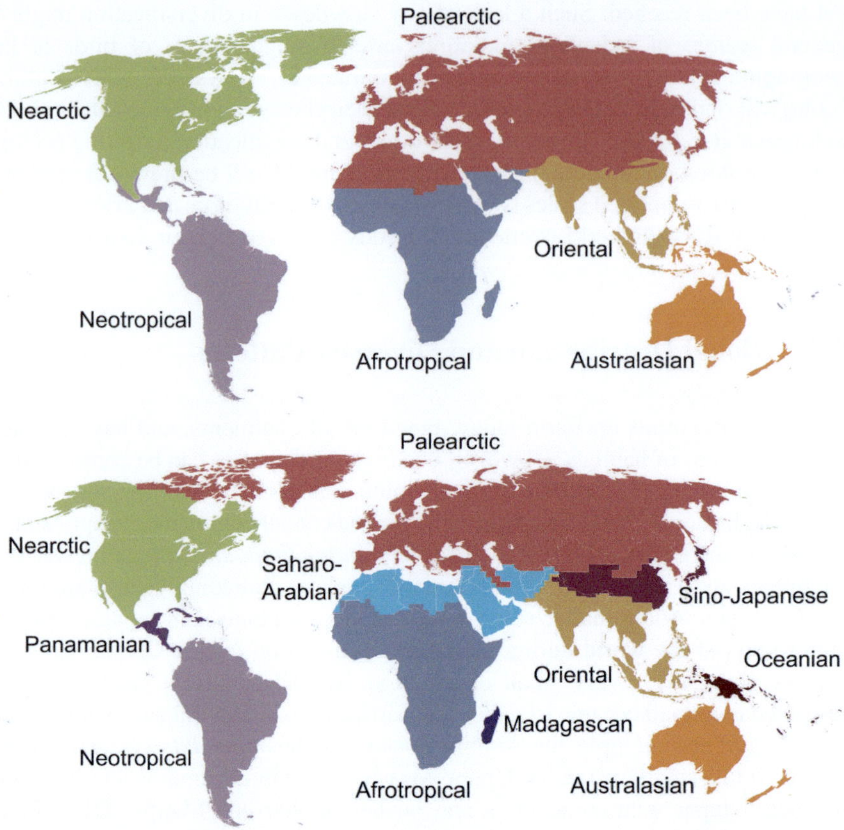

Fig. 8.2 Top: World map showing the six major biogeographic regions according to WWF (taken from https://en.wikipedia.org/wiki/Biogeographic_realm) basically following the original scheme by A.R. Wallace from 1876 (Wallace 1867). Bottom: An updated biogeographic regionalization by Holt et al. (2013) based on an integration of global distribution data and phylogenetic relationships of amphibian, non-pelagic birds, and nonmarine mammals. Redrawn after Holt et al. (2013)

these figures depend strongly on the taxonomic treatment applied. Species richness is also greatest in the Neotropics (Newton 2003), with the most diverse regions in Peru and Ecuador, where more than 600 species have been recorded within a few square kilometers (Bass et al. 2010; Greenberg 2016). The majority of avian diversity hotspots also lie in the Neotropical region. Orme et al. (2005) identified nine hotspots of species richness on Earth, defined as the richest 2.5% of grid cells at a resolution comparable to 1° latitude × 1° longitude. Six of these nine hotspots are located in the Neotropics (Andes, Amazon Basin, Guiana Highlands, Atlantic coastal forests, Mato Grosso Plateau, Panama, Costa Rica highlands), while two are in the Afrotropics (Western Great Rift Valley, Eastern Great Rift Valley) and the other in the Oriental region (the Himalaya). All hotspots are in tropical regions, and with the exception of the Amazon, all are associated with upland or montane regions that remained forested during glacial maxima. Mountains are often exceptionally diverse, although they cover just one eighth of the global land surface, excluding Antarctica; they harbor approximately one third of global terrestrial species diversity (Antonelli 2015).

As with most other biota, avian species diversity is generally greatest in tropical regions near the equator and decreases toward the poles. This so-called latitudinal diversity gradient is considered to be one of the most pervasive large-scale patterns in biology (Jablonski et al. 2017). It cannot be causally linked to a single mechanism and might be the product of diverse drivers. Generally, three different types of mechanisms have been invoked to explain this gradient: ecological, evolutionary, and historical hypotheses (Mittelbach et al. 2007; Pulido-Santacruz and Weir 2016). In the first case, geographical variation in resource availability should constrain diversity, with species carrying capacity decreasing as latitude increases. While evolutionary hypotheses imply that rates of speciation, extinction, and immigration should vary between different latitudes, historical hypotheses focus on the age of ecosystems and the time consequently available for species accumulation, with older ecosystems being considered generally more diverse.

In birds, high species diversity at low latitudes might be associated with low extinction rates (Pulido-Santacruz and Weir 2016). Higher extinction rates at higher latitudes might be related to a global temperature drop at the Eocene-Oligocene boundary (c. 34 Ma) and to climate perturbations during Plio-Pleistocene glacial cycles (over the last 5 million years). These regions were then repeatedly recolonized from the tropics, which were less affected by climate change, resulting in continuously lower extinction rates.

8.3 Geography of Speciation

Species richness within a given area is basically the result of speciation, extinction, and colonization. The latter process involves range expansion via dispersal of a species to an area where it has previously not occurred, followed by establishment of a viable breeding population (Bellemain and Ricklefs 2008). Speciation begins with the accumulation of genetically based divergence between populations and is completed by the formation of reproductive isolation among them. Different species concepts focus on different stages during this speciation continuum (Chap. 2), and different mechanisms can lead to reproductive isolation (Chap. 11). The process of speciation can be approached via different aspects. While one is concerned with processes and mechanisms leading to reproductive isolation (see Chap. 11), another considers speciation within a geographic context (Price 2008; Rundle and Nosil 2005). In birds, and many other organisms, there is usually a phase of geographic separation with no gene flow between diverging populations, termed allopatric speciation (Coyne and Orr 2004). It involves different steps, starting with range expansion, followed by range fragmentation, and the evolution of reproductive isolation between the geographically separated populations, and a further step can involve dispersal into a secondary contact zone leading to coexistence in sympatry (Price 2008; Price et al. 2014). Range fragmentation can proceed via vicariance, if the range is split by the formation of a physical barrier resulting from geological events such as montane uplift (Coyne and Orr 2004, see also below). Populations can also become separated by habitat fragmentation through climate change as occurred repeatedly during Pleistocene glacial cycles during the last 2.5 million and particularly last 1 million years when climate fluctuations were most intensive (Hewitt 2000). Allopatry can also be the result of the colonization of a previously unoccupied and isolated area. If it involves the colonization of a small new area such as an oceanic island due to a founder event by a subset of the ancestral population, it is often referred to as peripatric speciation (Coyne and Orr 2004; Lomolino et al. 2010). The lack of gene flow between separated populations during allopatric speciation per se promotes divergence, but different selection pressures caused by ecological differences in the separated populations might accelerate their divergence (Price 2008) (Chap. 11).

It is widely believed that most speciation events in birds involved some phases of allopatry with no gene flow, making allopatric speciation apparently the commonest route for speciation in birds (Phillimore et al. 2008; Price 2008).

Two geographically separated populations can also become reproductively isolated from each other, despite ongoing gene flow at least at the start of their divergence; this is termed parapatric speciation (Coyne and Orr 2004). Three different forms of parapatric speciation have been distinguished (Price 2008). In a model of clinal speciation, adaptation to local environmental conditions could lead to divergence within a population that occurs along an environmental gradient. Speciation can also simply occur as the result of geographic distance, with the population at either end of a continuous distribution area developing steadily greater differences, while intervening populations continue to exchange genes. In the island model of speciation, two populations are separated by a barrier, but the barrier is permeable, and they still exchange genes to some extent. Reproductive isolation can evolve when immigrants (and their offspring) possess low reproductive success or migration itself between adjacent populations is reduced (Price 2008).

A special case of speciation by distance is the ring species model. When an ancestral population spreads around a barrier, adjacent populations might continue to exchange genes, while gene flow between distant populations is gradually reduced. The populations at both ends of the expansion front increasingly diverge, and when they meet after having spread around the barrier, they might have achieved reproductive isolation from each other (cf. Coyne and Orr 2004; Mayr 1942). One avian example where such a process could have been relevant for the diversity currently observed is the Greenish Warbler *Phylloscopus trochiloides* species complex (Irwin et al. 2005). It has a ringlike distribution around the Tibetan Plateau with five described taxa and an additional, isolated, one in the Caucasus (Alcaide et al. 2014; Irwin et al. 2001, 2005) (Fig. 8.3). A break in morphology, song, and call is most evident between the taxa *viridanus* and *plumbeitarsus* north of the Tibetan Plateau. These forms are differentiated genome-wide, whereas the other populations display a genetic gradient around the ring. However, reproductive isolation is not complete in

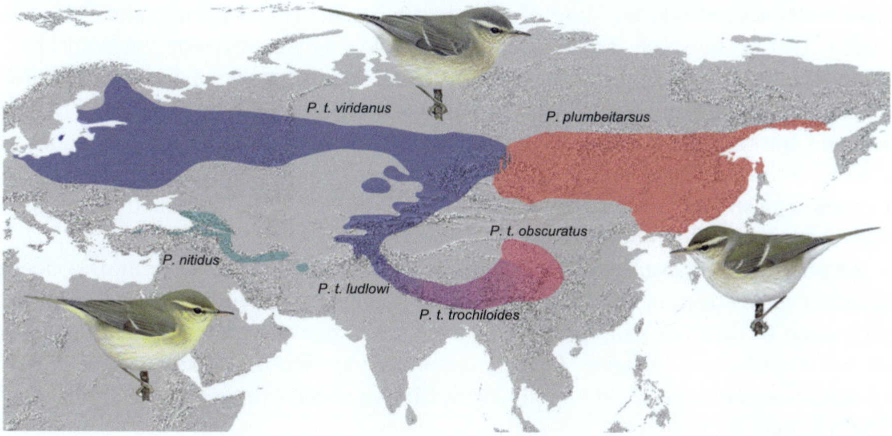

Fig. 8.3 Distribution of the different taxa of the Greenish Warbler *Phylloscopus trochiloides* species complex. It has a ringlike distribution around the Tibetan Plateau with five described taxa and an additional isolated one in the Caucasus. A break in morphology, song, and call is most evident between *viridanus* and *plumbeitarsus* north of the Tibetan Plateau. Distribution areas are modified from BirdLife International and Handbook of the Birds of the World (2016). Bird figures were taken with permission from del Hoyo et al. (2017)

the northern contact zone, and a genomic break in the Southwest can only be explained by secondary contact after divergence in allopatry (Alcaide et al. 2014). Hence, speciation by distance might not entirely explain the ringlike divergence of this complex and climate shifts during the last 1–2 million years which might have led to (repeated) phases of allopatry along the ringlike distribution. Speciation in a ring is indeed difficult to distinguish from secondary contact zones with ongoing gene flow. Such hybridization in secondary contact zones following allopatry have been widely documented in birds (Aliabadian et al. 2005; Weir and Price 2011) (Chap. 10) and might generally be difficult to distinguish from parapatric divergence.

In the case of sympatric speciation, reproductive isolation is achieved without any geographic separation. This model of speciation has been debated for decades, but it can clearly happen under particular, albeit reasonable, biological conditions, as demonstrated by mathematical models (Gavrilets 2014) and empirical works (Savolainen et al. 2006; Sorenson et al. 2003). However, it seems to occur very rarely, and only a few such events in birds can probably be referred to as having happened in sympatry. Most importantly, it must be shown that the divergence of two contemporary and sympatric sister taxa has not involved an allopatric stage. In other words, the initial divergence must have occurred sympatrically, and the possibility of geographic isolation must be eliminated. One of the apparently most plausible sympatric speciation cases in bird concerns African indigobirds of the genus *Vidua* (Sorenson et al. 2003). This group of songbirds is found in Western and Southern Africa, and they are all host-specific brood parasites, with nestlings reared by waxbills, munias, and allies (Estrildidae) in particular. Male indigobirds learn their songs from their host species, i.e., their foster parents. Females preferentially mate with males that mimic the song of their foster parents (Balakrishnan and Sorenson 2006). Unlike in some parasitic cuckoo species where different "host-specific" females mate freely with males, the assortative mating mechanism in indigobirds clearly results in strong cohesion between hosts and parasites. The switch by a female to a new host can then drive simultaneous sympatric divergence as a consequence of learned behavior. Her offspring learn the song from their new host and are most likely to mate with siblings or the offspring from another female that switched to the same new host species (Lovette 2016).

Other proposed examples of speciation without geographic separation concern allochronic speciation, wherein two sister populations diverge due to temporal separation of their breeding periods, as has been suggested on several archipelagos in the Pacific and Atlantic Oceans inhabited by Band-rumped Storm Petrels *Oceanodroma castro* (Bolton et al. 2008; Monteiro and Furness 1998; Smith et al. 2007) (Chap. 6).

8.4 Vicariance vs. Dispersal and the Dynamics of Range Evolution in Birds

Every bird species is restricted to a particular geographic area, i.e., its range, or it occurs in one particular habitat or climatic zone. Ranges differ considerably between different species (see above); nevertheless, strikingly similar distribution patterns can be detected among certain groups. For a long time, vicariance has been invoked to explain such global-scale biogeographic patterns in different bird groups, and vicariance has generally dominated historical biogeography. Vicariance is considered to be the spatial division of a species' range due to the formation of a barrier by historical events such as montane uplift or ocean formation as continents shift (tectonic rifting) (de Queiroz 2005; Lomolino et al. 2010). It was assumed that such historical events might cause congruent biogeographic patterns in different clades through synchronous allopatric speciation events (cf. Halas et al. 2005; Upchurch 2008). As several groups of organisms, including birds, possess distinct Southern Hemisphere distributions, it was hypothesized that vicariant evolution wrought by the breakup of the supercontinent Gondwana from c. 160–30 Ma might explain such biogeographic patterns (e.g. Cracraft 2001; Hedges et al. 1996; Upchurch 2008). This ancient supercontinent in the Southern Hemisphere included the modern-day landmasses of Africa and the Arabian Peninsula, Antarctica, Australia, the Indian subcontinent, Madagascar, and South America.

Palaeognathae have been considered a textbook example of vicariant evolution, and the continental breakup of Gondwana was thought to have shaped their diversification. With the exception of the tinamous of the Neotropical region, all members of the group are flightless, namely, the rheas of South America, the ostriches of Africa and formerly the Arabian Peninsula, the emus and cassowaries of Australasia, and the kiwis of New Zealand as well as the extinct elephant birds of Madagascar

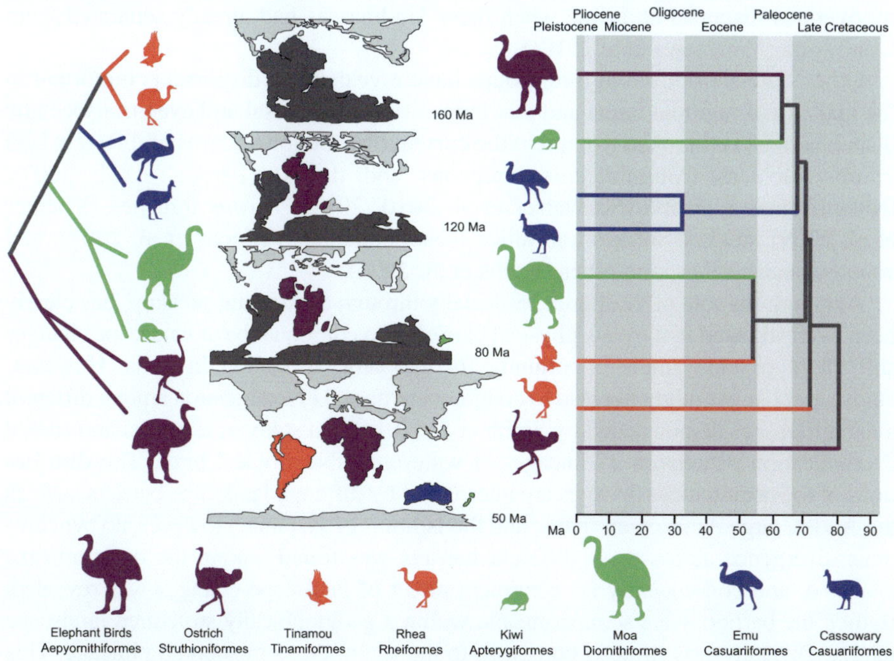

Fig. 8.4 Ratites (Palaeognathae) have for a long time been considered a textbook example of vicariant evolution, and the continental breakup of Gondwana was thought to have shaped their diversification. The breakup of Gondwana during the last 160 million years illustrating the relative position of the continents is shown in the middle, while the cladogram on the left shows the expected phylogenetic relationships of ratites under this scenario. Redrawn after Mitchell et al. (2014) with permission from AAAS. On the right, the most recent time-calibrated phylogenetic hypothesis of ratites based on genomic and morphological data redrawn after Yonezawa et al. (2017) with permission from Elsevier. There is an obvious conflict with the relationships predicted by vicariance through the Gondwana breakup. Moreover, the placement of the tinamous as the closest relatives of the moas meant that the ability to fly must have been lost in several lineages independently

and the extinct moas of New Zealand. However, with the increasing availability of dated molecular phylogenetic hypotheses in the last decade, it became explicitly testable whether branching events along a phylogenetic tree, and their temporality, are congruent with hypothesized drivers of vicariance such as the breakup of Gondwana. The integration in a phylogenetic context of ancient DNA data from the extinct elephant birds and moas revealed a conflict with the relationships predicted by vicariance through the Gondwana breakup (Mitchell et al. 2014) (Fig. 8.4). Moreover, placement of the tinamous as the closest relatives of moas meant that the ability to fly must have been independently lost in several lineages. A recent comprehensive analysis based on genomic and morphological data revealed that Palaeognathae originated in the late Cretaceous in the Northern Hemisphere, migrated to the Southern Hemisphere, and colonized New Zealand and Madagascar by over-sea dispersal at a time when these landmasses had already separated from Gondwana (Yonezawa et al. 2017).

Other studies of different bird groups have revealed that dispersal is important in the makeup of regional fauna and that long-distance dispersal and even transoceanic dispersal must be invoked to explain the current distribution patterns of different bird groups such as Columbiformes (pigeons and doves) (Pereira et al. 2007), Psittaciformes (parrots) (Schweizer et al. 2010, 2011), *Turdus* thrushes (Voelker et al. 2009), cuckoo shrikes and allies (Campephagidae) (Fuchs et al. 2007), and vangas, bush shrikes, and allies (Fuchs et al. 2006).

Although the role of vicariance, especially through continental breakup, has clearly been overestimated in the past, physical barriers as a consequence of landscape changes still might provide ample opportunity for allopatric speciation in birds. However, landscape changes might not lead to congruent temporal diversification among different avian groups, as demonstrated by Smith et al. (2014) in a study of temporal and spatial diversification patterns in 27 lineages of widespread Neotropical birds. The distributions of the taxa in these lineages are interrupted by different landscape barriers such as the Andes, large Amazonian rivers, and the Isthmus of Panama. However, no synchronous divergence across these different barriers was found among the different taxa analyzed, and consequently no congruent effect of landscape changes was revealed. Rather, the barriers were semipermeable within a geographically structured landscape matrix. Diversity was strongly correlated to the age of the corresponding lineage. This indicated that lineages that had persisted for a longer time period within a given landscape had more time to disperse across barriers and hence greater opportunity to differentiate across them. Additionally, ecology had a significant effect on lineage diversity. Lineages with low dispersal ability occupying the understory were significantly more diverse than more mobile groups inhabiting the canopy of tropical forests. Or, lineages with low dispersal ability seem to have accumulated genetic differences at a higher rate. Overall, diversification patterns in these Neotropical avian lineages are the result of interactions between the ecological properties of the different lineages and their environment as well as the age of a given lineage and are only indirectly influenced by landscape changes. This highlights the role of dispersal as a primary driver for allopatric speciation in birds (Smith et al. 2014).

In general, dispersal accelerates geographic range expansions in birds, and the rate at which sister lineages achieve sympatric ranges following speciation in spatial segregation is correlated with their dispersal ability (Pigot and Tobias 2015). Therefore, dispersal is of general importance for shaping species assemblages over space and time.

Physical barriers, time, and dispersal ability, however, are not the only factors that limit colonization of new areas and the dynamics of range expansions in birds. Geographic ranges are generally limited by a suite of biotic and abiotic factors (see also Chaps. 9 and 12) including climate, habitat constraints, food and nest sites, predators, and parasites and pathogens, as well as competition (Koenig 2016; Newton 2003). Among birds in general, ranges at higher latitudes might be strongly limited by winter temperatures, whereas competition might be more important in the tropics (Newton 2003). However, the relative importance of biotic factors in limiting species distributions is controversial (Pigot and Tobias 2013), not least because interspecific competition is difficult to detect in field studies and is not easily tested for in controlled experiments (Greenberg 2016).

However, phylogenies combined with ecological information can be used to test for the role of competition in limiting species ranges. In a study of ovenbirds (Furnariidae), including woodcreepers, a species-rich and ecologically diverse family of Neotropical birds, Pigot and Tobias (2013) showed that rates of secondary sympatry increase with time following diversification in pairs of sister species. This indicates that biotic interactions could be important in limiting range overlap between closely related species, assuming that lineages become more ecologically divergent the longer they are separated. In congruence, accelerated transition rates to sympatry were also revealed between species pairs that differ in ecomorphology related to resource use and foraging strategy. Biotic interactions also play a role in the growth of species diversity at the community level among songbirds in the Eastern Himalayas. Diversification appears to be limited by a failure of new species to expand their ranges and to colonize the local area, caused by competitive biotic interactions as a result of the available niche space being filled (Price et al. 2014).

Range expansion is not only an important first step in speciation; it also influences the number of coexisting species after lineages have diverged ecologically and/or achieved reproductive isolation in allopatry. It thus shapes the turnover of biodiversity in space and time. Understanding the evolutionary dynamics of range expansions and range limits is consequently one of the prerequisites for predicting range shifts and biodiversity patterns under ongoing human-induced climate change (Chap. 12).

Acknowledgments We thank Elisa Badia, Erina Balmer, Vanesa de Pietri, Jana Flury, Guy M. Kirwan, Stephan Lauper, and Christian Müller for their help in preparing this chapter.

References

Alcaide M, Scordato ESC, Price TD, Irwin DE (2014) Genomic divergence in a ring species complex. Nature 511:83–U433

Aliabadian M, Roselaar CS, Nijman V, Sluys R, Vences M (2005) Identifying contact zone hotspots of passerine birds in the Palaearctic region. Biol Lett 1:21–23

Alroy J (2008) Dynamics of origination and extinction in the marine fossil record. Proc Natl Acad Sci USA 105:11536–11542

Antonelli A (2015) Biodiversity: multiple origins of mountain life. Nature 524:300–301

Balakrishnan CN, Sorenson MD (2006) Song discrimination suggests premating isolation among sympatric indigobird species and host races. Behav Ecol 17:473–478

Bass MS, Finer M, Jenkins CN, Kreft H, Cisneros-Heredia DF, McCracken SF, Pitman NCA, English PH, Swing K, Villa G, Di Fiore A, Voigt CC, Kunz TH (2010) Global conservation significance of Ecuador's Yasuni National Park. PLoS One 5:e8767

Bellemain E, Ricklefs RE (2008) Are islands the end of the colonization road? Trends Ecol Evol 23:461–468

BirdLife International and Handbook of the Birds of the World (2016) Bird species distribution maps of the world. Version 6.0. Available at http://datazone.birdlife.org/species/requestdis

Bolton M, Smith AL, Gomez-Diaz E, Friesen VL, Medeiros R, Bried J, Roscales JL, Furness RW (2008) Monteiro's Storm-petrel *Oceanodroma monteiroi*: a new species from the Azores. IBIS 150:717–727

Brown JW, Rest JS, Garcia-Moreno J, Sorenson MD, Mindell DP (2008) Strong mitochondrial DNA support for a Cretaceous origin of modern avian lineages. BMC Biol 6:6

Bryson RW, Chaves J, Smith BT, Miller MJ, Winker K, Pérez-Emán JL, Klicka J (2014) Diversification across the New World within the 'blue' cardinalids (Aves: Cardinalidae). J Biogeogr 41:587–599

Collar N, Sharpe CJ (2017) Sombre Rockchat (*Oenanthe dubia*). In: del Hoyo J, Elliott A, Sargatal J, Christie DA, de Juana E (eds) Handbook of the birds of the world alive. Lynx Edicions, Barcelona (retrieved from http://www.hbw.com/node/58530 on 18 September 2017)

Cooper A, Penny D (1997) Mass survival of birds across the Cretaceous-Tertiary boundary: molecular evidence. Science 275:1109–1113

Coyne JA, Orr HA (2004) Speciation. Sinauer Associates, Sunderland

Cracraft J (2001) Avian evolution, Gondwana biogeography and the Cretaceous-Tertiary mass extinction event. Proc R Soc B Biol Sci 268:459–469

de Queiroz A (2005) The resurrection of oceanic dispersal in historical biogeography. Trends Ecol Evol 20:68–73

del Hoyo J, Collar NJ (2014) HBW and BirdLife International illustrated checklist of the birds of the world. Vol 1: No-passerines. Lynx Edicions, Barcelona

del Hoyo J, Elliott A, Sargatal J, Christie DA, de Juana E (eds) (2017) Handbook of the birds of the world alive. Lynx Edicions, Barcelona (retrieved from http://www.hbw.com/ on 14/02/2017)

Derryberry EP, Claramunt S, Derryberry G, Chesser RT, Cracraft J, Aleixo A, Pérez-Emán J, Remsen JJV, Brumfield RT (2011) Lineage diversification and morphological evolution in a large-scale continental radiation: the Neotropical ovenbirds and woodcreepers (Aves: Furnariidae). Evolution 65:2973–2986

Feduccia A (1995) Explosive evolution in Tertiary birds and mammals. Science 267:637–638

Feduccia A (2003) 'Big bang' for Tertiary birds? Trends Ecol Evol 18:172–176

Ficetola GF, Mazel F, Thuiller W (2017) Global determinants of zoogeographical boundaries. Nat Ecol Evol 1:00089

Fuchs J, Fjeldså J, Pasquet E (2006) An ancient African radiation of corvoid birds (Aves: Passeriformes) detected by mitochondrial and nuclear sequence data. Zool Scr 35:375–385

Fuchs J, Cruaud C, Couloux A, Pasquet E (2007) Complex biogeographic history of the cuckoo-shrikes and allies (Passeriformes: Campephagidae) revealed by mitochondrial and nuclear sequence data. Mol Phylogenet Evol 44:138–153

Gavrilets S (2014) Models of speciation: where are we now? J Hered 105:743–755

Greenberg RS (2016) Bird communities. In: Lovette IJ, Fitzpatrick JW (eds) The Cornell Lab of Ornithology handbook of bird biology, 3rd edn. Wiley, Cornell University, pp 537–577

Hackett SJ, Kimball RT, Reddy S, Bowie RCK, Braun EL, Braun MJ, Chojnowski JL, Cox WA, Han KL, Harshman J, Huddleston CJ, Marks BD, Miglia KJ, Moore WS, Sheldon FH, Steadman DW, Witt CC, Yuri T (2008) A phylogenomic study of birds reveals their evolutionary history. Science 320:1763–1768

Halas D, Zamparo D, Brooks DR (2005) A historical biogeographical protocol for studying biotic diversification by taxon pulses. J Biogeogr 32:249–260

Hedges SB, Parker PH, Sibley CG, Kumar S (1996) Continental breakup and the ordinal diversification of birds and mammals. Nature 381:226–229

Hewitt G (2000) The genetic legacy of the Quaternary ice ages. Nature 405:907–913

Holt B, Lessard JP, Borregaard MK, Fritz SA, Araujo MB, Dimitrov D, Fabre PH, Graham CH, Graves GR, Jonsson KA, Nogues-Bravo D, Wang ZH, Whittaker RJ, Fjeldså J, Rahbek C (2013) An update of Wallace's zoogeographic regions of the world. Science 339:74–78

Irwin DE, Bensch S, Price TD (2001) Speciation in a ring. Nature 409:333–337

Irwin DE, Bensch S, Irwin JH, Price TD (2005) Speciation by distance in a ring species. Science 307:414–416

Jablonski D, Huang S, Roy K, Valentine JW (2017) Shaping the latitudinal diversity gradient: new perspectives from a synthesis of paleobiology and biogeography. Am Nat 189:1–12

Jarvis ED, Mirarab S, Aberer AJ, Li B, Houde P, Li C, Ho SYW, Faircloth BC, Nabholz B, Howard JT, Suh A, Weber CC, da Fonseca RR, Li JW, Zhang F, Li H, Zhou L, Narula N, Liu L, Ganapathy G, Boussau B, Bayzid MS, Zavidovych V, Subramanian S, Gabaldon T, Capella-Gutierrez S, Huerta-Cepas J, Rekepalli B, Munch K, Schierup M, Lindow B, Warren WC, Ray D, Green RE, Bruford MW, Zhan XJ, Dixon A, Li SB, Li N, Huang YH, Derryberry EP, Bertelsen MF, Sheldon FH, Brumfield RT, Mello CV, Lovell PV, Wirthlin M, Schneider MPC, Prosdocimi F, Samaniego JA, Velazquez AMV, Alfaro-Nunez A, Campos PF, Petersen B, Sicheritz-Ponten T, Pas A, Bailey T, Scofield P, Bunce M, Lambert DM, Zhou Q, Perelman P, Driskell AC, Shapiro B, Xiong ZJ, Zeng YL, Liu SP, Li ZY, Liu BH, Wu K, Xiao J, Yinqi X, Zheng QM, Zhang Y, Yang HM, Wang J, Smeds L, Rheindt FE, Braun M, Fjeldså J, Orlando L, Barker FK, Jonsson KA, Johnson W, Koepfli KP, O'Brien S, Haussler D, Ryder OA, Rahbek C, Willerslev E, Graves GR, Glenn TC, McCormack J, Burt D, Ellegren H, Alström P, Edwards SV, Stamatakis A, Mindell DP, Cracraft J, Braun EL, Warnow T, Jun W, Gilbert MTP, Zhang GJ (2014) Whole-genome analyses resolve early branches in the tree of life of modern birds. Science 346:1320–1331

Jetz W, Thomas GH, Joy JB, Hartmann K, Mooers AO (2012) The global diversity of birds in space and time. Nature 491:444–448

Jonsson KA, Fabre PH, Fritz SA, Etienne RS, Ricklefs RE, Jorgensen TB, Fjeldså J, Rahbek C, Ericson PGP, Woog F, Pasquet E, Irestedt M (2012) Ecological and evolutionary determinants for the adaptive radiation of the Madagascan vangas. Proc Natl Acad Sci USA 109:6620–6625

Koenig WD (2016) Ecology of bird populations. In: Lovette IJ, Fitzpatrick JW (eds) The Cornell Lab of Ornithology handbook of bird biology, 3rd edn. Wiley, Cornell University, pp 495–534

Kreft H, Jetz W (2010) A framework for delineating biogeographical regions based on species distributions. J Biogeogr 37:2029–2053

Liu Y, Hu JH, Li SH, Duchen P, Wegmann D, Schweizer M (2016) Sino-Himalayan mountains act as cradles of diversity and immigration centres in the diversification of parrotbills (Paradoxornithidae). J Biogeogr 43:1488–1501

Livezey BC, Zusi RL (2007) Higher-order phylogeny of modern birds (Theropoda, Aves: Neornithes) based on comparative anatomy. II. Analysis and discussion. Zool J Linnean Soc 149:1–95

Lomolino MV, Riddle BR, Whittaker RJ, Brown JH (2010) Biogeography, 4th edn. Sinauer Associates, Sunderland

Lovette IJ (2016) How birds evolve. In: Lovett JC, Fitzpatrick JW (eds) The Cornell Lab of Ornithology handbook of bird biology, 3rd edn. Wiley, Cornell University, pp 63–99

Mayr E (1942) Systematics and the origin of species. Columbia University Press, New York

Mayr G (2011) Metaves, Mirandornithes, Strisores and other novelties – a critical review of the higher-level phylogeny of neornithine birds. J Zool Syst Evol Res 49:58–76

Mayr G (2017) Avian evolution: the fossil record of birds and its paleobiological significance. Wiley, Chichester, West Sussex

Mitchell KJ, Llamas B, Soubrier J, Rawlence NJ, Worthy TH, Wood J, Lee MSY, Cooper A (2014) Ancient DNA reveals elephant birds and kiwi are sister taxa and clarifies ratite bird evolution. Science 344:898–900

Mittelbach GG, Schemske DW, Cornell HV, Allen AP, Brown JM, Bush MB, Harrison SP, Hurlbert AH, Knowlton N, Lessios HA, McCain CM, McCune AR, McDade LA, McPeek MA, Near TJ, Price TD, Ricklefs RE, Roy K, Sax DF, Schluter D, Sobel JM, Turelli M (2007) Evolution and the latitudinal diversity gradient: speciation, extinction and biogeography. Ecol Lett 10:315–331

Monteiro LR, Furness RW (1998) Speciation through temporal segregation of Madeiran storm petrel (*Oceanodroma castro*) populations in the Azores? Philos Trans R Soc Lond Ser B Biol Sci 353:945–953

Newton I (2003) The speciation and biogeography of birds. Academic Press, London

Orme CDL, Davies RG, Burgess M, Eigenbrod F, Pickup N, Olson VA, Webster AJ, Ding TS, Rasmussen PC, Ridgely RS, Stattersfield AJ, Bennett PM, Blackburn TM, Gaston KJ, Owens IPF (2005) Global hotspots of species richness are not congruent with endemism or threat. Nature 436:1016–1019

Pacheco MA, Battistuzzi FU, Lentino M, Aguilar R, Kumar S, Escalante AA (2011) Evolution of modern birds revealed by mitogenomics: timing the radiation and origin of major orders. Mol Biol Evol 28:1927–1942

Pereira SL, Baker AJ (2006) A mitogenomic timescale for birds detects variable phylogenetic rates of molecular evolution and refutes the standard molecular clock. Mol Biol Evol 23:1731–1740

Pereira SL, Johnson KP, Clayton DH, Baker AJ (2007) Mitochondrial and nuclear DNA sequences support a Cretaceous origin of Columbiformes and a dispersal-driven radiation in the Paleogene. Syst Biol 56:656–672

Phillimore AB, Price TD (2008) Density-dependent cladogenesis in birds. PLoS Biol 6:483–489

Phillimore AB, Orme CDL, Thomas GH, Blackburn TM, Bennett PM, Gaston KJ, Owens IPF (2008) Sympatric speciation in birds is rare: insights from range data and simulations. Am Nat 171:646–657

Pigot AL, Tobias JA (2013) Species interactions constrain geographic range expansion over evolutionary time. Ecol Lett 16:330–338

Pigot AL, Tobias JA (2015) Dispersal and the transition to sympatry in vertebrates. Proc R Soc B Biol Sci 282:20141929

Pratt RC, Gibb GC, Morgan-Richards M, Phillips MJ, Hendy MD, Penny D (2009) Toward resolving deep Neoaves phylogeny: data, signal enhancement, and priors. Mol Biol Evol 26:313–326

Price T (2008) Speciation in birds. Roberts and Company, Greenwood Village, CO

Price TD, Hooper DM, Buchanan CD, Johansson US, Tietze DT, Alström P, Olsson U, Ghosh-Harihar M, Ishtiaq F, Gupta SK, Martens J, Harr B, Singh P, Mohan D (2014) Niche filling slows the diversification of Himalayan songbirds. Nature 509:222–225

Prum RO, Berv JS, Dornburg A, Field DJ, Townsend JP, Lemmon EM, Lemmon AR (2015) A comprehensive phylogeny of birds (Aves) using targeted next-generation DNA sequencing. Nature 526:569–U247

Pulido-Santacruz P, Weir JT (2016) Extinction as a driver of avian latitudinal diversity gradients. Evolution 70:860–872

Rabosky DL (2009) Ecological limits and diversification rate: alternative paradigms to explain the variation in species richness among clades and regions. Ecol Lett 12:735–743

Rabosky DL, Lovette IJ (2008) Density-dependent diversification in North American wood warblers. Proc R Soc B Biol Sci 275:2363–2371

Reddy S, Driskell A, Rabosky DL, Hackett SJ, Schulenberg TS (2012) Diversification and the adaptive radiation of the vangas of Madagascar. Proc R Soc B Biol Sci 279:2062–2071

Rundle HD, Nosil P (2005) Ecological speciation. Ecol Lett 8:336–352

Savolainen V, Anstett MC, Lexer C, Hutton I, Clarkson JJ, Norup MV, Powell MP, Springate D, Salamin N, Baker WJ (2006) Sympatric speciation in palms on an oceanic island. Nature 441:210–213

Schulte P, Alegret L, Arenillas I, Arz JA, Barton PJ, Bown PR, Bralower TJ, Christeson GL, Claeys P, Cockell CS, Collins GS, Deutsch A, Goldin TJ, Goto K, Grajales-Nishimura JM, Grieve RAF, Gulick SPS, Johnson KR, Kiessling W, Koeberl C, Kring DA, MacLeod KG, Matsui T, Melosh J, Montanari A, Morgan JV, Neal CR, Nichols DJ, Norris RD, Pierazzo E, Ravizza G, Rebolledo-Vieyra M, Reimold WU, Robin E, Salge T, Speijer RP, Sweet AR, Urrutia-Fucugauchi J, Vajda V, Whalen MT, Willumsen PS (2010) The Chicxulub asteroid impact and mass extinction at the Cretaceous-Paleogene boundary. Science 327:1214–1218

Schweizer M, Seehausen O, Güntert M, Hertwig ST (2010) The evolutionary diversification of parrots supports a taxon pulse model with multiple trans-oceanic dispersal events and local radiations. Mol Phylogenet Evol 54:984–994

Schweizer M, Seehausen O, Hertwig ST (2011) Macroevolutionary patterns in the diversification of parrots: effects of climate change, geological events and key innovations. J Biogeogr 38:2176–2194

Schweizer M, Hertwig ST, Seehausen O (2014) Diversity versus disparity and the role of ecological opportunity in a continental bird radiation. J Biogeogr 41:1301–1312

Slack KE, Jones CM, Ando T, Harrison GL, Fordyce RE, Arnason U, Penny D (2006) Early penguin fossils, plus mitochondrial genomes, calibrate avian evolution. Mol Biol Evol 23:1144–1155

Smith AL, Monteiro L, Hasegawa O, Friesen VL (2007) Global phylogeography of the band-rumped storm-petrel (*Oceanodroma castro*; Procellariiformes: Hydrobatidae). Mol Phylogenet Evol 43:755–773

Smith BT, McCormack JE, Cuervo AM, Hickerson MJ, Aleixo A, Cadena CD, Perez-Eman J, Burney CW, Xie X, Harvey MG, Faircloth BC, Glenn TC, Derryberry EP, Prejean J, Fields S, Brumfield RT (2014) The drivers of tropical speciation. Nature 515:406–409

Sorenson MD, Sefc KM, Payne RB (2003) Speciation by host switch in brood parasitic indigobirds. Nature 424:928–931

Suh A (2016) The phylogenomic forest of bird trees contains a hard polytomy at the root of Neoaves. Zool Scr 45:50–62

Upchurch P (2008) Gondwanan break-up: legacies of a lost world? Trends Ecol Evol 23:229–236

Vilhena DA, Antonelli A (2015) A network approach for identifying and delimiting biogeographical regions. Nat Commun 6:6848

Voelker G, Rohwer S, Outlaw DC, Bowie RCK (2009) Repeated trans-Atlantic dispersal catalysed a global songbird radiation. Glob Ecol Biogeogr 18:41–49

Wallace AR (1867) The geographical distribution of animals. Harper, New York

Weir JT (2006) Divergent timing and patterns of species accumulation in lowland and highland Neotropical birds. Evolution 60:842–855

Weir JT, Price TD (2011) Limits to speciation inferred from times to secondary sympatry and ages of hybridizing species along a latitudinal gradient. Am Nat 177:462–469

Winkler DW, Billerman SM, Lovette IJ (2015) Bird families of the world: an invitation to the spectacular diversity of birds. Lynx Edicions, Barcelona

Yonezawa T, Segawa T, Mori H, Campos PF, Hongoh Y, Endo H, Akiyoshi A, Kohno N, Nishida S, Wu JQ, Jin HF, Adachi J, Kishino H, Kurokawa K, Nogi Y, Tanabe H, Mukoyama H, Yoshida K, Rasoamiaramanana A, Yamagishi S, Hayashi Y, Yoshida A, Koike H, Akishinonomiya F, Willerslev E, Hasegawa M (2017) Phylogenomics and morphology of extinct Paleognaths reveal the origin and evolution of the ratites. Curr Biol 27:68–77

Chapter 9
Modeling Avian Distributions and Niches: Insights into Invasions and Speciation in Birds

Darius Stiels and Kathrin Schidelko

Abstract Avian evolutionary studies have recently benefited from a plethora of new techniques as well as conceptual progresses on the evolution of ecological niches. The so-called species distribution models (SDMs) allow for niche quantifications in a way that permits comparisons among species and populations. This review will introduce the theoretical background of niche concepts and niche conservatism, followed by an outline of popular methods for modeling and analyzing environmental niches. A comparison of ecological niches among native and non-native populations of invasive species can reveal niche shifts. They can point to evolutionary changes that evolved over comparatively short time scales of decades to a few centuries. On the other hand, ecological niches can also remain conserved over the invasion process. In a similar way, comparisons of ecological niches are also applicable among closely related taxa. Thereby, it is possible to infer changes of ecological niches over longer time scales and reveal otherwise hidden patterns and processes in the evolutionary history of avian clades. Finally, SDMs offer the potential to contribute to integrative taxonomic studies.

Keywords Niche evolution · Niche conservatism · Birds · Species distribution model · Ecological niche model · Invasive species · Phylogeography

9.1 Introduction

Birds have conquered nearly every terrestrial corner of the planet and even soar and feed over the oceans. However, bird species are not uniformly distributed and often differ in their ecological niches. Current individual ranges are the result of tightly linked evolutionary history processes, like speciation, and ecological factors, like climate. Today, these factors are complemented by human influences. Thus, the study of distributional patterns and their underlying processes is at the core of

D. Stiels (✉) · K. Schidelko
Section Ornithology, Zoological Research Museum A. Koenig, Bonn, Germany
e-mail: d.stiels@leibniz-zfmk.de

D. T. Tietze (ed.), *Bird Species*, Fascinating Life Sciences,
https://doi.org/10.1007/978-3-319-91689-7_9

fundamental ornithological research, but also not less than a paramount challenge under global change. Here, we review one particular tool in the study of avian distributions and ecological niches: species distribution models (SDMs). SDMs or ecological niche models (ENMs) have become a standard tool in various fields of biology including ecology, evolution, conservation biology, and related fields (e.g., Elith and Leathwick 2009; Elith et al. 2006; Graham et al. 2004a; Guisan and Thuiller 2005; Guisan et al. 2013; Jeschke and Strayer 2008; Pearman et al. 2007; Peterson et al. 2011). Both terms, SDMs and ENMs, might slightly differ in their emphasis—whether the geographical distribution or the niche is in focus—but are regularly used as synonyms (Peterson and Soberón 2012). Given the in-depth knowledge and the great interest in birds, SDMs are now widely used in ornithology. Only recently, avian niche models and their applications were extensively reviewed (Engler et al. 2017). In contrast to this exhaustive overview, we will limit our focus to applications for the study of avian niche evolution and niche conservatism and its implications for research on speciation and invasive species. Prior to this, we start with a few basic concepts for those who are not familiar with the general background of SDMs.

9.2 The Conceptual Background of SDMs or What Is a Niche?

SDMs are based on the niche concept originally dating back to the beginning of the twentieth century (Grinnell 1917). Joseph Grinnell, an American zoologist (1877–1939), was presumably the first who explicitly related the distribution of an animal, the California Thrasher *Toxostoma redivivum*, to the predominant climatic conditions in its range. These general relations were extended and specified by British zoologist George Evelyn Hutchinson (1903–1991). In 1957, he defined the species-specific niche as a multidimensional hypervolume of environmental conditions which allows a species to persist (Hutchinson 1957). Hypervolumes of high dimensions are nearly impossible to imagine. A simple special case of a hypervolume would be a two-dimensional environmental surface created by certain temperature and precipitation ranges that are occupied by a species.

It is worthwhile to distinguish between the so-called Grinnellian niche, applied in most SDM studies, and the Eltonian niche (Elton 1927; named after British ecologist and zoologist Charles Elton, 1900–1991). While the Grinnellian niche class defines the niche by "scenopoetic" variables (e.g., climate, habitat parameters) (Hutchinson 1978; Soberón 2007), Eltonian niches, in contrast, include resource variables interacting with consumers (e.g., Chase and Leibold 2003; Soberón 2007). The Eltonian noise hypothesis states that Eltonian niches are of predominant importance at smaller scales, while at large scales biotic factors become less relevant (Soberón and Nakamura 2009). Ecological niches were further divided into realized and fundamental niches (e.g., Hutchinson 1957; Soberón 2007). The set of abiotic and

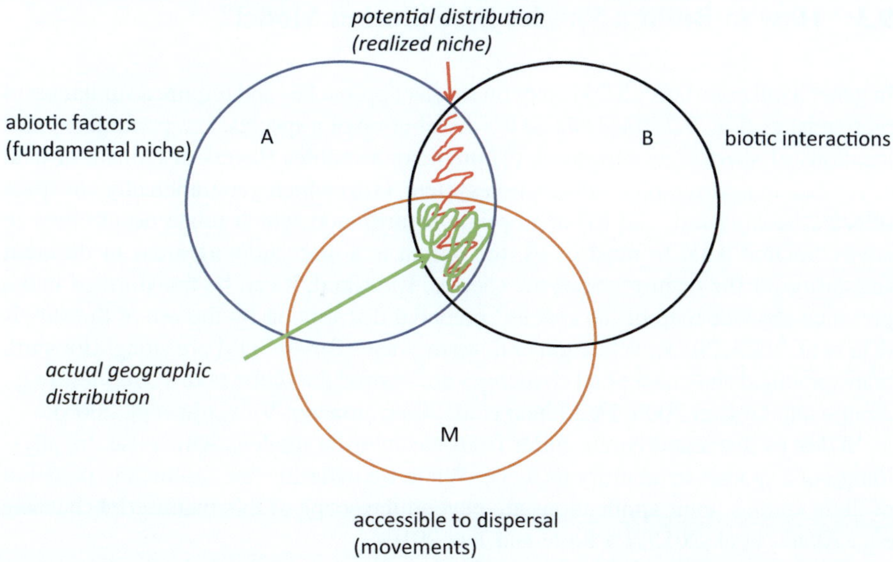

Fig. 9.1 BAM plot after Soberón and Peterson (2005). The geographic area with the right set of abiotic conditions for a species is represented by the blue circle A and is equivalent to the fundamental niche. Area B (black) contains a combination of interacting species. The overlapping area between A and B incorporates the potential distribution of the species, where abiotic and biotic conditions are suitable (red). The orange circle M represents areas where the species is able to move to. The intersection of all three circles stands for the actual geographic distribution of the species (green)

biotic conditions and resources in which an organism could survive and reproduce in the absence of competitors and other biotic interactions is called the fundamental niche. The part of the fundamental niche that is actually occupied by a species under prevailing locational factors is termed the realized niche. Interactions with other organisms such as competition, predation, or parasitism narrow the fundamental niche so that the realized niche is smaller than the niche that would be occupied without competitors. Soberón and Peterson (2005) gave a comprehensible explanation by presenting their so-called BAM plot (Fig. 9.1). Therein, the geographic distribution of a species is restricted by three main factors: (A) the abiotic conditions that allow a species to persist (this is equivalent to the fundamental niche sensu Hutchinson 1957) and (B) biotic factors, i.e., a combination of interacting species. The realized niche is an area where abiotic and biotic conditions are suitable for a species, but this is still not equal to the geographic range of a species. Some otherwise suitable areas might be simply inaccessible: (M) is the reachable area, which could be limited by barriers and species-specific dispersal capacities. As SDMs solely rely on occurrence data, they infer information on niche characteristics only from the geographic range. Moreover, in the case of invasive species, dispersal barriers are broken down by humans.

9.3 How to Build a Species Distribution Model?

In general, all correlative SDMs rely on similar approaches and require a similar set of prerequisites (Fig. 9.2): (1) Data on the distribution of a species, i.e., preferably exact locations of species' occurrences, (2) predictor variables (layers of climate or land cover data in a geographic information system, GIS) which geographically cover the selected background, and (3) appropriate algorithm(s) which relate occurrences to environmental data. In most cases, the output is a map showing areas of different suitability for the occurrence of the species. If desired, it can be transformed into a presence-absence map of the species' potential distribution by the use of thresholds (Liu et al. 2005, 2013). While general requirements for an SDM are straightforward, many practical and conceptual challenges are beyond the scope of this overview (e.g., Araújo and Guisan 2006; Heikkinen et al. 2006; Jiménez-Valverde et al. 2008).

SDMs as discussed herein differ from mechanistic models, which use the physiology of a species to quantify its niche. While we underline the fascinating potential of these studies, their applications are beyond the scope of this manuscript (but see, e.g., Khaliq et al. 2015; La Sorte and Jetz 2010).

9.3.1 Occurrence Data

Range maps are available for nearly every bird species (del Hoyo et al. 1992–2013, http://datazone.birdlife.org/home). However, polygons need to be transferred into occurrences with an appropriate resolution (e.g., Echarri et al. 2009; Schidelko et al. 2011). More exact occurrences are available through online databases like the Global Biodiversity Information Facility (GBIF, www.gbif.org) and mostly originate from museum specimens as well as citizen science data like those collected by birdwatchers, e.g., via eBird (www.ebird.org), underlining the tremendous importance of such data collections (Newbold 2010; Pyke and Ehrlich 2010). In addition, occurrence data come from tracking devices, which were only recently discovered as a data source for SDM studies (Jiguet et al. 2010; Quillfeldt et al. 2017). Most occurrence data show some kind of observer bias, e.g., specimens were collected along roads, close to settlements, or with different intensity across habitats or political territories (Araújo and Guisan 2006). Hence, data processing and filtering is pivotal in order to gain a reliable data source in every modeling approach (Boria et al. 2014; Ranc et al. 2017; Varela et al. 2014). However, even a limited number of data points might result in reasonable outputs (Pearson et al. 2007).

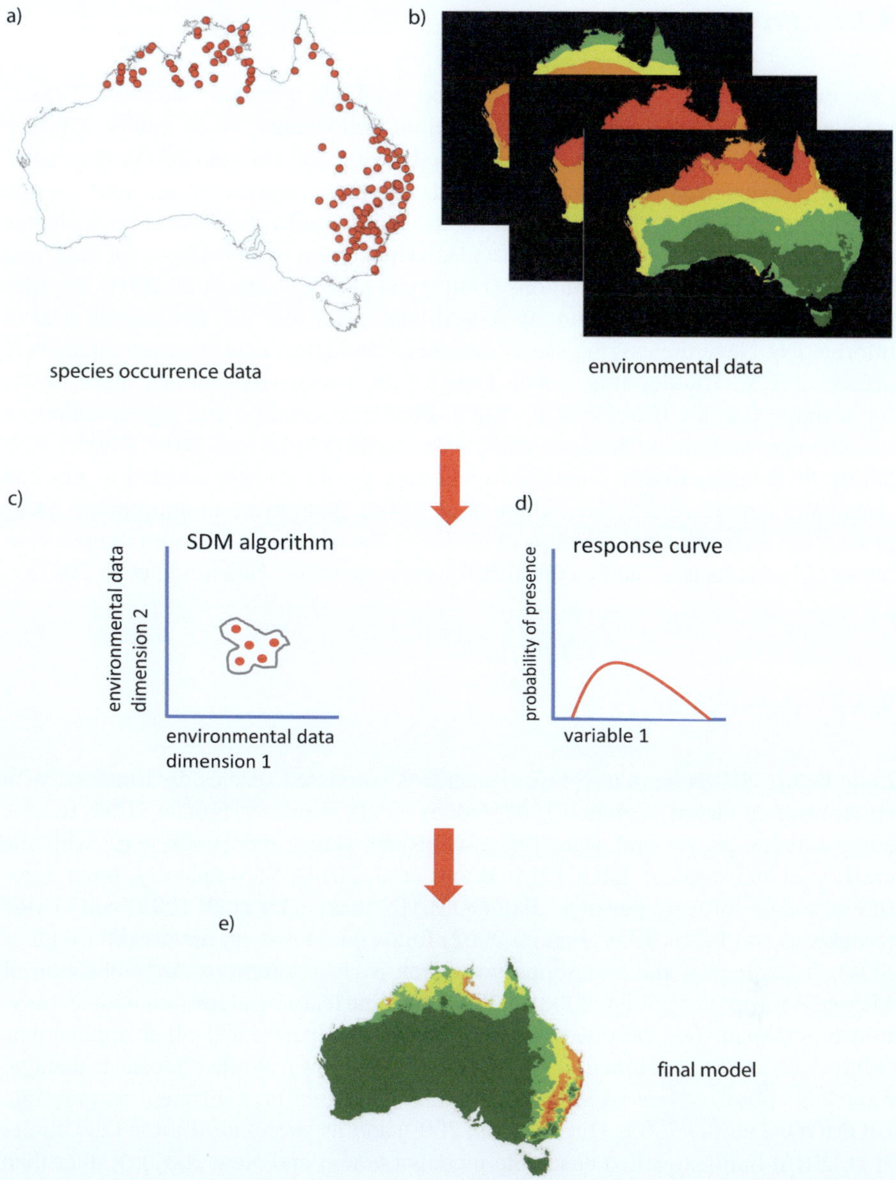

Fig. 9.2 Procedure of building a species distribution model: (**a**) species occurrence records are (**b**) combined with predictor variables (e.g., climate data like temperature or precipitation), (**c**) an appropriate modeling algorithm is selected, and (**d**) the model is fitted to the data. (**e**) Finally, the models' predictions are mapped onto geographical space. The model can also be projected beyond the training area onto other areas or time frames. By application of thresholds, the suitability map can be transformed into a binary presence-absence map of potential distribution

9.3.2 Predictor Variables

Like every model, SDMs are dependent on available predictor variables. Typical predictors are climate data like precipitation, temperature, or humidity. Climate possibly limits distributions at larger scales (Pearson and Dawson 2003). For example, the widely used 19 global BIOCLIM variables (Hijmans et al. 2005, www.worldclim.org) even include data sets for the future (under different climate change scenarios) and the past (e.g., for the last glacial maximum) which allows for hind- and forecasting of potential distributions (Huntley et al. 2007; Jetz et al. 2007; Nogués-Bravo 2009). In addition, remotely sensed land cover data can add a great deal of information. Nevertheless, the use of climate or land cover data has been intensively discussed (e.g., Thuiller et al. 2004). Data for the marine environment make SDMs for seabirds feasible (Engler et al. 2017). Predictor variables can highly influence model output and should be biologically meaningful (Engler et al. 2014; Rödder et al. 2009). Statistically highly correlated predictors should also be avoided as this can potentially tamper results (Heikkinen et al. 2006). Moreover, an appropriate background has to be selected (Phillips et al. 2009; Barve et al. 2011), and the incorporation of biotic factors can be considered (Anderson 2017; Heikkinen et al. 2007).

9.3.3 Algorithms

Long before SDMs were developed, scientists correlated species distributions with environmental factors (Grinnell 1917; Voous 1960). However, typical SDMs require computational power and were first established during the 1980s, e.g., with the creation of the software BIOCLIM (Booth et al. 2014). Subsequently, other algorithms and/or software packages like DOMAIN (Carpenter et al. 1993) and GARP (Stockwell and Peters 1999; Pereira 2002) followed. However, the breakthrough of SDMs in ecological and evolutionary research is closely tight to the publication of Maxent (Phillips et al. 2004, 2006, 2017), a machine learning algorithm with an easy-to-use graphical user interface that has regularly outperformed other applications (Elith et al. 2006; Heikkinen et al. 2006, but see below). While Maxent is a single modeling approach, one widespread alternative is based on a different assumption: BIOMOD (Thuiller 2003; Thuiller et al. 2009) and its successor (biomod2, Thuiller et al. 2016) build so-called ensemble models (Araújo and New 2007). Rather than relying on one single best model, outputs of different algorithms are summarized. The basic principle is that the potential weakness of single algorithms is outweighed by the model power of the majority of algorithms.

There is no consensus which modeling approach is superior, and this may almost certainly also depend on external preconditions (Aguirre-Gutiérrez et al. 2013; Araújo and New 2007; Elith and Graham 2009; Marmion et al. 2009; Qiao et al. 2015; Shabani et al. 2016; Zhu and Peterson 2017). There are different statistics to assess the quality of SDMs like AUC (area under the curve) or TSS (true skill statistic) among others, but it should be noted that their explanatory powers are limited (Lobo et al. 2008; Peterson et al. 2008). Despite great methodological and conceptual progress, SDMs remain a challenging field (e.g., Araújo and Guisan 2006).

9.3.4 Niche Comparisons

Ecological niches of different species or populations can be compared in geographical space based on model outputs. One straightforward approach is to project models from one species onto the range of another and to quantify its predictive power. Model comparisons can be based on statistics like Schoener's *D* inferred from model outputs and now implemented in software packages like ENMTools (Warren et al. 2010) or the approach by Broennimann et al. (2012). The so-called hypervolume algorithm directly compares *n*-dimensional environmental niche hypervolumes (Blonder et al. 2014, but see Qiao et al. 2017).

9.4 Niche Conservatism

One of the basic principles in evolution and ecology—niche conservatism—states that species retain their fundamental niche over time (Ricklefs and Latham 1992; Webb et al. 2002; Wiens and Graham 2005; Wiens et al. 2010). Moreover, given that ecological niches are more similar to each other than can be expected based on the phylogenetic relationships of the species, the term "phylogenetic niche conservatism" has been suggested (Losos 2008a). A vivid debate has risen about the biological meaning of niche conservatism as pattern or process (Losos 2008a, b, 2011; Wiens and Graham 2005; Wiens 2008; Pyron et al. 2015). However, it is widely accepted now that niche conservatism can be considered in the light of many different aspects rather than as a phenomenon per se. There are many obvious examples for niche conservatism in birds, e.g., the recent genomic studies that claim that many waterbirds are indeed closely related (Jarvis et al. 2014; Prum et al. 2015). In addition, feeding niches seem to be more conservative than climatic niches (Anciães and Peterson 2009; Pearman et al. 2014). Evidence for (or against) niche conservatism can be arranged along a time axis ranging from years or decades to millions of years (Peterson 2011; Engler et al. 2017). In this regard, the most recent events are species invasions, in which niches change or remain conservative during the invasion process (Guisan et al. 2014; Broennimann et al. 2007; Peterson 2011). In contrast, the analysis of speciation events, e.g., by comparing species pairs or phylogenies of closely related taxa, is usually related to much longer time periods. Herein, we will follow this temporal structure by having a closer look at avian invasions and review some aspects of niche evolution tied to speciation events.

9.5 Evaluating Avian Invasions

International trading and human transport have pushed the global spread of non-native species (Mack et al. 2000). While they are often discussed in terms of negative impacts, on a global scale, the impact of non-native birds is generally assumed to be small (Blackburn et al. 2009: Chap. 7.5 and references therein). Nevertheless, among invasive species, birds belong to the most ubiquitous and conspicuous non-native animals. Typical examples are Common Starlings *Sturnus vulgaris* in North America and Western European countryside birds in New Zealand or Common Mynas *Acridotheres tristis* from South and Southeast Asia, which are now distributed in various areas of the world.

Globally, about 420 bird species have established populations outside their indigenous range due to human transportation and subsequent deliberate or unintended release (Dyer et al. 2017). Hence, it is very surprising that there is only a very limited amount of SDM studies on invasive birds. In a recent systematic overview, Engler et al. (2017) found 27 studies dealing with SDMs of non-native birds. These studies only covered less than 10% of the species with non-native populations worldwide. There is also a strong bias in the geographic scope of SDM studies, as they mainly focus on few well-known non-native birds brought to Western Palearctic and Nearctic regions (Engler et al. 2017). In some parts of the world, intentional release was particularly important in the spread of non-native species. For example, in New Zealand so-called acclimatization societies propagated and conducted the introduction of British farmland birds, and in parts of Asia, birds are released for religious ceremonies (Blackburn et al. 2009; Severinghaus and Chi 1999; Sodhi et al. 2011). Engler et al. (2017) also found a taxonomic bias in published SDM studies. Only 11 families are represented, with ducks, geese, and swans (Anatidae); pheasants and allies (Phasianidae); waxbills, munias, and allies (Estrildidae); and African and New World parrots (Psittacidae) being particular prevalent among them. As a single prominent species, there are in-depth studies on Rose-ringed Parakeets *Psittacula krameri*, perhaps one of the most successful avian non-native species in Europe (Strubbe and Matthysen 2009).

The biological invasion of a non-native species is a multistage process. The so-called invasion pathway starts with a native species, which after transport, introduction, establishment, and spread becomes an invasive species or sometimes even a pest (Duncan et al. 2003; Sol et al. 2005; Blackburn et al. 2009; see Bauer and Woog 2011 for discussion on terms). While the success of invasive species depends on different factors, a predominant role of the pure number of individuals has become widely accepted (e.g., Blackburn et al. 2015 and references therein, but see Moulton et al. 2011, 2012a, b, 2013).

Biological invasions can be regarded as natural experiments and help to understand how species colonize new environments (e.g., Guisan et al. 2014; Sax et al. 2007 and dating back to Elton 1958). Over the last years, studies using SDMs have greatly contributed to the field of invasion biology, e.g., by identifying potential risk areas for invasions (Peterson and Viglais 2001). SDMs can also contribute to comprehensive studies of niche evolution (Guisan et al. 2014). The question in focus is whether non-native species retain their niche or whether they change it during the invasion process. Knowledge about niche shifts may also help to select appropriate models, e.g., based on all available records or only those from the native or non-native area, respectively (Broennimann and Guisan 2008; Peterson and Viglais 2001; Mau-Crimmins et al. 2006). Improved model predictions can subsequently lead to better assessments of current or future risk areas, although this target has only been rarely addressed for birds (Stiels et al. 2011).

Niche shifts between native and non-native ranges can also point to evolutionary processes during the invasion. In contrast, habitat or climate matching indicates niche conservatism. However, niche comparisons are a great challenge in non-native species (Elith et al. 2010). Invasive populations typically start their spread from a limited number of locations. Hence, there often was simply not enough time to reach equilibrium with the environment and to "fill" the niche. In this case, the niche occupied in the non-native range only represents a subset of the conditions found in the native range (e.g., Engler et al. 2017; Stiels et al. 2011). Guisan et al. (2014) classify niche changes into unfilling, stability, and expansion situations and stress the importance of accounting for available climatic conditions and climatic analogy. Hence, dissimilarity does not automatically point to evolutionary changes in the fundamental niche.

One of the few multispecies studies shows that niche conservatism is the dominant phenomenon observed among 28 non-native species in Europe (Strubbe et al. 2013). Therein, climatic niches between native and non-native ranges differ, but the non-native range can be regarded a subset of the native range instead of a shift outside the native niche space. Nevertheless, results may differ among species and might require species-specific approaches as highlighted for estrildid finches (Stiels et al. 2015). Intraspecific variation and subsequent genetic drift during the invasion process or interspecific interactions may allow species to expand their niche in the non-native range. For example, a species might extend its habitat preferences or its climate niche toward colder temperatures as shown for Alexandrine Parakeets *Psittacula eupatria* (Ancillotto et al. 2015). The great potential of SDM applications in the study of non-native birds is still not fully exploited, and further studies might easily explore neglected taxa or areas hitherto less in focus.

9.6 Speciation and Niche Evolution

Speciation and geographical distributions are closely tight. Since Ernst Mayr's epoch-making work (Mayr 1942), allopatric speciation has emerged as a fundamental evolutionary process which is closely linked to climate niche evolution (e.g., Wiens 2004; Hua and Wiens 2013). Hence, "niche conservatism as an emerging principle" (Wiens et al. 2010) is in focus of recent studies on speciation and phylogeography (see paragraph "niche conservatism" and Engler et al. 2017 for a current avian review). As already described for non-native species, SDMs of two or more different taxa of a clade can be analyzed in order to compare similarities and changes. The integration of phylogenetic analyses into recently developed statistical tools facilitates analyses of evolutionary changes in environmental niches. This even includes the reconstruction of so-called ancestral niches, i.e., the reconstructed niche space once potentially occupied by a common ancestor of recent taxa (e.g., Cooper et al. 2010; Evans et al. 2009; Graham et al. 2004b; Heibl and Calenge 2013; Revell 2012; Rödder et al. 2013). While methodological details are beyond the focus of this article, we stress their great potential for future avian evolutionary studies (Engler et al. 2017, see Gómez et al. 2016 for a recent avian example). We simultaneously call for very cautious interpretations, as similar diversification patterns can be caused by completely different processes (Warren et al. 2014).

A highly influential study on avian niche conservatism analyzed climatic niches of sister species: Peterson et al. (1999) conducted SDMs of allopatrically distributed sister species along a faunal divide—the Isthmus of Tehuantepec in Mexico. The basic assumption of their approach is that, if two taxa show conserved niches, SDMs of one species should be able to predict the distribution of the other and vice versa. Their prediction was confirmed, and their results point to a general pattern of conserved niches. Similar methods can also be transferred to intraspecific populations like non-native species (see above) or subspecies (Peterson and Holt 2003). However, their results were not as homogeneous as for the analyses of sister species, and predictive power varied among studied taxa. SDMs also offer opportunities for the study of sister species with overlapping ranges and hybrid zones like those of two European *Hippolais* warblers (Engler et al. 2013).

While these examples refer to current distributional patterns, SDMs also allow for a closer look at the influence of historical events on present patterns like range contractions to refugial areas. Hence, paleodistribution models can greatly help to understand current diversity and its underlying evolutionary processes (Nogués-Bravo 2009). For instance, the North American Painted Bunting *Passerina ciris* shows a disjunct breeding range. Shipley et al. (2013) conducted current SDMs as well as paleodistribution models in order to test whether the distributional gap is of bioclimatic origin. As current SDMs showed suitable climatic conditions in this gap, this hypothesis was refused, while a relation to migratory constraints that connect breeding and wintering ranges was preferred. In addition, this explanation can be associated with past distributions as inferred from hindcasting, given that the species might follow past expansion paths after being restricted to southerly areas during the Pleistocene (see also Ruegg et al. 2006). Another example stems from the Neotropics. An analysis of potential Pleistocene refugial areas for members of the thrush-like *Schiffornis* complex revealed broad congruence with current phylogroups identified by molecular methods (Peterson and Nyári 2007). Recently, SDMs were also applied to test the ring-species concept in Greenish Warblers *Phylloscopus trochiloides* (Peterson and Anamza 2017). The range of this species forms a ring around the Tibetan plateau, but the existence of Pleistocene refugia revealed by paleodistribution models questions the validity of the ring-species concept (Peterson and Anamza 2017). Hypotheses on former refugia can also be addressed by multispecies approach. Stacked SDMs of many estrildid finch species were used to test refugial hypotheses for forest and savanna species in Africa and Australia (Schidelko et al. 2011, 2013). They revealed a pattern of relative stability of biodiversity hotspots but also point to the importance of current shelf zones as historical retreat areas.

Many of the aforementioned examples mainly relied on single species or species pairs; however, applications of SDMs are equally transferable to multispecies phylogenies. The question, whether climatic niches are conservative, was intensively studied in North and Central American jays of the genus *Aphelocoma* (Rice et al. 2003; McCormack et al. 2009). These studies are particular informative, because originally a strong deviation from niche conservatism was found (Rice et al. 2003). This was rejected later, mainly based on recent sophisticated statistical approaches like the use of null models (McCormack et al. 2009). While we restrict our overview to moderate time scales up to the genus level, it is well worth noting that in some studies longer time scales were taken into account and analyses up to the family level were conducted (reviewed by Engler et al. 2017). Examples of analyzed genera also include the Asiatic genus *Pomatorhinus*, for which results about niche conservatism not only differ for different subclades but also for different predictor variables (Nyári and Reddy 2013).

The migratory behavior of many birds makes it possible to look at so-called seasonal niches (Joseph 1996; Joseph and Stockwell 2000; Martínez-Meyer et al. 2004; Engler et al. 2017). Birds are exposed to different environmental conditions during and outside the breeding season, often combined with a different distribution—the breeding and the wintering range. Hence, it is not only possible to analyze, whether niches remain stable throughout the year, but also to ask, whether realized breeding and wintering niches might evolve differently in a clade over evolutionary time scales (Gómez et al. 2016; Martínez-Meyer et al. 2004; Laube et al. 2015).

9.7 Assisting Taxonomy

Newly described taxa are often only known from a handful of locations, and SDMs can help to gain information about their potential range and its environmental correlates. Moreover, even taxonomic implications might be supported from SDMs as mainly outlined by Wiens (2004) and Wiens and Graham (2005): A newly found population, e.g., in a mountain range, might be isolated from another well-known population. SDMs are able to reveal whether the intervening habitat (e.g., lowland areas with a different vegetation type and climate) might represent a potential barrier to gene flow (Wiens and Graham 2005). If SDMs show a corridor of suitable habitat between both areas, then gene flow is more likely, and genetic discreteness of the new population is questionable. Highly valuable interpretations can also be inferred from direct comparisons of niches between both populations (see Niche Comparisons above). If the ecological niche of the new population is not a subset of the well-known population, this can point to evolutionary changes that happened in the past (and subsequently conservative niches that prevented further spread) (Wiens and Graham 2005). One avian case study analyzed a disjunct population of a hummingbird, the Blossomcrown *Anthocephala floriceps* in

Colombia, where the population from the Sierra de Santa Marta is disjunct from the Andean population. Both live under different climatic conditions as cross-projections of population-specific SDMs show (Lozana-Jaramillo et al. 2014). In addition, SDMs were applied for recently described taxa; a prominent example is different members of the Neotropical tapaculos (Rhinocryptidae) (Avendaño and Donagan 2015; Avendaño et al. 2015; Nemésio et al. 2013).

In a nutshell, SDMs offer the potential to become a central part in integrative taxonomic studies. They already tremendously contributed to invasion biology, phylogeography, and research related to climatic niche evolution. In combination with recent methodological advances and an increasing knowledge about avian systematics and genomics, they will likely play an important role in further ornithological studies (Engler et al. 2017).

References

Aguirre-Gutiérrez J, Carvalheiro LG, Polce C, van Loon EE, Raes N, Reemer M, Biesmeijer JC (2013) Fit for purpose: species distribution model performance depends on evaluation criteria – Dutch hoverflies as a case study. PLoS One 8(3):e63708. https://doi.org/10.1371/journal.pone.0063708

Anciães M, Peterson AT (2009) Ecological niches and their evolution among Neotropical manakins (Aves: Pipridae). J Avian Biol 40:591–605

Ancillotto L, Strubbe D, Menchetti M, Mori E (2015) An overlooked invader? Ecological niche, invasion success and range dynamics of the Alexandrine parakeet in the invaded range. Biol Invasions 18(2):583–595. https://doi.org/10.1007/s10530-015-1032-y

Anderson RP (2017) When and how should biotic interactions be considered in models of species niches and distributions. J Biogeogr 44:8–17

Araújo MB, Guisan A (2006) Five (or so) challenges for species distribution modelling. J Biogeogr 33:1677–1688

Araújo MB, New M (2007) Ensemble forecasting of species distributions. Trends Ecol Evol 22:42–47

Avendaño JE, Donagan TM (2015) A distinctive new subspecies of *Scytalopus griseicollis* (Aves, Passeriformes, Rhinocryptidae) from the northern Eastern Cordillera of Colombia and Venezuela. ZooKeys 506:137–153

Avendaño JE, Cuervo AM, López-O JPm Gutiérrez-Pinto N, Cortés-Diago A, Cadena CD (2015) A new species of tapaculo (Rhinocryptidae: *Scytalopus*) from the Serranía de Perijá of Colombia and Venezuela. Auk 132:450–466

Barve N, Barve V, Jiménez-Valverde A, Lira-Noriega A, Maher SP, Peterson AT, Soberón J, Villalobos F (2011) The crucial role of the accessible area in ecological niche modelling and species distribution modelling. Ecol Model 222:1810–1819

Bauer H-G, Woog F (2011) On the 'invasiveness' of non-native bird species. IBIS 153:204–206

Blackburn TM, Lockwood JL, Cassey P (2009) Avian invasions. The ecology and evolution of exotic birds. Oxford University Press, Oxford, New York

Blackburn TM, Lockwood JL, Cassey P (2015) The influence of numbers on invasion success. Mol Ecol 24:1942–1953

Blonder B, Lamanna C, Violle C, Enquist BJ (2014) The n-dimensional hypervolume. Glob Ecol Biogeogr 23:595–609

Booth TH, Nix HA, Busby JR, Hutchinson MF (2014) BIOCLIM: the first species distribution modelling package, its early applications and relevance to most current MAXENT studies. Divers Distrib 20:1–9

Boria RA, Olson LE, Goodman SM, Anderson RP (2014) Spatial filtering to reduce sampling bias can improve the performance of ecological niche models. Ecol Model 275:73–77

Broennimann O, Guisan A (2008) Predicting current and future biological invasions: both native and invaded ranges matter. Biol Lett 4:585–589

Broennimann O, Treier UA, Müller-Schärer H, Thuiller W, Peterson AT, Guisan A (2007) Evidence of climatic niche shift during biological invasion. Ecol Lett 10:701–709

Broennimann O, Fitzpatrick MC, Pearman PB, Petitpierre B, Pellissier L, Yoccoz NG, Thuiller W, Fortin M-J, Randin C, Zimmermann NE, Graham CH, Guisan A (2012) Measuring ecological niche overlap from occurrence and spatial environmental data. Glob Ecol Biogeogr 21:481–497

Carpenter G, Gillison AN, Winter J (1993) DOMAIN: a flexible modelling procedure for mapping potential distributions of plants and animals. Biodivers Conserv 2:667–680

Chase J, Leibold M (2003) Ecological niches: linking classical and contemporary approaches. University of Chicago Press, Chicago

Cooper N, Jetz W, Freckleton RP (2010) Phylogenetic comparative approaches for studying niche conservatism. J Evol Biol 23:2529–2539

del Hoyo J, Elliott A, Sargatal J (eds) (1992–2013) Handbook of the birds of the world, 16 vol and special volume. Lynx Edicions, Barcelona

Duncan RP, Blackburn TM, Sol D (2003) The ecology of bird introductions. Annu Rev Ecol Evol Syst 34:71–98

Dyer EE, Cassey P, Redding DW, Collen B, Franks V, Gaston KJ, Jones KE, Kark S, Orme CDL, Blackburn TM (2017) The global distribution and drivers of alien bird species richness. PLoS Biol 15(1):e2000942. https://doi.org/10.1371/journal.pbio.2000942

Echarri F, Tambussi C, Hospitaleche CA (2009) Predicting the distribution of the crested tinamous, *Eudromia spp.* (Aves, Tinamiformes). J Ornithol 150:75–84

Elith J, Graham CH (2009) Do they? How do they? WHY do they differ? On finding reasons for differing performances of species distribution models. Ecography 32:66–77

Elith J, Leathwick JR (2009) Species distribution models: ecological explanation and prediction across space and time. Annu Rev Ecol Evol Syst 40:677–697

Elith J, Graham CH, Anderson RP, Dudik M, Ferrier S, Guisan A, Hijmans RJ, Huettmann F, Leathwick JR, Lehmann A, Li J, Lohmann LG, Loiselle BA, Manion G, Moritz C, Nakamura M, Nakazawa Y, Overton JMM, Peterson AT, Phillips SJ, Richardson K, Scachetti-Pereira R, Schapire RE, Soberón J, Williams S, Wisz MS, Zimmermann NE (2006) Novel methods improve prediction of species' distributions from occurrence data. Ecography 29:129–151

Elith J, Kearney M, Phillips S (2010) The art of modelling range-shifting species. Methods Ecol Evol 1:330–342

Elton CS (1927) Animal ecology. Sidgwick and Jackson, London

Elton CS (1958) The ecology of invasions by animals and plants. Springer, Boston

Engler JO, Rödder D, Elle O, Hochkirch A, Secondi J (2013) Species distribution models contribute to determine the effect of climate and interspecific interactions in moving hybrid zones. J Evol Biol 26:2487–2496

Engler JO, Rödder D, Stiels D, Förschler MI (2014) Suitable, reachable but not colonised: seasonal niche duality in an endemic mountainous songbird. J Ornithol 155:657–669

Engler JO, Stiels D, Schidelko K, Strubbe D, Quillfeldt P, Brambilla M (2017) Avian SDMs: current state, challenges, and opportunities. J Avian Biol. https://doi.org/10.1111/jav.01248

Evans MEK, Smith SA, Flynn RS, Donoghue MJ (2009) Climate, niche evolution, and diversification of the "bird-cage" evening primroses (*Oenothera*, sections *Anogra* and *Kleinia*). Am Nat 173:225–240

Gómez C, Tenorio EA, Montoya P, Cadena CD (2016) Niche-tracking migrants and niche-switching residents: evolution of climatic niches in New World warblers (Parulidae). Proc R Soc B 283:20152458

Graham CH, Ferrier S, Huettmann F, Moritz C, Peterson AT (2004a) New developments in museum-based informatics and applications in biodiversity analysis. Trends Ecol Evol 19:497–503

Graham CH, Ron SR, Santos JC, Schneider CJ, Moritz C (2004b) Integrating phylogenetics and environmental niche models to explore speciation mechanisms in dendrobatid frogs. Evolution 58:1781–1793

Grinnell J (1917) The niche-relationships of the California thrasher. Auk 34:427–433

Guisan A, Thuiller W (2005) Predicting species distribution: offering more than simple habitat models. Ecol Lett 8:993–1009

Guisan A, Tingley R, Baumgartner JB, Naujokaitis-Lewis I, Sutcliffe PR, Tulloch AIT, Regan TJ, Brotons L, McDonald-Madden E, Mantyka-Pringle C, Martin TG, Rhodes JR, Maggini R, Setterfield SA, Elith J, Schwartz MW, Wintle BA, Broennimann O, Austin M, Ferrier S, Kearney MR, Possingham HP, Buckley YM (2013) Predicting species distributions for conservation decisions. Ecol Lett 16:1424–1435

Guisan A, Petitpierre B, Broennimann O, Daehler C, Kueffer C (2014) Unifying niche shift studies: insights from biological invasions. Trends Ecol Evol 29:260–269

Heibl C, Calenge C (2013) PHYLOCLIM: integrating phylogenetics and climatic niche modelling. Available from https://cran.r-project.org/web/packages/phyloclim/index.html

Heikkinen RK, Luoto M, Araújo MB, Virkkala R, Thuiller W, Sykes MT (2006) Methods and uncertainties in bioclimatic envelope modelling under climate change. Prog Phys Geogr 30:751–777

Heikkinen RK, Luoto M, Virkkala R, Pearson RG, Körber J-H (2007) Biotic interactions improve prediction of boreal bird distributions at macro-scales. Gobal Ecol Biogeogr 16:754–763

Hijmans RJ, Cameron SE, Parra JL, Jones PG, Jarvis A (2005) Very high resolution interpolated climate surfaces for global land areas. Int J Climatol 25:1965–1978

Hua X, Wiens JJ (2013) How does climate influence speciation. Am Nat 182:1–12. https://doi.org/10.1086/670690

Huntley B, Green RE, Collingham YC, Willis SG (2007) A climatic atlas of European breeding birds. Lynx Edicions, Barcelona

Hutchinson GE (1957) Concluding remarks. Cold Spring Harb Symp Quant Biol 22:415–427

Hutchinson GE (1978) An introduction to population biology. Yale University Press, New Haven, CT

Jarvis ED et al (2014) Whole-genome analyses resolve early branches in the tree of life of modern birds. Science 346:1320–1331

Jeschke JM, Strayer DL (2008) Usefulness of bioclimatic models for studying climate change and invasive species. Ann N Y Acad Sci 1134:1–24

Jetz W, Wilcove DS, Dobson AP (2007) Projected impacts of climate and land-use change on global diversity of birds. PLoS Biol 5:e157. https://doi.org/10.1371/journal.pbio.0050157

Jiguet F, Barbet-Massin M, Chevallier D (2010) Predictive distribution models applied to satellite tracks: modelling the western African winter range of European migrant Black Storks *Ciconia nigra*. J Ornithol 152:111–118

Jiménez-Valverde A, Lobo JM, Hortal J (2008) Not as good as they seem: the importance of concepts in species distribution modelling. Divers Distrib 14:885–890

Joseph L (1996) Preliminary climatic overview of migration patterns in South American Austral migrant passerines. Ecotropica 2:185–193

Joseph L, Stockwell D (2000) Temperature-based models of the migration of Swainson's Flycatcher (*Myiarchus swainsoni*) across South America: a new use for museum specimens of migratory birds. Proc Acad Natl Sci Phila 150:293–300

Khaliq I, Fritz SA, Prinzinger R, Pfenninger M, Böhning-Gaese K, Hof C (2015) Global variation in thermal physiology of birds and mammals: evidence for phylogenetic niche conservatism only in the tropics. J Biogeogr 42:2187–2196

La Sorte FA, Jetz W (2010) Avian distributions under climate change: towards improved projections. J Exp Biol 213:862–869

Laube I, Graham CH, Böhning-Gaese K (2015) Niche availability in space and time: migration in *Sylvia* warblers. J Biogeogr 42:1896–1906

Liu C, Berry PM, Dawson TP, Pearson RG (2005) Selecting thresholds of occurrence in the prediction of species distributions. Ecography 28:385–393

Liu C, White M, Newell G (2013) Selecting thresholds for the prediction of species occurrence with presence-only data. J Biogeogr 40:778–789

Lobo JM, Jiménez-Valverde A, Real R (2008) AUC: a misleading measure of the performance of predictive distribution models. Glob Ecol Biogeogr 17:145–151

Losos JB (2008a) Phylogenetic niche conservatism, phylogenetic signal and the relationship between phylogenetic relatedness and ecological similarity among species. Ecol Lett 11:995–1007

Losos JB (2008b) Rejoinder to Wiens (2008): phylogenetic niche conservatism, its occurrence and importance. Ecol Lett 11:1005–1007

Losos JB (2011) Seeing the forest for the trees: the limitations of phylogenies in comparative biology. Am Nat 177:709–727

Lozana-Jaramillo M, Rico-Guevara A, Cadena CD (2014) Genetic differentiation, niche divergence, and the origin and maintenance of the disjunct distribution in the Blossomcrown *Anthocephala floriceps* (Trochilidae). PLoS One 9(9):e108345. https://doi.org/10.1371/journal.pone.0108345

Mack RN, Simberloff D, Lonsdale WM, Evans H, Clout M, Bazzaz FA (2000) Biotic invasions: causes, epidemiology, global consequences, and control. Ecol Appl 10:689–710

Marmion M, Parviainen M, Luoto M, Heikkinen RK, Thuiller W (2009) Evaluation of consensus methods in predictive species distribution modelling. Divers Distrib 15:59–69

Martínez-Meyer E, Peterson AT, Navarro-Sigüenza A (2004) Evolution of seasonal ecological niches in the *Passerina* buntings (Aves: Cardinalidae). Proc R Soc B 271:1151–1157

Mau-Crimmins TM, Schussmann HR, Geiger EL (2006) Can the invaded range of a species be predicted using only native-range data? Lehmann lovegrass (*Eragrostis lehmanniana*) in the southwestern United States. Ecol Model 193:736–746

Mayr E (1942) Systematics and the origin of species. Columbia University Press, New York

McCormack JE, Zellmer AJ, Knowles LL (2009) Does niche divergence accompany allopatric divergence in *Aphelocoma* jays as predicted under ecological speciation? Insights from tests with niche models. Evolution 64:1231–1244

Moulton MP, Cropper WP Jr, Avery ML (2011) A reassessment of the role of propagule pressure in influencing fates of passerine introductions to New Zealand. Biodivers Conserv 20:605–623

Moulton MP, Cropper WP Jr, Moulton LE, Avery ML, Peacock D (2012a) A reassessment of historical records of avian introductions to Australia: no case for propagule pressure. Biodivers Conserv 21:155–174

Moulton MP, Cropper WP Jr, Avery ML (2012b) Historical records of passerine introductions to New Zealand fail to support the propagule pressure hypothesis. Biodivers Conserv 21:297–307

Moulton MP, Cropper WP Jr, Avery ML (2013) Is propagule size the critical factor in predicting introductions outcomes in passeriform birds? Biol Invasions 15:1449–1458

Nemésio A, Rasmussen C, Aguiar AP, Pombal JP Jr, Dubois A (2013) Nomenclatural issues in ornithology: the incredible controversy on the identity of a long overlooked Brazilian bird. Zootaxa 3734:241–258

Newbold T (2010) Applications and limitations of museum data for conservation and ecology, with particular attention to species distribution models. Prog Phys Geogr 34:3–22

Nogués-Bravo D (2009) Predicting the past distribution of species climatic niches. Glob Ecol Biogeogr 18:521–531

Nyári ÁS, Reddy S (2013) Comparative phyloclimatic analysis and evolution of ecological niches in the scimitar babblers (Aves: Timaliidae: *Pomatorhinus*). PLoS One 8(2):e55629. https://doi.org/10.1371/journal.pone.0055629

Pearman PB, Guisan A, Broennimann O, Randin CF (2007) Niche dynamics in space and time. Trends Ecol Evol 23:149–158

Pearman PB, Lavergne S, Roquet C, Wüest R, Zimmermann NE, Thuiller W (2014) Phylogenetic patterns of climatic, habitat and trophic niches in a European avian assemblage. Glob Ecol Biogeogr 23:414–424

Pearson RG, Dawson TP (2003) Predicting the impacts of climate change on the distribution of species: are bioclimate envelope models useful? Glob Ecol Biogeogr 12:361–371

Pearson RG, Raxworthy CJ, Nakamura M, Peterson AT (2007) Predicting species distributions from small numbers of occurrence records: a test case using cryptic geckos in Madagascar. J Biogeogr 34:102–117

Pereira RS (2002) Desktop GARP. Kansas: University of Kansas Center for Research. Available from http://www.nhm.ku.edu/desktopgarp/index.html

Peterson AT (2011) Ecological niche conservatism: a time-structured review of evidence. J Biogeogr 38:817–827

Peterson AT, Anamza T (2017) Reexamining *Phylloscopus trochiloides* as a ring species: a refugial counter hypothesis. J Avian Biol 48(12):1608–1613. https://doi.org/10.1111/jav.01167

Peterson AT, Holt RD (2003) Niche differentiation in Mexican birds: using point occurrences to detect ecological innovation. Ecol Lett 6:774–782

Peterson AT, Nyári ÁS (2007) Ecological niche conservatism and Pleistocene refugia in the thrush-like mourner, *Schiffornis* sp., in the Neotropics. Evolution 62–1:173–183

Peterson AT, Soberón J (2012) Species distribution modelling and ecological niche modelling: getting the concepts right. Natureza & Conservação 10:102–107

Peterson TA, Viglais DA (2001) Predicting species invasions using ecological niche modeling: new approaches from bioinformatics attack a pressing problem. Bioscience 51:363–371

Peterson AT, Soberón J, Sánchez-Cordero V (1999) Conservatism of ecological niches in evolutionary time. Science 285:1265–1267

Peterson AT, Papes M, Soberón J (2008) Rethinking receiver operating characteristic analysis applications in ecologinal niche modeling. Ecol Model 213:63–72

Peterson AT, Soberón J, Pearson RG, Anderson RP, Martínez-Meyer E, Nakamura M, Araújo MB (2011) Ecological niches and geographic distributions. Princeton University Press, Princeton, Oxford

Phillips SJ, Dudík M, Schapire RE (2004) A maximum entropy approach to species distribution modeling. Proceedings of the 21st international conference on machine learning. Banff, Canada

Phillips SJ, Anderson RP, Schapire RE (2006) Maximum entropy modeling of species geographic distributions. Ecol Model 190:231–259

Phillips SJ, Dudík M, Elith J, Graham CH, Lehmann A, Leathwick J, Ferrier S (2009) Sample selection bias and presence-only distribution models: implications for background and pseudoabsence data. Ecol Appl 19:181–197

Phillips SJ, Anderson RP, Dudík M, Schapire RE, Blair ME (2017) Opening the black-box: an open-source release of Maxent. Ecography 40:887–893

Prum RO, Berv JS, Dornburg A, Field DJ, Townsend JP, Moriarty Lemmon E, Lemmon AR (2015) A comprehensive phylogeny of birds (Aves) using targeted next-generation DNA sequencing. Nature 526:569–573

Pyke GH, Ehrlich PR (2010) Biological collections and ecological/environmental research: a review, some observations and a look to the future. Biol Rev 85:247–266

Pyron RA, Costa GC, Patten MA, Burbrink FT (2015) Phylogenetic niche conservatism and the evolutionary basis of ecological speciation. Biol Rev 90:1248–1262

Qiao H, Soberón J, Peterson AT (2015) No silver bullets in correlative ecological niche modelling: insights from testing among many potential algorithms for niche estimation. Methods Ecol Evol 6:1126–1136

Qiao H, Escobar LE, Saupe EE, Ji L, Soberón J (2017) A cautionary note on the use of hypervolume kernel density estimators in ecological niche modelling. Glob Ecol Biogeogr 26:1066–1070

Quillfeldt P, Engler JO, Silk JRD, Phillips RA (2017) Influence of device accuracy and choice of algorithm for species distribution modelling of seabirds: a case study using black-browed albatrosses. J Avian Biol 48(12):1549–1555. https://doi.org/10.1111/jav.01238

Ranc N, Santini L, Rondinini C, Boitani L, Poitevin F, Angerbjörn A, Maiorano L (2017) Performance tradeoffs in target-group bias correction for species distribution models. Ecography 40:1076–1087

Revell LJ (2012) Phytools: an R package for phylogenetic comparative biology (and other things). Methods Ecol Evol 3:217–223

Rice NH, Martínez-Meyer E, Peterson AT (2003) Ecological niche differentiation in the *Aphelocoma* jays: a phylogenetic perspective. Biol J Linn Soc 80:369–383

Ricklefs RE, Latham RE (1992) Intercontinental correlation of geographical ranges suggests stasis in ecological traits of relict genera of temperate perennial herbs. Am Nat 139:1305–1321

Rödder D, Schmidtlein S, Veith M, Lötters S (2009) Alien invasive Slider Turtle in unpredicted habitat: a matter of niche shift or predictors studied. PLoS One 4(11):e7843. https://doi.org/10.1371/journal.pone.0007843

Rödder D, Lawing AM, Flecks M, Ahmadzadeh F, Dambach J, Engler JO, Habel JC, Hartmann T, Hörnes D, Ihlow F, Schidelko K, Stiels D, Polly PD (2013) Evaluating the significance of paleophylogeographic species distribution models in reconstructing quaternary range-shifts of Nearctic chelonians. PLoS One 8(10):e72855. https://doi.org/10.1371/journal.pone.0072855

Ruegg KC, Hijmans RJ, Moritz C (2006) Climate change and the origin of migratory pathways in the Swainson's thrush *Catharus ustulatus*. J Biogeogr 33:1172–1182

Sax DF, Stachowicz JJ, Brown JH, Bruno JF, Dawson MN, Gaines SD, Grosberg RK, Hastings A, Holt RD, Mayfield MM, O'Connor MI, Rice WR (2007) Ecological and evolutionary insights from species invasions. Trends Ecol Evol 22:465–471

Schidelko K, Stiels D, Rödder D (2011) Historical stability of diversity patterns in African estrildid finches (Aves: Estrildidae)? Biol J Linn Soc 102:455–470

Schidelko K, Wüstenhagen N, Stiels D, van den Elzen R, Rödder D (2013) Continental shelf as potential retreat areas for Austral-Asian estrildid finches (Passeriformes: Estrildidae) during the Pleistocene. J Avian Biol 44:121–132

Severinghaus LL, Chi L (1999) Prayer animal release in Taiwan. Biol Conserv 89:301–304

Shabani F, Kumar L, Ahmadi M (2016) A comparison of absolute performance of different correlative and mechanistic species distribution models in an independent area. Ecol Evol 6:5973–5986

Shipley JR, Contina A, Batbayar N, Bridge ES, Peterson AT, Kelly JF (2013) Niche conservatism and disjunct populations: a case study with painted buntings (*Passerina ciris*). Auk 130:476–486

Soberón J (2007) Grinnellian and Eltonian niches and geographic distributions of species. Ecol Lett 10:1115–1123

Soberón J, Nakamura M (2009) Niches and distributional areas: concepts, methods, and assumptions. Proc Natl Acad Sci USA 106:19644–19650

Soberón J, Peterson AT (2005) Interpretation of models of fundamental ecological niches and species' distributional areas. Biodivers Inform 2:1–10

Sodhi NS, Şekercioğlu ÇH, Barlow J, Robinson SK (2011) Conservation of tropical birds. Wiley-Blackwell, Oxford

Sol D, Blackburn TM, Cassey P, Duncan RP, Clavell J (2005) The ecology and impact of non-indigenous birds. In: del Hoyo J, Elliott A, Sargatal J (eds) Handbook of the birds of the world. X. Cuckoo-shrikes to thrushes. Lynx Edicions, Barcelona, pp 13–35

Stiels D, Schidelko K, Engler JO, van den Elzen R, Rödder D (2011) Predicting the potential distribution of the invasive Common Waxbill *Estrilda astrild* (Passeriformes: Estrildidae). J Ornithol 152:769–780

Stiels D, Gaißer B, Schidelko K, Engler JO, Rödder D (2015) Niche shift in four non-native estrildid finches and implications for species distribution models. Ibis 157:75–90

Stockwell DRB, Peters DP (1999) The GARP modelling system: problems and solutions to automated spatial prediction. Int J Geogr Inf Syst 13:143–158

Strubbe D, Matthysen E (2009) Establishment success of invasive ring-necked and monk parakeets in Europe. J Biogeogr 36:2264–2278

Strubbe D, Broennimann O, Chiron F, Matthysen E (2013) Niche conservatism in non-native birds in Europe: niche-unfilling rather than niche-expansion. Glob Ecol Biogeogr 22(8):962–970. https://doi.org/10.1111/geb.12050

Thuiller W (2003) Biomod – optimizing predictions of species distributions and projecting potential future shifts under global change. Glob Chang Biol 9:1353–1362

Thuiller W, Araújo MB, Lavorel S (2004) Do we need land-cover data to model species distributions in Europe? J Biogeogr 31:353–361

Thuiller W, Lafourcade B, Engler R, Araújo MB (2009) BIOMOD – a platform for ensemble forecasting of species distributions. Ecography 32:369–373

Thuiller W, Georges D, Engler R, Breiner F (2016) biomod2: ensemble platform for species distribution modeling. R Core Team. https://cran.r-project.org/web/packages/biomod2/biomod2.pdf

Varela S, Anderson RP, García-Valdés R, Fernández-González F (2014) Environmental filters reduce the effects of sampling bias and improve predictions of ecological niche models. Ecography 37:1084–1091

Voous KH (1960) Atlas van de Europese vogels. Elsevier, Amsterdam

Warren DL, Glor RE, Turelli M (2010) ENMTools: a toolbox for comparative studies of environmental niche models. Ecography 33:607–611

Warren DL, Cardillo M, Rosaur DF, Bolnick DI (2014) Mistaking geography for biology: inferring processes from species distributions. Trends Ecol Evol 29(10):572–580

Webb CO, Ackerly DD, McPeek MA, Donoghue MJ (2002) Phylogenies and community ecology. Annu Rev Ecol Syst 33:475–505

Wiens JJ (2004) Speciation and ecology revisited: phylogenetic niche conservatism and the origin of species. Evolution 58:193–197

Wiens JJ (2008) Commentary on Losos (2008): Niche conservatism déjà vu. Ecol Lett 11:1004–1005

Wiens JJ, Graham CH (2005) Niche conservatism: integrating evolution, ecology, and conservation biology. Annu Rev Ecol Evol Syst 36:519–539

Wiens JJ, Ackerly DD, Allen AP, Anacker BL, Buckley LB, Cornell HV, Damschen EI, Davies TJ, Grytnes J-A, Harrison SP, Hawkins BA, Holt RD, McCain CM, Stephens PR (2010) Niche conservatism as an emerging principle in ecology and conservation biology. Ecol Lett 13:1310–1324

Zhu G-P, Peterson AT (2017) Do consensus models outperform individual models? Transferability evaluations of diverse modeling approaches for an invasive moth. Biol Invasions 19(9):2519–2532. https://doi.org/10.1007/s10530-017-1460-y

Chapter 10
Phylogeography and the Role of Hybridization in Speciation

Leo Joseph

Abstract Human beings have a strong, innate desire to classify and name things. We like things to be clear-cut. The way we approach classification of birds is as good an example as any of this. So it always comes as something of a surprise to non-ornithologists to learn that how we classify birds at the level of the species around us is still subject of so much at times fiery debate. Various chapters in this book approach this from different perspectives. In this chapter, the focus is on reminding us that evolution is an ongoing, dynamic process and that appreciating this evolution can help us make sense of why it is sometimes so complicated to pin names on birds and indeed many other organisms. This will take us into a few particular aspects of bird evolution. One will be the process of hybridization between populations that may or may not be of the same species or between species that may or may not be each other's closest relatives. Another will concern the study of genetic diversity that exists within a species. In particular, we will examine what we have learned from the way that that diversity has come to be apportioned and distributed across the geographical range and landscapes inhabited by a species. These two areas have opened windows into the dynamics of evolution that give us new understanding of bird species. Genetic boundaries between species and sub-species are frequently very "leaky." Only certain parts of the genome, the entire complement of genetic material in a species, may be contributing to the differences that we can see between bird species. If the chapter can convey to the reader that we must learn to think of birds as continually evolving evolutionary lineages, then it will have had some success.

Keywords Birds · Speciation · Hybrid zones · Phylogeography · Genomics

---

L. Joseph (✉)
Australian National Wildlife Collection, National Research Collections Australia, CSIRO, Canberra, ACT, Australia
e-mail: Leo.Joseph@csiro.au

D. T. Tietze (ed.), *Bird Species*, Fascinating Life Sciences,
https://doi.org/10.1007/978-3-319-91689-7_10

10.1 Introduction

Phylogeography, hybridization, and speciation are inextricably intertwined. To help explain this and to chart a course for this chapter, some definitions and discussion are needed at the outset.

Phylogeography was originally conceived as a bridge between population genetics and systematics formed by analyses of mitochondrial DNA (mtDNA) (Avise et al. 1987). Its goals have always been to describe patterns of genetic diversity within a species across its geographical distribution and to understand the processes driving those patterns. Its molecular scope grew to include multilocus data, a term that in practice describes inclusion of up to 30 single regions of the nuclear genome. Currently, the tools of genomics enable genetic diversity to be studied across the entire genome, often including thousands of single sites that vary within and between species—single nucleotide polymorphisms or SNPs. Phylogeographic data are usually characterized by extensive sampling of individuals across the range of a species. This contrasts with the smaller number of individuals and higher number of loci that can suffice when estimating phylogenetic relationships among species and higher taxa (Nei and Roychoudhury 1974). Phylogeography and phylogeny are increasingly less discrete in scope (Smith et al. 2014a) but are reasonably seen as different parts of a continuum (Edwards et al. 2016).

Phylogeographic data enable tests of biogeographic hypotheses, description of the evolution of reproductive isolation, inference of processes underlying the origin, distribution and maintenance of biodiversity, and inferences about temporal changes in a population's physical and biotic environments (Beheregaray 2008). Revisions of species boundaries and taxonomic changes are prompted by phylogeographic data (especially when integrated with other data sets). Examples in birds abound [e.g., from Australia (Dolman and Joseph 2015, 2016; McLean et al. 2017a, b); from Eurasia (Olsson et al. 2013); from the Americas (Cadena and Cuervo 2010; Harvey and Brumfield 2015; Oswald et al. 2016); from Africa (Bowie et al. 2009; Voelker et al. 2013)]. Often, phylogeographic studies reveal much previously unknown (cryptic) but strongly geographically structured genetic diversity within a species. When researchers understandably try to understand the drivers of that diversity (Smith et al. 2014b; Harvey et al. 2017), one often has a sense of unfinished taxonomic business. Alternatively, cryptic population structure revealed by phylogeography can open very different, non-taxonomic windows of research into how natural selection has operated at the molecular level (Ribeiro et al. 2011; Pavlova et al. 2013). I will examine some cases in this chapter.

Hybridization has been defined many ways: reproduction between different species (Rheindt and Edwards 2011) or between genetically distinct populations producing offspring of mixed ancestry (Abbott et al. 2013), for example. In molecular terms, hybridization can be seen as an invasion of one genome by another (Mallet 2005), and genomic approaches to hybridization are now commonplace (Payseur and Rieseberg 2016). Perhaps not so often appreciated, among ornithologists at least, is that hybridization can *promote* rather than impede differentiation and speciation. Hybridization can select for reinforcement of reproductive isolation and cause hybrid speciation (Barton 2013; see Italian Sparrow *Passer italiae* example below). Also important is what hybridization is *not*. It is not something that "goes wrong" long after speciation is complete nor is it some "weakness" in how two species maintain separation or reproductive isolation (see Mallet 2005).

Genomic data in phylogeography reveal even higher frequencies of movements of genes between species than had been appreciated (Edwards et al. 2016; Mallet et al. 2016): species boundaries are far more porous than supposed (Chan and Levin 2005). Introgression, the one-way movement of genes from one population into another after hybridization, and two-way gene flow explain this. The result is known as reticulation and is a rearrangement of genetic diversity within and among species as they diverge from their common ancestor. Reticulation at different taxonomic levels will recur in this chapter.

Ornithologists have long studied hybridization and the regions where it is thought to occur—hybrid zones (Sibley 1957; Short 1969; Ford 1974). Descriptions of species hybridizing are common in the literature. Books and review papers have been written about it in birds (Grant and Grant 1992; McCarthy 2006).

Speciation is the study of how two or more species evolve from a common ancestor. In birds, it too has been the subject of books (Price 2008; Newton 2004) and reviews (Friesen 2015 for marine birds; Toews et al. 2016a). Reproductive isolation between two populations is perhaps the most unequivocal criterion by which researchers will accept that two species are involved. As with hybridization, we need to stress what speciation is *not*. It is not the same as species delimitation, but the process by which populations diverge from their common ancestor is a component of speciation that may or may not lead to its completion. Different lenses of data, say from phylogeography, plumage variation, or ecology and behavior, coupled with different ways of defining species, can mean that different workers consider speciation complete (and so worthy of species delimitation) at different stages of the process of diverging from a common ancestor. This concept has been captured well in de Queiroz's (2007) well-known figure (Fig. 10.1). The "gray zone" described by de Queiroz (2007) is very familiar to ornithologists faced with taxonomically interpreting cases where genetic distances between two populations under study and calculated from DNA sequences are between 0.5% and 2% and few phenotypic differences are discernible (see Roux et al. 2016).

How reproductive isolation evolves is a contentious and active area of study. What it offers to species delimitation in contrast to details of how it evolves likely have no simple demarcation as this book will show. Similarly, different evolutionary and environmental histories of birds within and among different regions mean that there is no single level of genetic divergence between two populations that indicates they have completed speciation. Against that, empirical and theoretical work is showing that the process of divergence from a common ancestor can reach a rapid transition to a "tipping point" resulting in reproductive isolation (Gompert et al. 2017; Nosil et al. 2017; Yamaguchi and Iwasa 2017).

Speciation research can only benefit from case-by-case phylogeographic study of how genetic diversity is geographically distributed within and among species whose

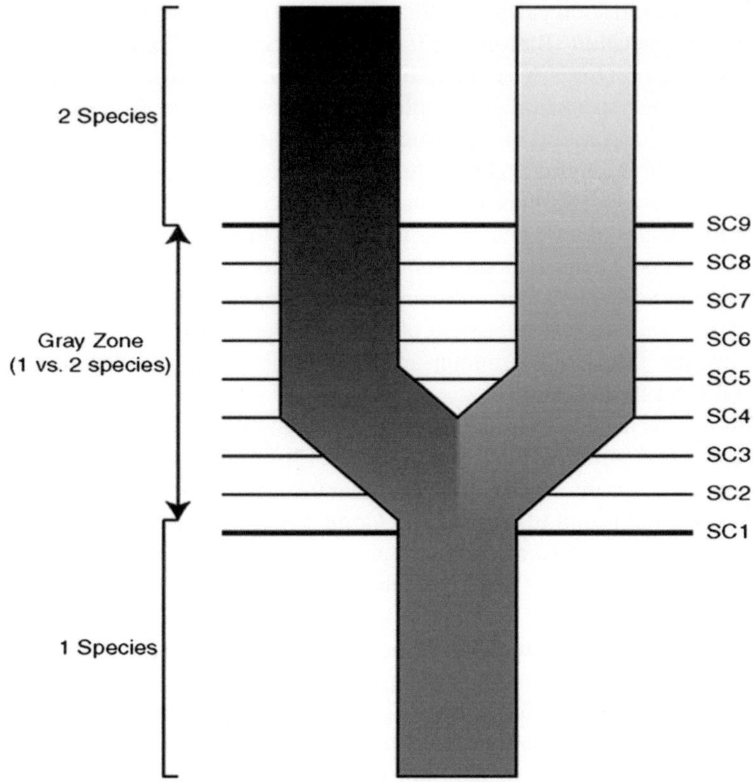

Fig. 10.1 Well-known and extraordinarily helpful figure from de Queiroz (2007) to describe how conflicts arise among different researchers when recognizing that speciation is or is not complete and when viewed through different lenses of data and species concepts (SCs)

origins we seek to understand along with consequences of hybridization (cf. Price 2008). The concepts by which we try to recognize the limits between species and so apply names to them are covered more explicitly in other chapters. One point will recur: It is often hard, if not foolhardy, to compartmentalize ongoing evolution by placing names too rigidly on many populations. Inevitably, we need to try and do so, because we need to communicate about species and their component populations, many of which are plainly and unquestioningly distinct. Yet in more and more cases to which we apply modern techniques, we learn that populations are evolutionarily more ephemeral than we perceive (Harvey et al. 2017) and nomenclature captures one phase of this change.

Tying these somewhat anticipatory threads together to form a foundation for the rest of this chapter, I note that hybridization revealed in phylogeographic studies is increasingly being stressed as a process *that simply will occur almost inevitably* during speciation. Gene flow happens *during* divergence (Smadja and Butlin 2011). Divergence and speciation in birds are not instantaneous events but can happen more easily and quickly than often thought (Mallet 2008; see Friis et al. 2016 for juncos).

The footprint of that gene flow is detectable by analytical methods now in use, but opinions differ about its *effect* on and potential role in the process of speciation (Butlin and Ritchie 2013). Barton (2013), for example, considers it negligible noting that gene flow of favorable alleles across a broad, two-dimensional habitat may not slow divergence; speciation between populations in geographic contact is then almost as easy as between those completely isolated from each other.

I now move to closer examination of just a few current topics in phylogeography, hybridization, and speciation in birds. The choice of topics discussed, while perhaps eclectic, is designed as an entrée to the explosively burgeoning literature. Cited examples will illustrate research already done or how different findings from across the globe often make similar or even sometimes contrasting points and I hope they will stimulate further research.

10.2 Some General Observations from Avian Phylogeography: Historical Population Size Changes and Introgression

A recurrent finding in many phylogeographic studies has been the genetic footprint of major reductions in population size, i.e., to refugia, at the Last Glacial Maximum (LGM) ~20,000 years ago followed by population (and presumably geographical range expansions) since then. This has been particularly common in European (Hewitt 1999) and North American studies (Boulet and Gibbs 2006; Friis et al. 2016) where widespread glaciation was the driver and even in Antarctica (Younger et al. 2015). Similar patterns in Australia are commonly found, but due to widespread cold, windy aridity at the LGM, not glaciation (Joseph and Omland 2009; Toon et al. 2010; Kearns et al. 2011). It is not a universal result, however. The Australian Grey Butcherbird *Cracticus torquatus* evidently expanded at the LGM and hybridized with a non-sister species, the Silver-backed Butcherbird *C. argenteus* (Kearns et al. 2014; see also Dong et al. 2017). Still other studies show important changes in population size having occurred before the LGM (in Brazil, Batalha-Filho et al. 2012; in Africa, Bowie et al. 2006; in Australia, Chan et al. 2014). A distinction to make here is that the species involved usually evolved earlier, in the Pliocene at least, whereas the Pleistocene molded their present-day diversity. This has been stressed in Africa (Voelker et al. 2010) and in the lowland Neotropical rainforests (Smith et al. 2014a, b) where dispersal and differentiation superimposed on earlier, large-scale landscape events such as the uplift of the Andes has been a major driver of avian speciation (see also Ribas et al. 2012). To complicate things, very recent speciation (middle Pleistocene onward) appears to have occurred too in North America, Europe, and Australia (Toews and Irwin 2008; Dolman and Joseph 2016; Friis et al. 2016). The point here is that the interplay between speciation before and sometimes during the Pleistocene relative to the role of the Pleistocene in shaping present-day genetic diversity within species is complex and idiosyncratic.

Introgression repeatedly emerges from phylogeographic data. Introgression during divergence of populations and species from their common ancestor is clearly a pervasive phenomenon (reviews in Rheindt and Edwards 2011; Edwards et al. 2016; Ottenburghs et al. 2017). The application of coalescent theory in population genetics has been instrumental here. By this approach, genetic diversity within a species is traced back in time to where it coalesces in a common ancestor. This is contrasted with the gene frequency-based approach of classical population genetics. Peters et al. (2007) were perhaps the first to use coalescence in detecting ancient introgression (~14,000 years ago) in birds. They detected the widespread occurrence and high frequency of mtDNA more typical of one species, the Asian Falcated Duck *Anas falcata*, in another more widespread species, the Gadwall *A. strepera* of the broader Holarctic. Peters et al. (2007) were perhaps the first to use coalescence theory in birds to carefully distinguish between the two reasons why two or more present-day species can share genetic diversity. One reason is that the different species have hybridized. The other reason, which they rejected, but which is often the favored explanation of such data, is that they still share much of their common ancestor's genetic variation. That is, unlike say plumage that we can readily see as diagnostic between two species and which can evolve rapidly under natural selection, the genetic marker being studied may evolve at a slower and probably more clock-like rate and so has not yet had time to "catch up" with the more visible species-specific markers of plumage. This phenomenon is termed incomplete lineage sorting (Joseph et al. 2006, e.g., in *Artamus* woodswallows). In a further study of ducks (Peters et al. 2014) as well as in many other papers in avian phylogeography (Degnan 1993), it has become clear that while mtDNA will always be a useful marker to include in a phylogeographic study, it can mislead, if not carefully interpreted. We will see further examples of this elsewhere in the chapter.

Phylogeographic studies also can detect the difference between recent and current hybridization and past gene flow. Recent hybridization will be characterized by first-generation (F1) hybrids and backcrosses between them and the parental types. Examples come from Manthey and Robbins (2016) showing hybridization likely due to recent anthropogenic disturbance. Older hybridization will lead to more complicated patterns of admixture (e.g., Manthey et al. 2012) as we will see elsewhere in the chapter. Warranting mention here is the "ghost of gene flow past." An example comes from Australia's Western Grasswren *Amytornis textilis*. Most of this species' populations in the west of the Australian continent are extinct but known from museum specimens. Austin et al. (2013) detected a divergent outlier in the mtDNA diversity of the eastern Australian subspecies *A. t. myall*. This result would have remained enigmatic were it not for the availability of museum specimens of the extinct western populations, which shared this form of mtDNA. Coupled with subfossil evidence of the species having been more continuously distributed in the past and indeed ecological data, Austin et al. (2013) concluded that they had detected evidence of past gene flow prior to the extinction of the western populations.

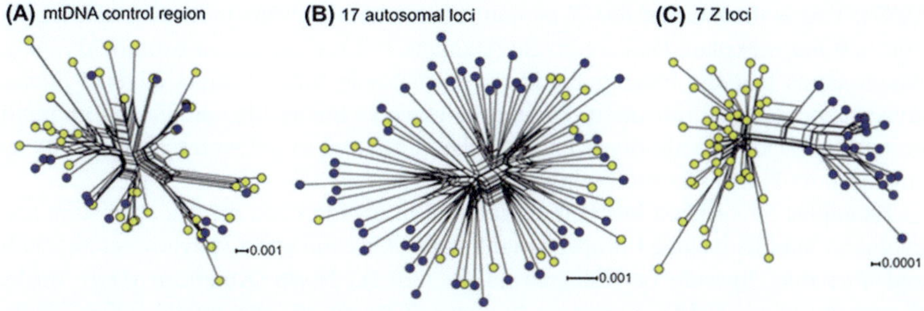

Fig. 10.2 Networks of DNA sequence diversity from Dhami et al. (2016) showing contrasting patterns in Grey Teal *Anas gracilis* (each yellow circle is one individual) and Chestnut Teal *A. castanea* (blue circles). The networks are derived from DNA sequences obtained from (**a**) the mtDNA control region, (**b**) 17 autosomal loci, and (**c**) 7 Z loci. Note how only in the Z chromosome sequences do the two species "pull apart"

10.3 Phylogeography, Sex Chromosomes, and Speciation

Phylogeographic studies of two closely related Australian ducks, the sexually monomorphic Grey Teal *Anas gracilis* and sexually dimorphic Chestnut Teal *A. castanea*, lead us neatly, if surprisingly, to a consideration of sex chromosomes in speciation. Initial phylogeographic and systematic studies of genetic diversity within and between these two species found no differences between them, whether studied with allozymes (Sraml et al. 1996), mtDNA (Joseph et al. 2009), or multilocus DNA (Dhami et al. 2013). Geographic structure in one species, *A. castanea*, was apparent between eastern and western Australia. Insights into the birds' evolution and speciation were gained nonetheless. For example, the species had only recently diverged from their common ancestor, divergent selection best explaining their obvious differences in ecology and life history. Incomplete lineage sorting was the favored explanation of their shared diversity. But where in the genome were the genetic differences between the two species of teal?

An answer came when loci on the Z sex chromosome of the two teal were sequenced (in birds, males are ZZ and females ZW; Dhami et al. 2016; Fig. 10.2). Differences between the two species were immediately apparent: All Chestnut Teal Z chromosome sequences were more closely related to each other than any were to the Grey Teal Z chromosome sequences and *vice versa*. The Z chromosome has repeatedly emerged as the site of many and sometimes key important genetic differences especially those controlling reproductive isolation between closely related species. We briefly address why this is so while acknowledging that the Z chromosome will not necessarily reveal all in every case.

Novel mutations on sex chromosomes are directly exposed to selection in the sex having only one of each sex chromosome—XY males in mammals such as humans and ZW females in birds. Selection is therefore expected to act faster on sex chromosomes and to fix some types of beneficial mutations (Charlesworth et al.

1987). This is the basis of fast-X evolution or fast-Z evolution in birds (see Ellegren 2009). It helps explain Haldane's rule (Haldane 1922; Carling and Brumfield 2008), which states that the heterogametic sex (females in birds) suffers adverse fitness consequences in hybridization. Population genetics theory also predicts more rapid accumulation of genetic divergence on the Z relative to autosomes because of its smaller population size promoting genetic drift.

Examples of Z-linked loci being important in tracking and driving speciation and evolution in birds include European nightingales *Luscinia* spp. (Storchová et al. 2009) and *Ficedula* flycatchers (Ellegren et al. 2012), North American *Anas* ducks (Lavretsky et al. 2015), *Passerina* buntings (Carling and Brumfield 2008, 2009), *Certhia* treecreepers (Manthey and Spellman 2014; Manthey et al. 2016), and in a comparative analysis of meliphagoid birds across hybrid zones in northeastern Australia (Peñalba et al. 2017) to name just a few. We should also consider physical rearrangements of entire blocks of chromosomes called chromosomal inversions. These involve an entire section of a chromosome being physically inverted. The genes in these inversions are generally inherited intact. This is because inversions hinder efficient recombination or crossing over between members of a pair of chromosomes during meiosis, which forms gametes. Inversions therefore lock up favorable combinations of genes. Chromosomal rearrangements have been more extensive on the Z chromosome relative to the rest of the genome (with notable exceptions such as one part of Chromosome 1A). This has been elegantly shown by Kawakami et al. (2014) in a comparison between two species with reasonable genomic data, the Australian Zebra Finch *Taeniopygia guttata* and the European-breeding Collared Flycatcher *Ficedula albicollis* (Fig. 10.3; see also van Doren et al. 2017).

Hooper and Price (2015, 2017) found that the number of inversion differences between closely related species is consistently predicted by whether their geographical ranges overlap and the extent of that overlap and illustrated this neatly in estrildid finches. Geographical range overlap in and of itself is not an absolute predictor of inversion differences, however. For example, there are no such differences between Collared and Pied Flycatchers *Ficedula albicollis* and *F. hypoleuca*, respectively (Backström et al. 2010a). Nonetheless, those two species have diverged at seven Z-linked loci significantly more than expected under neutrality; two of the detected candidate regions contain genes that are associated with plumage coloration in birds (Backström et al. 2010b). Indeed, the four black-and-white *Ficedula* flycatcher species (the other two being *F. speculigera* and *F. semitorquata*) show greater genetic divergence on the Z chromosome than on the autosomes. This has been attributed most simply to a fast-Z phenomenon (Sætre and Sæther 2010; Hogner et al. 2012).

Hooper and Price (2015, 2017) also reported variable rates of fixation of inversions across the autosomes, but inversions are more likely to be fixed on the Z chromosome than the average autosome. A role for gene flow in divergence is seen here. Gene flow on secondary contact between partially reproductively isolated forms may promote the spread of an inversion and its favorable combination of genes. Hooper and Price (2015) note that, because Z-linked genes diverge in function more rapidly than autosomal genes, so at any time before reproductive isolation is complete, an inversion on the Z chromosome should be more likely to

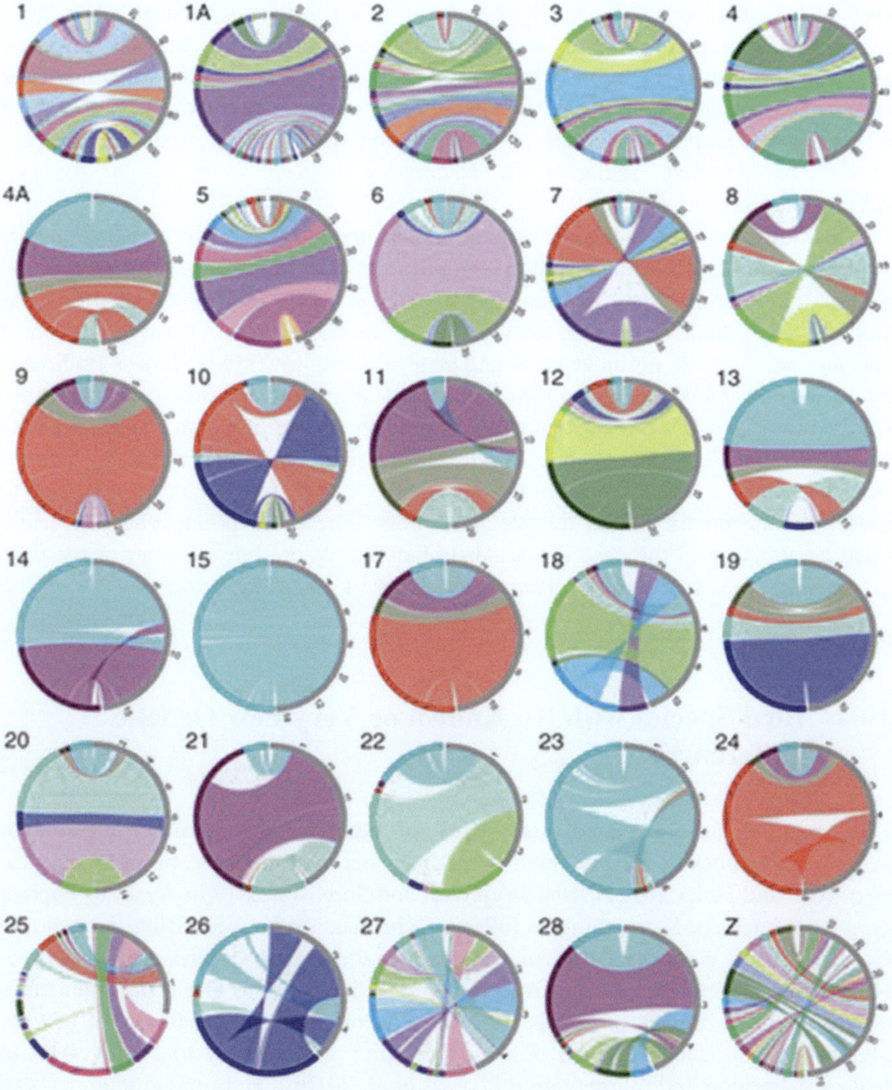

Fig. 10.3 From Kawakami et al. (2014) showing comparative circular visualization of the organization of blocks of loci on homologous chromosomes in the Collared Flycatcher *Ficedula albicollis* and Zebra Finch *Taeniopygia guttata*. Each circle represents a chromosome as numbered. Each circle's left half represents the Collared Flycatcher and the right half represents the Zebra Finch. Very little structural rearrangement has happened on some chromosomes since these two species diverged from their common ancestor, e.g., chromosomes 4 and 15, whereas far greater rearrangements have occurred in one region of chromosome 1A and along the entire Z chromosome. Scale is indicated on the Zebra Finch side of plots, in megabases

capture two or more alleles locally adapted—either to that population's habitat or genomic background—than an inversion on an autosome. They reason that inversions on the Z may be more strongly selected for, if gene flow is an important mechanism driving their selective advantage. Ornithologists should heed a most informative example of the role of hybridization spreading an inversion and promoting an adaptive radiation in *Heliconius* butterflies of South America (Dasmahapatra et al. 2012).

This brings us back to the two Australian teal and why the Z chromosome and its genes warrant attention in speciation genomics. At least for *Ficedula* flycatchers, the reasons have been neatly summarized by Sæther et al. (2007), Qvarnström and Bailey (2009), and Nater et al. (2015). The Z chromosome is a known location of species-specific male plumage traits and genes causing low hybrid fitness (indicated by an intronic marker in *Passerina* buntings; Carling and Brumfield 2008, 2009). Limited introgression of genes on it and limited recombination along its length keep blocks of genes together despite hybridization, suggesting that it is a hotspot for sexual traits in adaptive speciation where male signals and female preferences to those signals remain linked despite gene flow. These are points one can easily imagine applying in the case of the Australian teal where one member of the pair, *A. castanea*, is sexually dimorphic and the other, *A. gracilis*, is not.

10.4 Bird Species with No Known or Very Few Genetic Differences

Other pairs of closely related birds stand out either for still unknown location of genetic differences or for the small number of loci that appear to be involved (see also Nosil and Feder 2012; Faust-Stryjewski and Sorenson 2017 in Australo-Papuan *Lonchura* finches). Key examples follow. Hybridization between Blue-winged and Golden-winged Warblers *Vermivora cyanoptera* and *V. chrysoptera*, respectively, of North America, has been extensively studied with a view to clarifying their taxonomic status as species or subspecies. They seem to differ genetically at just six regions in their genome, each of which contains genes likely to be involved in feather patterning and pigmentation (Toews et al. 2016b). Hybridization appears to be a long-term feature of their evolutionary history, as also reported in an endangered subspecies of the Australian Yellow-tufted Honeyeater *Lichenostomus melanops cassidix* and the more common parapatric subspecies *L. melanops gippslandicus* (Pavlova et al. 2014). In another thoroughly studied case of hybridizing European birds, the Carrion and Hooded Crows *Corvus corone* and *C. cornix*, respectively, genomic differences appear to be concentrated in one relatively small region of one chromosome. This is likely a chromosomal inversion and involves genes controlling plumage differences (Poelstra et al. 2014), the details of which in terms of gene expression and biochemical pathways have been elucidated (Poelstra et al. 2015). Finally, the genomes of three Holarctic redpoll finches *Acanthis* spp. appear largely undifferentiated (Mason and Taylor 2015), and speciation in isolation followed by secondary contact can confidently be eliminated. Differences between the apparent species may be due to differences in gene expression related to the phenotypic differences in what is still one variable lineage.

10.5 Hybrid Zones: A Closer Look

10.5.1 Suture Zones and Multiple Hybrid Zones

Hybrid zones are regions where two well-differentiated taxa interbreed or hybridize (Barton and Hewitt 1981). Having had a similarly long history of research in ornithology (Ford 1974), they have been argued to indicate, at a minimum, that the speciation process is incomplete or may never be complete (Coyne and Orr 2004). Increasingly, porous, "leaky" genetic boundaries between clearly well-differentiated taxa, which nonetheless remain well-differentiated despite such porosity, appear to be the norm (Chan and Levin 2005; Bay and Ruegg 2017; Chattopadhyay et al. 2017; Griffith and Hooper 2017; McLean et al. 2017a, b; Slender et al. 2017). Worth stressing here is that hybrid zones are not black holes from which genes never emerge; they are evolutionary conduits through which adaptive and neutral markers can differentially move (Brumfield et al. 2001; Gompert et al. 2017).

A special opportunity to study speciation is provided by suture zones (Remington 1968, but see Swenson and Howard 2004), regions where species pairs variously either have hybrid zones, or come into close contact, or replace each other on either side of the zone while not occurring within it. Typically, there will be a range of taxonomic divergence associated with these zones from undifferentiated populations on either side through to clearly differentiated allopatric species. These zones are ideal natural experiments in which to apply genomic tools in the study of how divergence and reproductive isolation evolve. Essentially a maturing of comparative rather than single species phylogeography (Bermingham and Moritz 1998), genomic study of these zones informs how gene flow affects divergence of populations that have evolved under shared environmental history (Winger 2017; Winger and Bates 2015; Peñalba et al. 2017; Fuchs and Bowie 2015). Although much has been learned from studies of birds in several such zones using earlier mtDNA and multilocus approaches (e.g., Carpentarian, Nullarbor and Eyrean Barrier regions in Australia – Balakrishnan et al. 2010; Dolman and Joseph 2012, 2016), I acknowledge the surge of this kind of genomic work in birds and in many nonavian groups and regions (e.g., Edwards et al. 2016; Gompert et al. 2017). They show how modern genomic tools are bringing entirely new levels of resolution to our understanding of the details of speciation in allopatry, parapatry, and sympatry. By studying different parts of the genome (autosomes vs. sex chromosomes vs. mtDNA), they also build on some seminal earlier work showing differential patterns of introgression among loci in North American *Passerina* buntings (Carling et al. 2010) and identifying genomic regions contributing to speciation or adaptive introgression (Borge et al. 2005; Gompert et al. 2017).

At the time of writing, foundational results from two such studies are emerging. One is from the Marañon Valley in Peru (Winger 2017; Winger and Bates 2015) and the other from tropical northeastern Australia where an especially complex set of three zones occurs (Peñalba et al. 2017; Ford 1986, 1987). I briefly review them and then examine the finer detail emerging from single species studies.

Both studies found that stronger plumage divergence (measured spectrophotometrically in Peru or through a proxy of taxonomic rank from Schodde and Mason's 1999 review in Australia) is associated with deeper genetic divergence. In the Peruvian case, this was further argued to be likely due to geographic isolation for up to two million years (Winger and Bates 2015), consistent with conventional allopatric speciation theory. Z chromosome loci were most divergent in the Australian case at least. This further supports the role of sex chromosomes in the study of speciation and the utility of using different marker sources for reconstructing evolutionary history. Conversely, lineages lacking plumage divergence across the same geographic barriers have been more recently isolated or exhibit a signature of genetic introgression after formerly isolated populations came into contact. This indicates that gene flow will impede divergence in phenotypic traits important to speciation and that evolutionary outcomes of cycles of isolation and divergence are indeed sensitive to how long gene flow is stopped. Morphometric trait evolution showed greater idiosyncrasy in both studies and so was a poor predictor of genetic divergence.

These comparative studies within regions both indicate which phenotypic traits can be predicted to diverge. In the Australian study, an additional variable of habitat (rainforest vs. mangrove vs. woodland) was included. After taxonomic rank, habitat was the next strongest predictor of genetic divergence (Peñalba et al. 2017). Together these studies undoubtedly affirm some long held if informally stated convictions among ornithologists that plumage divergence predicts genetic divergence. The full analyses of the data sets reveal that this is far from absolute, both studies finding several major divergences between essentially undifferentiated populations regarded as consubspecific. Peñalba et al.'s (2017) case of the Dusky Myzomela *Myzomela obscura* especially warrants further study.

10.5.2 Detail Emerging from Single Species and Hybrid Zones: Three Case Studies

Work on birds has been instrumental in showing how genomic tools can vastly refine our knowledge of how hybridization, hybrid zones, and phylogeography inform our understanding of speciation. The hybrid zone between two Central American manakins, White-collared *Manacus candei* and Golden-collared Manakin *M. vitellinus*, has long been a landmark example. Earlier work established discordance among geographical clines in molecular markers and plumage traits. Differential introgression of the

M. vitellinus yellow collar across the hybrid zone was driven by sexual selection (Parsons et al. 1993; Brumfield et al. 2001; McDonald et al. 2001). Parchman et al. (2013) took the study further. They tested alternative models concerning whether adaptively important genes involved in reproductive isolation cluster in discrete "islands" in the genome or are scattered through the genome (see Cruickshank and Hahn 2014). Their data comprised 59,100 SNPs, single sites in the genome where there is variation in the DNA sequence between two populations under study. These sites were then mapped to a draft of the *Manacus* genome to determine in which genes they fell. Finding that genetic regions involved in adaptive divergence and reproductive isolation are scattered throughout the genome, they also concluded that many relevant loci had signatures of strong genetic differentiation and introgression. This is consistent with the hypothesis that loci involved in isolation are often characterized by a history of divergent selection and have not just drifted to differentiation.

A further case is especially valuable, because it integrates so much biological and genetic data to link a species' deeper history to biological drivers of more recent patterns in gene flow. It epitomizes how phylogeography, hybridization, speciation, and ongoing divergence are intertwined as claimed at this chapter's outset. The Red-backed Fairywren *Malurus melanocephalus* occurs in subtropical and tropical Australia. Two subspecies are diagnosed essentially by whether their dorsal color is red or orange (Schodde and Mason 1999; Fig. 10.4). The red-backed subspecies *M. m. cruentatus* occurs across Australia's tropical north, the orange-backed *M. m. melanocephalus* being in more southeastern parts of the range south of Cape York Peninsula (Fig. 10.4). Morphometric traits have likely been subject to ecological selection and simply track local environmental variation regardless of subspecies identity; ecological selection appears to have also influenced the evolution of tail length as an intrasexual signal used primarily among competing males (Baldassarre et al. 2013). The species' phylogeography was first assayed from 29 anonymous nuclear loci, six introns, and one mtDNA locus, amounting to over 15,000 base pairs per individual. The major phylogeographic disjunction in the species is *within* the red-backed subspecies' range west of Cape York Peninsula at the Carpentarian Barrier (Lee and Edwards 2008; CB in Fig. 10.4), and this was confirmed by later work with 2702 SNPs (Baldassarre et al. 2013, 2014; Fig. 10.5). The same later study showed that alleles for red plumage color have introgressed east across the hybrid zone following secondary contact and into the genomic background of the orange subspecies. That is, the plumage cline is displaced significantly east of the major phylogeographic break and the vast majority of the individual SNP clines (Fig. 10.5). Previous experimental work demonstrated an extra-pair mating advantage for red males (Baldassarre and Webster 2013). So it appears that, as in the manakins, sexual selection is driving asymmetrical introgression of red plumage alleles across the hybrid zone. This integration of phylogeography and superb field biology shows how sexual selection can complicate definition of taxonomic boundaries and promote gene flow, particularly at an intermediate stage of divergence.

The third hybrid zone to discuss here will lead us to an entirely new approach to speciation. It concerns three Australian rosella parrots, the Northern, Pale-headed and Eastern Rosellas *Platycercus venustus*, *P. adscitus*, and *P. eximius*, respectively.

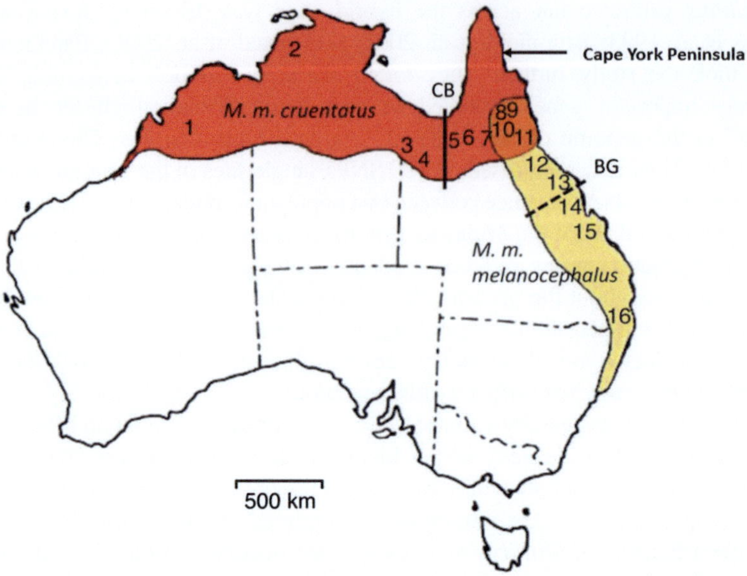

Fig. 10.4 Map modified from Baldassarre et al. (2014), showing the currently recognized distribution of the red-backed *M. m. cruentatus* subspecies in the west and the orange-backed *M. m. melanocephalus* subspecies in the east. The area of overlap in the northeast represents the hypothesized region of plumage overlap (Schodde and Mason 1999). The solid vertical line represents the Carpentarian Barrier (CB), and the dashed line represents the Burdekin Gap (BG). Numbers refer to sampling locations

A hybrid zone around the middle latitudes of Australia's east coast had long been recognized between *P. adscitus* and *P. eximius*. Few data and specimens were available concerning it, however, the original basis having been field observations (review in Schodde 1997). Aside from such scant information forming a basis for considering these two taxa conspecific, it was a surprise and contrary to earlier mtDNA work (Ovenden et al. 1987) when multilocus data showed *P. adscitus* and *P. venustus* to be sister species east and west, respectively, of the Carpentarian Barrier (Fig. 10.4) and that *P. eximius* was in turn their sister (Shipham et al. 2015). The latter study had a key twist, however. Phylogenetic analysis using mtDNA only from the Australian mainland populations of *P. eximius* leads to a finding of *P. eximius* and *P. adscitus* being sisters. Alternatively, when mtDNA from the geographically isolated Tasmanian populations of *P. eximius* is sampled, a result emerges consistent with multilocus data. Why did mtDNA within one species yield such different results?

Shipham et al. (2017) then used genomic methods to test alternative hypotheses to explain this observation. They favor the view that hybridization between *P. eximius* and *P. adscitus* has occurred and that mtDNA of *P. adscitus* has so completely introgressed southward into mainland southeast Australian *P. eximius* that it has completely "captured" the latter. The Tasmanian populations of *P. eximius*

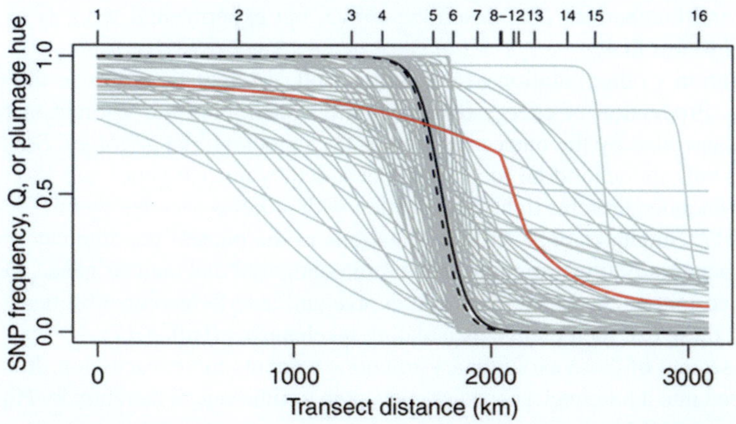

Fig. 10.5 Example from Baldassarre et al. (2014) of population genomic data increasingly being seen in studies of genomic clines across hybrid zones. Numbers along the top refer to sampling locations from west to east across northern Australia as shown in Fig. 10.4. The solid black and dashed black lines are summary geographic clines for all 2702 loci or from all 102 diagnostic loci, respectively. Clines for each of the 102 diagnostic SNP loci are shown in gray and that for plumage hue is shown in red. Higher values for hue correspond to redder dorsal plumage color in more western populations. All SNP loci have allele frequency greater than 80% on the western end of the cline and less than 20% on the eastern end, even those with very wide-fitted clines that appear to deviate from this criterion

isolated from the mainland by Bass Strait retain that species' original mtDNA. Kearns et al. (2014) found a similar case of mtDNA capture likely still progressing in Australian butcherbirds, and Dong et al. (2014) document a similar case in Taiwanese scimitar babblers *Pomatorhinus* spp. The problem to address next is this: Given that there are so many cases of hybridization leading to localized gene flow within and across hybrid zones, why is it that in some cases there is such an extreme outcome of the mtDNA of one form completely capturing that of another?

10.6 Mitonuclear Incompatibility, Hybridization, and Speciation

Mitochondria, so often called powerhouses of cells, are vital to an organism's survival. Sloan et al. (2017) note that deleterious mutations in mtDNA adversely affect energy production and that evolution has devised two solutions to this. One amounts to the mitochondrial genome of one species or population "rescuing" that of another by completely replacing it (e.g., Drovetski et al. 2015), through hybridization, with more selectively fitter, foreign mtDNA. This is adaptive introgression (Borge et al. 2005; Bonnet et al. 2017) and recalls the idea in my Introduction that hybridization involves an invasion of one genome by another. It likely explains the rosella and butcherbird data described above, but experimental work is of course required to test that.

Evolution's other solution to deleterious mtDNA mutations lies in the nuclear genome. Production of energy by a smoothly functioning mitochondrion requires proteins encoded by the nuclear as well as the mitochondrial genomes. Some 1500 proteins with mitochondrial function are encoded by nuclear genes and 180 of these nuclear-encoded proteins tightly cofunction with proteins encoded by mtDNA genes (Hill 2017; Sunnucks et al. 2017). Mutations in the nuclear genome can therefore lead to potentially fitter combinations of mitochondrial and nuclear genes. Different coadapted mitonuclear genotypes can evolve and lead to incompatibilities within a species; these can then evolve as isolating mechanisms (Hill 2017).

The second of these evolutionary solutions warrants more discussion. It has been developed into a mitonuclear species concept in ornithological literature by Hill (2017 and references therein). The process that generates energy, oxidative phosphorylation, cannot function, and an organism cannot survive, unless the cofunctioning mitochondrial and nuclear gene products needed for oxidative phosphorylation are in step. Tight cofunctioning of nuclear and mitochondrially encoded genes underpins the mitonuclear species concept. As mtDNA evolves faster than the nuclear genome, there must be perpetual coevolution of the two genomes. One can think of species having uniquely coadapted mitonuclear genotypes. Once populations diverge for this genotype, there will be reduced fitness of offspring due to mitonuclear incompatibilities. That mitochondrial genotype, Hill (2017) argues, becomes the best current method for diagnosing species. Does this mean that Eastern and Pale-headed Rosellas are the same species? Hill's (2017) concept may be taken to argue "yes," but recall that Northern and Pale-headed Rosellas are closest relatives. One would have to treat all three as conspecific. One could, but is it a useful taxonomy? This shows that different ways of diagnosing species usually always have some merit but can be in conflict with each other.

Whatever the mitonuclear concept's strengths and weaknesses, and Hill's (2017) discussion is well balanced with examples challenging the concept, it is a refreshingly compelling one. It prompts us to think differently about existing datasets. For example, the Eastern Yellow Robin *Eopsaltria australis* of Australia exhibits a geographically structured intraspecific mtDNA divergence of 6%, a magnitude usually seen between *genera* of birds, not within a species (see Dai et al. 2017 for another example in China). This mtDNA divergence is geographically organized east–west across the bird's range (Pavlova et al. 2013). Nuclear DNA markers are organized north–south, perpendicular to the mtDNA structure. Conventional explanations of genetic drift and neutral evolution have been eliminated in explaining this pattern (Pavlova et al. 2013). A case has been built arguing that natural selection at the level of proteins encoded by the mtDNA and driven by adaptation to different climates has determined the mtDNA structure (Morales et al. 2015, 2017a). Further, adaptive mitochondrial introgression and selection against incompatible mitonuclear combinations are likely involved (Morales et al. 2017b). Finally, genomic scans (Morales et al. 2017c) show that genetic differentiation between the two adjacent but climatically divergent mtDNA lineages in this species is mostly limited, not only to the mitochondrial genome but also to a large part of chromosome 1A, very possibly in an inversion. This part of chromosome 1A contains tightly linked genes having mitochondrial functions. Notably, a second region of divergence is on the Z chromosome, suggesting that nuclear gene flow occurs primarily via male hybrids, in accordance with Haldane's rule. These findings suggest not so much that there are two species within the Eastern Yellow Robin but that mitonuclear coevolution has been critical in climatic adaptation during population divergence within the species.

Before leaving mitonuclear incompatibility, I revisit the Z chromosome's role in speciation. The mitonuclear concept predicts that at least some nuclear-encoded genes for proteins of mitochondrial function should be located on the Z chromosome. This is because in female birds the paternal Z chromosome must cofunction with the maternal mtDNA. The prediction is supported in the Eastern Yellow Robin and in other species such as Gouldian Finches *Erythrura gouldiae* (Pryke and Griffith 2009).

10.7 Ring Species as a Special Case of Divergence with Gene Flow: Are There Any Surviving Examples?

A theme of current speciation research emphasized in this chapter is that populations and species can diverge despite the reticulation caused by ongoing gene flow. A specific example is Mayr's (1942) ring species concept (Fig. 10.6). This posits that an ancestral population spreads in a geographical ring away from its origin. Eventually, populations at the beginning and end of the ring are in contact and behave as separate species reproductively isolated from each other in sympatry. The hypothesis makes very specific predictions about patterns of genetic diversity that should be recoverable. In ornithology at least, these predictions have not survived such testing with genetic data or even with retrodictive modeling of paleodistributions. Consequently, more conventional alternative explanations of speciation in allopatry followed by range expansion leading to secondary contact cannot be rejected. They explain the data as well if not better than the ring species hypothesis. Examples are the Herring Gull *Larus argentatus* complex of Northern Hemisphere circumpolar regions (Liebers et al. 2004), the Crimson Rosella *Platycercus elegans* complex of southeastern Australia (Joseph et al. 2008; Fig. 10.6), and the Greenish Warbler *Phylloscopus trochiloides* complex encircling Tibet (Alcaide et al. 2014; Peterson and Anamza 2017). The point here is not that the ring species hypothesis should be abandoned but that it continues to engender stimulating evolutionary insights into the details of speciation in each case (e.g., Mihailova et al. 2014; Irwin et al. 2016).

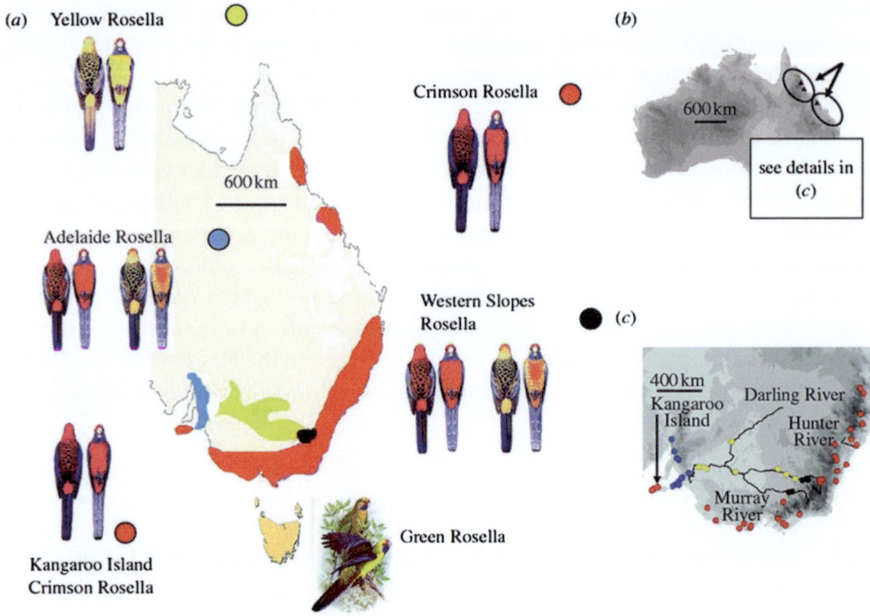

Fig. 10.6 Example of a putative ring species, not supported as such by genetic analysis from Joseph et al. (2008): (**a**) Distribution and plumage phenotype variation in parrots of the Crimson Rosella complex in eastern Australia (from Forshaw and Cooper (2002) in this and subsequent figures). Note the narrow zone of unsuitable habitat currently separating Adelaide and Yellow. (**b**, **c**) Sampling scheme of vouchered museum specimens for mtDNA analyses of the full mainland distribution of the complex

10.8 Hybrid Species

Can hybridization lead to the evolution of new species in birds? A now well-studied case is that of the Italian Sparrow *Passer italiae*. Long hypothesized to be of hybrid origin because of male plumage traits intermediate between House *Passer domesticus* and Spanish Sparrows *P. hispaniolensis*, its hybrid origin is confirmed by extensive nuclear and mitochondrial genetic work as well as ecological and behavioral studies (Elgvin et al. 2011; Hermansen et al. 2011; Trier et al. 2014; Sætre et al. 2017). Z-linked and mitonuclear reproductive barriers limit gene flow and maintain the integrity of the three populations (Trier et al. 2014). Hill (2017) notes it as a case in which a novel but adaptive combination of genes for mitochondrial and nuclear-encoded proteins arose through hybridization. Postzygotic selection against mitonuclear incompatibilities now limits gene flow between the three populations (Trier et al. 2014). Italian Sparrows have come into secondary contact with Spanish Sparrows from which they have diverged significantly across 81 protein-coding genes (Sætre et al. 2017). Six of these genes showing the greatest divergence are associated with learning and neural development in other bird species suggesting a role for behavioral isolating mechanisms. Prezygotic assortative mating may also contribute to the maintenance of coadapted mitochondrial and nuclear genes (Hermansen et al. 2011).

House and Spanish Sparrows have also hybridized across the Mediterranean in North Africa but with outcomes differing in terms of which mtDNA alleles are most frequent in hybrids (Belkacem et al. 2017). Clearly, much remains to be learned (see also Brelsford et al. 2011 for the case of Audubon's Warbler *Setophaga auduboni*). Importantly, other cases where two species interact differently at different zones of contact offer rich potential for study. Prime examples have been studied in European populations of Barn Swallows *Hirundo rustica* (Scordato et al. 2017) and North American Mountain Chickadees *Poecile gambeli* (Manthey et al. 2012). Much remains to be learned in Australia in the Crimson Rosella group (Joseph et al. 2008) and Brown and Inland Thornbills *Acanthiza pusilla* and *A. apicalis* (Black et al. 2015). A special case warranting an entire chapter is the role of hybridization in driving the history of one of the most famous examples of adaptive evolution, Darwin's finches (see Grant and Grant 2016 and references therein and for alternative views McKay and Zink 2014).

10.8.1 Hybrid Zones Sometimes Move

Although on evolutionarily short time scales of centuries and decades, data are now available to show that some hybrid zones have moved and some have not (Morales-Rozo et al. 2017). Insights gained from this phenomenon vary. Long-term study of North American birds provides the richest examples. In *Sphyrapicus* sapsuckers, Billerman et al. (2016) used climatic models to accurately recover the hybrid zone's shift, which was itself documented with museum specimens collected over the last 100 years. They predicted future trends in the ranges of the respective parental taxa and that the climatic niche of the hybrids will disappear under climate change. Similarly in *Passerina* buntings, Carling and Zuckerberg (2011) suggest that continued spread of the hybrid zone threatens one of the parental species. In *Poecile* chickadees, Taylor et al. (2014) found a set of loci consistently linked to genomic regions likely under selection and linked to reproductive isolation. These loci showed patterns of elevated divergence and reduced introgression regardless of when samples were collected and despite multiple generations of admixture between the parental species. In contrast, Mettler and Spellman (2009) argued that a *Pheucticus* grosbeak hybrid zone had not moved in 40 years and constituted a "tension zone" maintained by a balance between dispersal into the hybrid zone and selection against hybrids. In Australia, it is unclear whether a well-studied hybrid zone between forms of the Australian Magpie *Gymnorhina tibicen* maintained by selection (Hughes et al. 2001) is stationary or moving (Burton and Martin 1976).

10.9 A View to the Future

This chapter has skimmed the surface of the vast body of published work accruing almost daily on the genetic and genomic details of phylogeography, hybridization, and speciation of the world's birds. Genomic data are bringing much to our understanding of evolutionary history within and among species. As with any mature field, our understanding has reached a point where we appreciate well the theoretical problems still to be addressed and how data now obtainable from genomic methods may help. Disentangling effects of selection and drift in different parts of the genome as populations diverge is one such challenge (Southcott and Kronforst 2017). In contrast, our understanding of the genetics of adaptation at the molecular level, though developing, is still a frontier. Much has been learned from certain species such as climate-driven selection in Australia's Eastern Yellow Robin (discussed above) and about the role individual genes play [*MC1R* in plumage coloration (San-Jose et al. 2017; Faust-Stryjewski and Sorenson 2017); *ALX1* and *HMGA2* in bill morphology (Lamichhaney et al. 2015, 2016); *DRD3* in evolution of migration (Chap. 7; Delmore et al. 2015)]. Floodgates are likely to open in the coming years, however, as more and more genomes are sequenced. At present, regions of the genome showing interest for one aspect or another of a bird's evolution can often be mapped to a reference genome in one of a few phylogenetically distant species for which reasonable genomic assemblies are available (e.g., the non-passerine chicken or the two oscine passerines Collared Flycatcher and Zebra Finch). This is like trying to determine one's location 3 days into a drive across a continent using a map showing approximate location of one road and no provincial boundaries or towns other than the journey's start and end points. As more annotated genomes (i.e., with genes mapped and named) come online, more and more genomic reference maps will be available across the avian tree of life. The mapping analogy would be that one knows one's whereabouts, because all roads, towns, and geographical features are shown accurately to scale. Then we will better understand either the function of genes or their proximity to genes of known function in a given study species (consider the clines in Fig. 10.5 for the Red-backed Fairywren) and so develop an understanding of how speciation occurs by truly integrating history and selection. The importance of this is neatly illustrated in the comparison of Ruegg et al. (2014) and Delmore et al. (2015) who both tried to elucidate the genetic basis to the evolution of migration in Swainson's Thrush *Catharus ustulatus*, the former mapping data to the Zebra Finch genome and the latter mapping to a draft assembly for Swainson's Thrush itself. This explains their different conclusions about the genes involved in the evolution of migration and how they are organized across the genome.

Byers et al. (2017) have neatly summarized these challenges. Figure 10.7, here reproduced from their paper, says it all! At the time of writing, the floodgates of genomic resources with which to study birds are about to open. Research into the evolution of birds through understanding the interplay among phylogeography, hybridization, and speciation has a bright future.

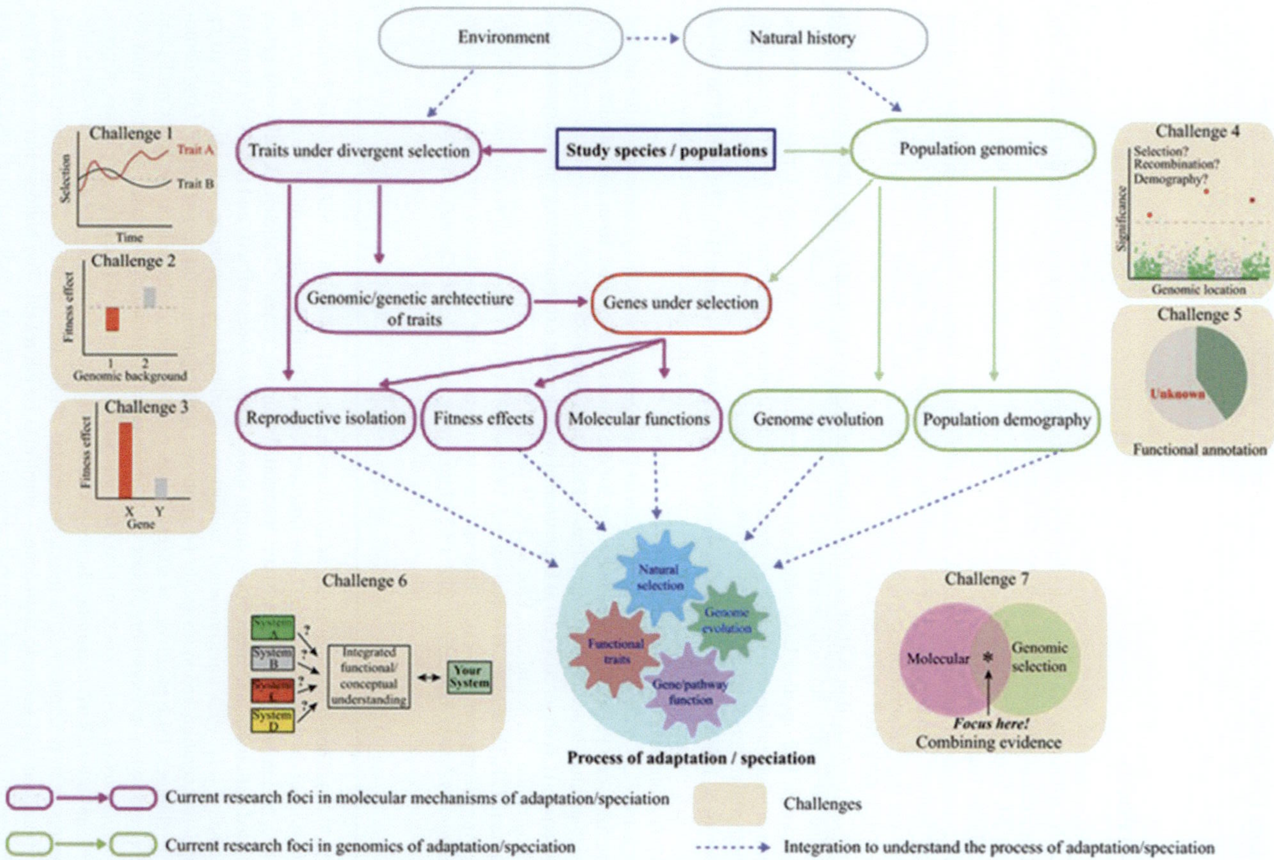

Fig. 10.7 Summary figure from Byers et al. (2017) arguing the need for integrating genomics and biology in speciation research. Current research foci and challenges in studying adaptation and speciation using molecular (magenta arrows) and genomic approaches (green arrows). Light red boxes highlight the challenges in these

Acknowledgments I am grateful to Dieter Thomas Tietze and Darren Irwin for the opportunity to write this chapter. Joshua Peñalba and Craig Moritz have helped me find and, I hope, understand some literature that I might not have found otherwise. Andrew Black offered helpfully incisive comments to improve my expression. The manuscript was closed on 22 December 2017.

References

Abbott R, Albach D, Ansell S et al (2013) Hybridization and speciation. J Evol Biol 26:229–246. https://doi.org/10.1111/j.1420-9101.2012.02599.x

Alcaide M, Scordato ESC, Price TD, Irwin DE (2014) Genomic divergence in a ring species complex. Nature 511:83–85

Austin J, Joseph L, Pedler L, Black A (2013) Uncovering cryptic evolutionary diversity in extant and extinct populations of the southern Australian arid zone Western and Thick-billed Grasswrens (Passeriformes: Maluridae: *Amytornis*). Conserv Genet 14:173–1184. https://doi.org/10.1007/s10592-013-0504-9

Avise JC, Arnold J, Ball RM et al (1987) Intraspecific phylogeography: the mitochondrial DNA bridge between population genetics and systematics. Annu Rev Ecol Syst 18:489–522

Backström N, Palkopoulou E, Qvarnström A, Ellegren H (2010a) No evidence for Z-chromosome rearrangements between the pied flycatcher and the collared flycatcher as judged by gene-based comparative genetic maps. Mol Ecol 19:3394–3405

Backström N, Lindell J, Zhang Y, Palkopoulou E, Qvarnström A, Sætre G-P, Ellegren H (2010b) A high-density scan of the Z chromosome in *Ficedula* flycatchers reveals candidate loci for diversifying selection. Evolution 64:3461–3475

Balakrishnan CN, Lee JY, Edwards SV (2010) Phylogeography and phylogenetics in the nuclear age. In: Grant P, Grant R (eds) Searching for the causes of evolution: from field observations to mechanisms. Princeton University Press, Princeton, pp 65–88

Baldassarre DT, Webster MS (2013) Experimental evidence that extrapair mating drives asymmetrical introgression of a sexual trait. Proc R Soc B Biol Sci 280:20132175

Baldassarre DT, Thomassen HA, Karubian J, Webster MS (2013) The role of ecological variation in driving divergence of sexual and non-sexual traits in the red-backed fairy-wren (*Malurus melanocephalus*). BMC Evol Biol 13:75 http://www.biomedcentral.com/1471-2148/13/75

Fig. 10.7 (continued) research areas, as follows: Challenge 1, selection on traits might be dynamic and dependent on time; Challenge 2, the fitness effects of individual genes might be dependent on genomic background, such as epistatic interactions among genes; Challenge 3, different genes associated with traits that affect fitness might have different fitness effects, and these effects might additionally vary with environmental conditions; Challenge 4, candidate loci identified using adaptation/speciation genomics might be spurious, with significant outliers resulting from demographic changes, false positives, or selective sweeps, rather than selection on the locus itself; and Challenge 5, even in model organisms (e.g., thale cress *Arabidopsis thaliana*, common fruit fly *Drosophila melanogaster*, house mouse *Mus musculus*), the function of a large proportion of genes remains unknown. Two remaining challenges of an integrative approach—Challenge 6, synthesizing results from multiple systems into a conceptual whole, and Challenge 7, combining evidence from molecular and genomic approaches—must also be addressed. The bottom circle depicts the integration of molecular and genomic approaches to studying adaptation and speciation and suggests some future integrative approaches (blue dashed arrows). See Byers et al. (2017) for further discussion

Baldassarre DT, White TA, Karubian J, Webster MS (2014) Genomic and morphological analysis of a semipermeable avian hybrid zone suggests asymmetrical introgression of a sexual signal. Evolution 68:2644–2657

Barton NH (2013) Does hybridization influence speciation? J Evol Biol 26:267–269

Barton NH, Hewitt GM (1981) Hybrid zones and speciation. In: Atchley WR, Woodruff D (eds) Evolution and speciation. Cambridge University Press, Cambridge, pp 109–145

Batalha-Filho H, Cabanne GS, Miyaki CY (2012) Phylogeography of an Atlantic forest passerine reveals demographic stability through the last glacial maximum. Mol Phylogenet Evol 65:892–902

Bay RA, Ruegg K (2017) Genomic islands of divergence or opportunities for introgression? Proc R Soc B 284:20162414. https://doi.org/10.1098/rspb.2016.2414

Beheregaray L (2008) Twenty years of phylogeography: the state of the field and the challenges for the Southern Hemisphere. Mol Ecol 17:3754–3774. https://doi.org/10.1111/j.1365-294X.2008.03857.x

Belkacem AA, Gast O, Stuckas H, Canal D, LoValvo M, Giacalone G, Päckert M (2017) North African hybrid sparrows (*Passer domesticus, P. hispaniolensis*) back from oblivion – ecological segregation and asymmetric mitochondrial introgression between parental species. Ecol Evol 6:5190–5206. https://doi.org/10.1002/ece3.2274

Bermingham E, Moritz C (1998) Comparative phylogeography: concepts and applications. Mol Ecol 7:367–369

Billerman SM, Murphy MA, Carling MD (2016) Changing climate mediates sapsucker (Aves: *Sphyrapicus*) hybrid zone movement. Ecol Evol 6:7976–7990

Black AB, Horton P, Joseph L (2015) Diverse interactions, including hybridisation, between Brown and Inland Thornbills in South Australia. S Aust Ornithol 41:18–34

Bonnet T, Leblois R, Rousset F, Crochet P-A (2017) A reassessment of explanations for discordant introgressions of mitochondrial and nuclear genomes. Evolution 71:2140–2158

Borge T, Lindroos K, Nádvorník P, Syvänen AC, Sætre GP (2005) Amount of introgression in flycatcher hybrid zones reflects regional differences in pre- and post-zygotic barriers to gene exchange. J Evol Biol 18:1416–1424

Boulet M, Gibbs HL (2006) Lineage origin and expansion of a Neotropical migrant songbird after recent glaciation events. Mol Ecol 15:2505–2525

Bowie RCK, Fjeldså J, Hackett SJ, Bates JM, Crowe TM (2006) Coalescent models reveal the relative roles of ancestral polymorphism, vicariance, and dispersal in shaping phylogeographical structure of an African montane forest robin. Mol Phylogenet Evol 38:171–188. https://doi.org/10.1016/j.ympev.2005.06.001

Bowie RCK, Fjeldså J, Kiure J (2009) Multilocus molecular DNA variation in Winifred's Warbler *Scepomycter winifredae* suggests cryptic speciation and the existence of a threatened species in the Rubeho–Ukaguru Mountains of Tanzania. Ibis 151:709–719

Brelsford A, Mila B, Irwin DE (2011) Hybrid origin of Audubon's warbler. Mol Ecol 20:2380–2389. https://doi.org/10.1111/j.1365-294X.2011.05055.x

Brumfield R, Jernigan JW, McDonald DB, Braun MJ (2001) Evolutionary implications of divergent clines in an avian (*Manacus*: Aves) hybrid zone. Evolution 55:2070–2087

Burton TC, Martin AA (1976) Analysis of hybridization between Black-backed and White-backed Magpies in south-eastern Australia. Emu 76:30–36

Butlin RK, Ritchie MG (2013) Pulling together or pulling apart: hybridization in theory and practice. J Evol Biol 26:294–298

Byers KJRP, Xu S, Schlüter PM (2017) Molecular mechanisms of adaptation and speciation: why do we need an integrative approach? Mol Ecol 26:277–290

Cadena CD, Cuervo AM (2010) Molecules, ecology, morphology, and songs in concert: how many species is *Arremon torquatus* (Aves: Emberizidae)? Biol J Linn Soc 99:152–176

Carling MD, Brumfield RT (2008) Haldane's rule in an avian system: using cline theory and divergence population genetics to test for differential introgression of mitochondrial, autosomal and sex-linked loci across the *Passerina* bunting hybrid zone. Evolution 62:2600–2615

Carling MD, Brumfield RT (2009) Speciation in *Passerina* buntings: introgression patterns of sex-linked loci identify a candidate gene region for reproductive isolation. Mol Ecol 18:834–847

Carling MD, Zuckerberg B (2011) Spatio-temporal changes in the genetic structure of the *Passerina* bunting hybrid zone. Mol Ecol 20:1166–1175

Carling MD, Lovette IJ, Brumfield RT (2010) Historical divergence and gene flow: coalescent analyses of mitochondrial, autosomal and sex-linked loci in *Passerina* buntings. Evolution 64:1762–1772

Chan KMA, Levin SA (2005) Leaky prezygotic isolation and porous genomes: rapid introgression of maternally inherited DNA. Evolution 59:720–729

Chan YL, Schanzenbach D, Hickerson MJ (2014) Detecting concerted demographic response across community assemblages using hierarchical approximate Bayesian computation. Mol Biol Evol 31:2501–2515. https://doi.org/10.1093/molbev/msu187

Charlesworth B, Coyne JA, Barton NH (1987) The relative rates of evolution of sex chromosomes and autosomes. Am Nat 130:113–146

Chattopadhyay B, Garg KM, Gwee CY, Edwards SV, Rheindt FE (2017) Gene flow during glacial habitat shifts facilitates character displacement in a Neotropical flycatcher radiation. BMC Evol Biol 17:210. https://doi.org/10.1186/s12862-017-1047-3

Coyne JA, Orr HA (2004) Speciation. Sinauer Associates, Sunderland, MA

Cruickshank TE, Hahn MW (2014) Reanalysis suggests that genomic islands of speciation are due to reduced diversity, not reduced gene flow. Mol Ecol 23:3133–3157

Dai C, Hao Y, Lei F (2017) The absence of reproductive isolation between non-sister and deeply diverged mitochondrial lineages of the black-throated tit (*Aegithalos concinnus*) revealed by a multilocus genetic analysis in a contact zone. BMC Evol Biol 17:266. https://doi.org/10.1186/s12862-017-1114-9

Dasmahapatra KK, Walters JR, Briscoe AD et al (*Heliconius* Genome Consortium) (2012) Butterfly genome reveals promiscuous exchange of mimicry adaptations among species. Nature 487:94–98

de Queiroz K (2007) Species concepts and species delimitation. Syst Biol 56:879–886

Degnan SM (1993) The perils of single gene trees – mitochondria1 versus single-copy nuclear DNA variation in white-eyes (Aves: Zosteropidae). Mol Ecol 2:219–225

Delmore KE, Hübner S, Kane NC, Schuster R, Andrew RL, Câómara F, Guigo´ R, Irwin DE (2015) Genomic analysis of a migratory divide reveals candidate genes for migration and implicates selective sweeps in generating islands of differentiation. Mol Ecol 24:1873–1888. https://doi.org/10.1111/mec.13150

Dhami KK, Joseph L, Roshier DA, Heinsohn R, Peters JL (2013) Multilocus phylogeography of Australian teals (*Anas* spp.): a case study of the relationship between vagility and genetic structure. J Avian Biol 44:169–178. https://doi.org/10.1111/j.1600-048X.2012.05826.x

Dhami KK, Joseph L, Roshier DA, Peters JL (2016) Recent speciation and elevated Z-chromosome differentiation between a sexually dichromatic and monochromatic pair of Australian teals (Anatidae: *Anas*). J Avian Biol 47:92–102. https://doi.org/10.1111/jav.00693

Dolman G, Joseph L (2012) A species assemblage approach to comparative phylogeography of birds in southern Australia. Ecol Evol 2:354–369. https://doi.org/10.1002/ece3.87

Dolman G, Joseph L (2015) Evolutionary history of birds across southern Australia: structure, history and taxonomic implications of mitochondrial DNA diversity in an ecologically diverse suite of species. Emu 115:35–48. https://doi.org/10.1071/MU14047

Dolman G, Joseph L (2016) Multi-locus sequence data reveal Pleistocene speciation in semi-arid southern Australian birds (*Cinclosoma* spp.) was associated with increased genetic drift. BMC Evol Biol 16:226. https://doi.org/10.1186/s12862-016-0798-6

Dong F, Zou F-S, Lei F-M, Liang W, Li S-H, Yang X-J (2014) Testing hypotheses of mitochondrial gene-tree paraphyly: unravelling mitochondrial capture of the Streak-breasted Scimitar Babbler (*Pomatorhinus ruficollis*) by the Taiwan Scimitar Babbler (*Pomatorhinus musicus*). Mol Ecol 23:5855–5867

Dong F, Hung C-M, Li X-L, Gao J-Y, Zhang Q, Wu F, Lei F-M, Li S-H, Yang X-J (2017) Ice age unfrozen: severe effect of the last interglacial, not glacial, climate change on East Asian avifauna. BMC Evol Biol 17:244. https://doi.org/10.1186/s12862-017-1100-2

Drovetski SV, Semenov G, Red'kin YA, Sotnikov VN, Fadeev IV, Koblik EA (2015) Effects of asymmetric nuclear introgression, introgressive mitochondrial sweep, and purifying selection on phylogenetic reconstruction and divergence estimates in the pacific clade of *Locustella* warblers. PLoS One 10(4):e0122590. https://doi.org/10.1371/journal.pone.0122590

Edwards SV, Potter S, Schmitt CJ, Bragg JC, Moritz C (2016) Reticulation, divergence, and the phylogeography-phylogenetics continuum. Proc Natl Acad Sci USA 113:8025–8032

Elgvin TO, Hermansen JS, Fijarczyk A, Bonnet T, Borge T et al (2011) Hybrid speciation in sparrows II: a role for sex chromosomes? Mol Ecol 20:3823–3837

Ellegren H (2009) Genomic evidence for a large-Z effect. Proc R Soc Lond B 276:361–366

Ellegren H, Smeds L, Burri R et al (2012) The genomic landscape of species divergence in *Ficedula* flycatchers. Nature 491:756–760

Faust-Stryjewski K, Sorenson M (2017) Mosaic genome evolution in a recent and rapid avian radiation. Nat Ecol Evol 1:1912–1922. https://doi.org/10.1038/s41559-017-0364-7

Ford J (1974) Concepts of subspecies and hybrid zones and their application in Australian ornithology. Emu 74:113–123

Ford J (1986) Avian hybridization and allopatry in the region of the Einasleigh Uplands and Burdekin-Lynd Divide, north-eastern Queensland. Emu 86:87–110

Ford J (1987) Hybrid zones in Australian birds. Emu 87:158–178

Forshaw J, Cooper WT (2002) Australian parrots, 3rd edn. Robina Press, Queensland

Friesen V (2015) Speciation in seabirds: why are there so many species. . .and why aren't there more? J Ornithol 156(Suppl 1):S27–S39. https://doi.org/10.1007/s10336-015-1235-0

Friis G, Aleixandre P, Rodríguez-Estrella R, Navarro-Sigüenza AG, Milá B (2016) Rapid postglacial diversification and long-term stasis within the songbird genus *Junco*: phylogeographic and phylogenomic evidence. Mol Ecol 25:6175–6195

Fuchs J, Bowie RCK (2015) Concordant genetic structure in two species of woodpecker distributed across the primary West African biogeographic barriers. Mol Phylogenet Evol 88:64–74

Gompert Z, Mandeville EG, Buerkle CA (2017) Analysis of population genomic data from hybrid zones. Annu Rev Ecol Evol Syst 48:207–229

Grant PR, Grant BR (1992) Hybridization of bird species. Science 256:193–197

Grant PR, Grant BR (2016) Introgressive hybridization and natural selection in Darwin's finches. Biol J Linn Soc 117:812–822

Griffith SC, Hooper DM (2017) Geographical variation in bill colour in the long-tailed finch: evidence for a narrow zone of admixture between sub-species. Emu 117:141–150

Haldane JBS (1922) Sex-ratio and unisexual sterility in hybrid animals. J Genet 12:101–109

Harvey MG, Brumfield RT (2015) Genomic variation in a widespread Neotropical bird (*Xenops minutus*) reveals divergence, population expansion, and gene flow. Mol Phylogenet Evol 83:305–316

Harvey MG, Seeholzer GF, Smitha BT, Rabosky DL, Cuervo AM, Brumfield RT (2017) Positive association between population genetic differentiation and speciation rates in New World birds. Proc Natl Acad Sci USA 114:6328–6333

Hermansen JS, Sæther SA, Elgvin TO, Borge T, Hjelle E et al (2011) Hybrid speciation in sparrows I: phenotypic intermediacy, genetic admixture and barriers to gene flow. Mol Ecol 20:3812–3822

Hewitt GM (1999) Post-glacial re-colonization of European biota. Biol J Linn Soc 68:87–112

Hill GE (2017) The mitonuclear compatibility species concept. Auk 134:393–409

Hogner S, Sæther SA, Borge T, Bruvik T, Johnsen A, Sætre G-P (2012) Increased divergence but reduced variation on the Z chromosome relative to autosomes in *Ficedula* flycatchers: differential introgression or the faster-Z effect? Ecol Evol 2:379–397

Hooper DM, Price TD (2015) Rates of karyotypic evolution in estrildid finches differ between island and continental clades. Evolution 69:890–903

Hooper DM, Price TD (2017) Chromosomal inversion differences correlate with range overlap in passerine birds. Nat Ecol Evol 1:1526–1534

Hughes JM, Baker AM, de Zylva G, Mather PB (2001) A phylogeographic analysis of southern and eastern populations of the Australian magpie: evidence for selection in maintenance of the distribution of two plumage morphs. Biol J Linn Soc 74:25–34. https://doi.org/10.1006/bijl.2001.0564

Irwin DE, Alcaide M, Delmore KE, Irwin JH, Owens GL (2016) Recurrent selection explains parallel evolution of genomic regions of high relative but low absolute differentiation in a ring species. Mol Ecol 25:4488–4507

Joseph L, Omland K (2009) Phylogeography: its development and impact in Australo-Papuan ornithology with special reference to paraphyly in Australian birds. Emu 109:1–23

Joseph L, Wilke T, Ten Have J, Chesser RT (2006) Implications of mitochondrial DNA polyphyly in two ecologically undifferentiated but morphologically distinct migratory birds, the masked and white-browed Woodswallows *Artamus* spp. of inland Australia. J Avian Biol 37:625–636

Joseph L, Dolman G, Saint K, Donnellan S, Berg M, Bennett A (2008) Where and when does a ring start and end? Testing the ring species hypothesis in a species complex of Australian parrots. Proc R Soc B Biol Sci 275:2431–2440. https://doi.org/10.1098/rspb.2008.0765

Joseph L, Adcock G, Linde C, Omland K, Heinsohn R, Roshier D (2009) A tangled, rampantly paraphyletic tale of two teal: population history of the Grey and Chestnut Teal of Australia. J Avian Biol 40:430–439. https://doi.org/10.1111/j.1600-048X.2008.04652.x

Kawakami T, Smeds L, Backström AN, Husby A, Qvarnström A, Mugal CF, Olason P, Ellegren HS (2014) A high-density linkage map enables a second-generation collared flycatcher genome assembly and reveals the patterns of avian recombination rate variation and chromosomal evolution. Mol Ecol 23:4035–4058. https://doi.org/10.1111/mec.12810

Kearns A, Joseph L, Omland K, Cook L (2011) Testing the impact of transient Plio-Pleistocene barriers in monsoonal Australo-Papua: did mangrove habitats maintain genetic connectivity in the Black Butcherbird? Mol Ecol 20:5042–5059. https://doi.org/10.1111/j.1365-294X.2011.05330.x

Kearns A, Joseph L, Toon A, Cook L (2014) Australia's arid-adapted butcherbirds experienced range expansions during Pleistocene glacial maxima. Nat Commun 5:3994. https://doi.org/10.1038/ncomms4994

Lamichhaney S, Berglund J, Almén MS et al (2015) Evolution of Darwin's finches and their beaks revealed by genome sequencing. Nature 518:371–375

Lamichhaney S, Han F, Berglund J et al (2016) A beak size locus in Darwin's finches facilitated character displacement during a drought. Science 352:470–474

Lavretsky P, Dacosta JM, Hernández-Baños BE, Engilis A, Sorenson MD, Peters JL (2015) Speciation genomics and a role for the Z chromosome in the early stages of divergence between Mexican ducks and mallards. Mol Ecol 24:5364–5378

Lee JY, Edwards SV (2008) Divergence across Australia's Carpentarian barrier: statistical phylogeography of the red-backed fairy wren (*Malurus melanocephalus*). Evolution 62:3117–3134

Liebers D, de Knijff P, Helbig AJ (2004) The herring gull complex is not a ring species. Proc R Soc B 271:893–901. https://doi.org/10.1098/rspb.2004.2679

Mallet J (2005) Hybridization as an invasion of the genome. Trends Ecol Evol 20:229–237

Mallet J (2008) Hybridization, ecological races and the nature of species: empirical evidence for the ease of speciation. Philos Trans R Soc B 363:2971–2986

Mallet J, Besansky N, Hahn MW (2016) How reticulated are species? BioEssays 38:140–149

Manthey JD, Robbins MB (2016) Genomic insights into hybridization in a localized region of sympatry between pewee sister species (*Contopus sordidulus* × *C. virens*) and their chromosomal patterns of differentiation. Avian Res 7:6. https://doi.org/10.1186/s40657-016-0040-7

Manthey JD, Spellman GM (2014) Increased differentiation and reduced gene flow in sex chromosomes relative to autosomes between lineages of the brown creeper *Certhia americana*. J Avian Biol 45:149–156. https://doi.org/10.1111/j.1600-048X.2013.00233.x

Manthey JD, Klicka J, Spellman GM (2012) Is gene flow promoting the reversal of Pleistocene divergence in the Mountain Chickadee (*Poecile gambeli*)? PLoS One 7(11):e49218. https://doi.org/10.1371/journal.pone.0049218

Manthey JD, Robbins MB, Moyle RG (2016) A genomic investigation of the putative contact zone between divergent Brown Creeper (*Certhia americana*) lineages: chromosomal patterns of genetic differentiation. Genome 59:115–125. https://doi.org/10.1139/gen-2015-0093

Mason NA, Taylor SA (2015) Differentially expressed genes match bill morphology and plumage despite largely undifferentiated genomes in a Holarctic songbird. Mol Ecol 24:3009–3025

Mayr E (1942) Systematics and the origin of species. Columbia University Press, New York

McCarthy EM (2006) Handbook of avian hybrids of the world. Oxford University Press, Oxford

McDonald DB, Clay RP, Brumfield RT, Braun MJ (2001) Sexual selection on plumage and behavior in an avian hybrid zone: experimental tests of male-male interactions. Evolution 55:1443–1451

McKay BD, Zink RM (2014) Sisyphean evolution in Darwin's finches. Biol Rev 90:689–698. https://doi.org/10.1111/brv.12127

McLean A, Toon A, Schmidt D, Hughes J, Joseph L (2017a) Phylogeography and geno-phenotypic discordance in a widespread Australian bird, the variegated fairy-wren, *Malurus lamberti* (Aves: Maluridae). Biol J Linn Soc 121:655–669. https://doi.org/10.1093/biolinnean/blx004

McLean A, Joseph L, Toon A, Schmidt D, Drew A, Mason IJ, Hughes J (2017b) Reassessment of a possible case of intraspecific gene flow across Australia's Great Dividing Range in the variegated fairy-wren, *Malurus lamberti* (Aves: Maluridae) and its systematic consequences. Biol J Linn Soc 122:210–223. https://doi.org/10.1093/biolinnean/blx054

Mettler RD, Spellman GM (2009) A hybrid zone revisited: molecular and morphological analysis of the maintenance, movement, and evolution of a Great Plains avian (Cardinalidae: *Pheucticus*) hybrid zone. Mol Ecol 18:3256–3267

Mihailova M, Berg ML, Buchanan KL, Bennett ATD (2014) Odour-based discrimination of subspecies, species and sexes in an avian species complex, the crimson rosella. Anim Behav 95:155–164

Morales HE, Pavlova A, Joseph L, Sunnucks P (2015) Positive and purifying selection in mitochondrial genomes of a bird with mitonuclear discordance. Mol Ecol 24:2820–2837

Morales HE, Sunnucks P, Joseph L, Pavlova A (2017a) Perpendicular axes of differentiation generated by mitochondrial introgression. Mol Ecol 26:3241–3255

Morales HE, Pavlova A, Sunnucks P, Major R, Amos N, Joseph L, Wang B, Lemmon AR, Endler JA Delhey K (2017b) Neutral and selective drivers of colour evolution in a widespread Australian passerine. J Biogeogr 44:522–536

Morales HE, Pavlova A, Amos N, Major R, Bragg J, Kilian A, Greening C, Sunnucks P (2017c) Mitochondrial-nuclear interactions maintain a deep mitochondrial split in the face of nuclear gene flow. bioRxiv preprint first posted online Dec. 20, 2016. https://doi.org/10.1101/095596

Morales-Rozo A, Tenorio EA, Carling MD, Cadena CD (2017) Origin and cross-century dynamics of an avian hybrid zone. BMC Evol Biol 17:257. https://doi.org/10.1186/s12862-017-1096-7

Nater A, Burri R, Kawakami T, Smeds L, Ellegren H (2015) Resolving evolutionary relationships in closely related species with whole-genome sequencing data. Syst Biol 64:1000–1017

Nei M, Roychoudhury AK (1974) Sampling variances of heterozygosity and genetic distance. Genetics 76:379–390

Newton I (2004) Speciation and biogeography of birds. Academic Press, London

Nosil P, Feder JL (2012) Genomic divergence during speciation: causes and consequences. Philos Trans R Soc B Biol Sci 367:332–342

Nosil P, Feder JL, Flaxman SM, Gompert Z (2017) Tipping points in the dynamics of speciation. Nat Ecol Evol 1:1–8. https://doi.org/10.1038/s41559-016-0001

Olsson U, Leader PJ, Carey GJ, Khan AA, Svensson L, Alström P (2013) New insights into the intricate taxonomy and phylogeny of the *Sylvia curruca* complex. Mol Phylogenet Evol 67:72–85

Oswald JA, Harvey MG, Remsen RC, Foxworth DU, Cardiff SW et al (2016) Willet be one species or two? A genomic view of the evolutionary history of *Tringa semipalmata*. Auk 133:593–614

Ottenburghs J, Kraus RHS, van Hooft P, van Wieren SE, Ydenberg RC, Prins HHT (2017) Avian introgression in the genomic era. Avian Res 8:30. https://doi.org/10.1186/s40657-017-0088-z

Ovenden JR, Mackinlay AG, Crozier RH (1987) Systematics and mitochondrial genome evolution of Australian rosellas (Aves, Platycercidae). Mol Biol Evol 4:526–543

Parchman TL, Gompert Z, Braun MJ, Brumfield RT, McDonald DB, Uy JAC, Zhang G, Jarvis ED, Schlinger BA, Buerkle CA (2013) The genomic consequences of adaptive divergence and reproductive isolation between species of manakins. Mol Ecol 22:3304–3317

Parsons TJ, Olson SL, Braun MJ (1993) Unidirectional spread of secondary sexual plumage traits across an avian hybrid zone. Science 260:1643–1646

Pavlova A, Amos JN, Joseph L, Loynes K, Austin JJ, Keogh JS, Stone GN, Nicholls JA, Sunnucks P (2013) Perched at the mito-nuclear crossroads: divergent mitochondrial lineages correlate with environment in the face of ongoing nuclear gene flow in an Australian bird. Evolution 67:3412–3428

Pavlova A, Selwood P, Harrisson KA, Murray N, Quin B, Menkhorst P, Smales I, Sunnucks P (2014) Integrating phylogeography and morphometrics to assess conservation merits and inform conservation strategies for an endangered subspecies of a common bird species. Biol Conserv 174:136–146

Payseur BA, Rieseberg LH (2016) A genomic perspective on hybridization and speciation. Mol Ecol 25:2337–2360

Peñalba JV, Mason IJ, Schodde R, Moritz C, Joseph L (2017) Characterizing divergence through three adjacent Australian avian transition zones. J Biogeogr 44:2247–2258. https://doi.org/10.1111/jbi.13048

Peters JL, Zhuravlev Y, Fefelov I, Logie A, Omland KE (2007) Nuclear loci and coalescent method support ancient hybridization as cause of mitochondrial paraphyly between Gadwall and Falcated Duck (*Anas* spp.). Evolution 61:1992–2006

Peters JL, Winker K, Millam KC, Lavretsky P, Kulikova I, Wilson RE, Zhuravlev YN, McCracken KG (2014) Mito-nuclear discord in six congeneric lineages of Holarctic ducks (genus *Anas*). Mol Ecol 23:2961–2974

Peterson AT, Anamza T (2017) Reexamining *Phylloscopus trochiloides* complex as a ring species: a refugial counter-hypothesis. J Avian Biol 48(12):1608–1613. https://doi.org/10.1111/jav.01167

Poelstra JW, Vijay N, Bossu CM, Lantz H, Ryll B, Müller I, Baglione V, Unneberg P, Wikelski M, Grabherr MG, Wolf JBW (2014) The genomic landscape underlying phenotypic integrity in the face of gene flow in crows. Science 344:1410–1414

Poelstra JW, Vijay N, Hoeppner MP, Wolf JBW (2015) Transcriptomics of colour patterning and coloration shifts in crows. Mol Ecol 24:4617–4628

Price TD (2008) Speciation in birds. Roberts and Company, Greenwood Village

Pryke SR, Griffith SC (2009) Postzygotic genetic incompatibility between sympatric color morphs. Evolution 63:793–798

Qvarnström A, Bailey RI (2009) Speciation through evolution of sex-linked genes. Heredity 102:4–15

Remington CL (1968) Suture-zones of hybrid interaction between recently joined biotas. In: Dobzhansky T, Hecht MK, Steere WC (eds) Evolutionary biology. Springer, New York, pp 321–428

Rheindt FE, Edwards SV (2011) Genetic introgression: an integral but neglected component of speciation in birds. Auk 128:620–632

Ribas CC, Maldonado-Coelho M, Smith BT, Cabanne GS, d'Horta FM, Naka LN (2012) Towards an integrated historical biogeography of the Neotropical lowland avifauna: combining diversification analysis and landscape evolution. Ornitol Neotrop 23:187–206

Ribeiro AM, Lloyd P, Bowie RCK (2011) A tight balance between natural selection and gene flow in a southern African arid-zone endemic bird. Evolution 65:3499–3514

Roux C, Fraïsse C, Romiguier J, Anciaux Y GN, Bierne N (2016) Shedding light on the grey zone of speciation along a continuum of genomic divergence. PLoS Biol 14:e2000234. https://doi.org/10.1371/journal.pbio.2000234
Ruegg K, Anderson EC, Boone J, Pouls J, Smith TB (2014) A role for migration-linked genes and genomic islands in divergence of a songbird. Mol Ecol 23:4757–4769. https://doi.org/10.1111/mec.12842
Sæther SA, Sætre G-P, Borge T et al (2007) Sex chromosome-linked species recognition and evolution of reproductive isolation in flycatchers. Science 318:95–97
Sætre G-P, Sæther SA (2010) Ecology and genetics of speciation in *Ficedula* flycatchers. Mol Ecol 19:1091–1106
Sætre G-P, Cuevas A, Hermansen JS, Elgvin TO, Fernández LP, Sæther SA, Cascio Sætre CL, Eroukhmanoff F (2017) Rapid polygenic response to secondary contact in a hybrid species. Proc R Soc B 284:20170365. https://doi.org/10.1098/rspb.2017.0365
San-Jose LM, Ducrest A-L, Ducret V, Simon C, Richter H, Wakamatsu K, Roulin A (2017) MC1R variants affect the expression of melanocortin and melanogenic genes and the association between melanocortin genes and coloration. Mol Ecol 26:259–276
Schodde R (1997) Psittacidae. In: Houston WMK, Wells A (eds) Zoological catalogue of Australia, vol 37.2. Aves. (Columbidae to Coraciidae). CSIRO Publishing, Melbourne, pp 109–218
Schodde R, Mason IJ (1999) The directory of Australian birds. CSIRO Publishing, Canberra
Scordato ESC, Wilkins MR, Semenov G, Rubtsov AS, Kane NC, Safran RJ (2017) Genomic variation across two barn swallow hybrid zones reveals traits associated with divergence in sympatry and allopatry. Mol Ecol 26(20):5676–5691. https://doi.org/10.1111/mec.14276
Shipham A, Schmidt D, Joseph L, Hughes J (2015) Phylogenetic analysis of the Australian rosella parrots (*Platycercus*) reveals discordance among molecules and plumage. Mol Phylogenenet Evol 91:150–159. https://doi.org/10.1016/j.ympev.2015.05.012
Shipham A, Joseph L, Schmidt D, Hughes JH (2017) A genomic approach reinforces a hypothesis of mitochondrial capture in eastern Australian rosellas. Auk 134:181–192. https://doi.org/10.1642/AUK-16-31.1
Short L (1969) Taxonomic aspects of avian hybridization. Auk 86:84–105
Sibley CG (1957) The evolutionary and taxonomic significance of sexual selection and hybridization in birds. Condor 59:166–191
Slender AL, Louter M, Gardner MG, Kleindorfer S (2017) Patterns of morphological and mitochondrial diversity in parapatric subspecies of the Thick-billed Grasswren (*Amytornis modestus*). Emu – Aust Ornithol 117:264–275. https://doi.org/10.1080/01584197.2017.1313686
Sloan DB, Havird JC, Sharborough J (2017) The on-again, off-again relationship between mitochondrial genomes and species boundaries. Mol Ecol 26:2212–2236
Smadja C, Butlin RK (2011) A framework for comparing processes of speciation in the presence of gene flow. Mol Ecol 20:5123–5140
Smith BT, Harvey MG, Faircloth BC, Glenn TC, Brumfield RT (2014a) Target capture and massively parallel sequencing of ultraconserved elements for comparative studies at shallow evolutionary time scales. Syst Biol 63:83–95
Smith BT, McCormack JE, Cuervo AM, Hickerson MJ et al (2014b) The drivers of tropical speciation. Nature 515:406–409. https://doi.org/10.1038/nature13687
Southcott L, Kronforst MR (2017) A neutral view of the evolving genomic architecture of speciation. Ecol Evol 7:6358–6366. https://doi.org/10.1002/ece3.3190
Sraml M, Christidis L, Easteal S, Horn P, Collet C (1996) Molecular relationships within Australasian waterfowl (Anseriformes). Aust J Zool 44:47–58
Storchová R, Reif J, Nachman MW (2009) Female heterogamety and speciation: reduced introgression of the Z chromosome between two species of nightingales. Evolution 64:456–471
Sunnucks P, Morales HE, Lamb AM, Pavlova A, Greening C (2017) Integrative approaches for studying mitochondrial and nuclear genome co-evolution in oxidative phosphorylation. Front Genet 8:25. https://doi.org/10.3389/fgene.2017.00025
Swenson NG, Howard DJ (2004) Do suture zones exist? Evolution 58:2391–2397

Taylor SA, Curry RL, White TA, Ferretti V, Lovette I (2014) Spatiotemporally consistent genomic signatures of reproductive isolation in a moving hybrid zone. Evolution 68:3066–3081
Toews DPL, Irwin DE (2008) Cryptic speciation in a Holarctic passerine revealed by genetic and bioacoustic analyses. Mol Ecol 17:2691–2705
Toews DPL, Campagna L, Taylor SA et al (2016a) Genomic approaches to understanding population divergence and speciation in birds. Auk 133:13–30. https://doi.org/10.1642/AUK-15-51.1
Toews DPL, Taylor SA, Vallender R, Brelsford A, Butcher BG, Messer PW, Lovette IJ (2016b) Plumage genes and little else distinguish the genomes of hybridizing warblers. Curr Biol 26:2313–2318
Toon A, Hughes J, Joseph L (2010) Multilocus analysis of honeyeaters (Aves: Meliphagidae) highlights the spatio-temporal heterogeneity in the influence of biogeographic barriers in the Australian monsoonal zone. Mol Ecol 19:2980–2994. https://doi.org/10.1111/j.1365-294X.2010.04730.x
Trier CN, Hermansen JS, Sætre G-P, Bailey RI (2014) Evidence for mito-nuclear and sex-linked reproductive barriers between the hybrid Italian Sparrow and its parent species. PLoS Genet 10 (1):e1004075. https://doi.org/10.1371/journal.pgen.1004075
van Doren BM, Campagna L, Helm B, Illera JC, Lovette IJ, Liedvogel M. (2017) Correlated patterns of genetic diversity and differentiation across an avian family. Mol Ecol 26:3982–3997
Voelker G, Outlaw RK, Bowie RCK (2010) Pliocene forest dynamics as a primary driver of African bird speciation. Glob Ecol Biogeogr 19:111–121
Voelker G, Marks BD, Kahindo C et al (2013) River barriers and cryptic biodiversity in an evolutionary museum. Ecol Evol 3:536–545
Winger BM (2017) Consequences of divergence and introgression for speciation in Andean cloud forest birds. Evolution 71:1815–1831
Winger BM, Bates JM (2015) The tempo of trait divergence in geographic isolation: Avian speciation across the Marañon Valley of Peru. Evolution 69:772–787
Yamaguchi R, Iwasa Y (2017) A tipping point in parapatric speciation. J Theor Biol 421:81–92
Younger JL, Clucas GV, Kooyman G, Wienecke B, Rogers AD, Trathan PN, Hart T, Miller KJ (2015) Too much of a good thing: sea ice extent may have forced emperor penguins into refugia during the last glacial maximum. Glob Chang Biol 21:2215–2226. https://doi.org/10.1111/gcb.12882

Chapter 11
Ecological Speciation: When and How Variation Among Environments Can Drive Population Divergence

Pim Edelaar

Abstract Speciation is a complex process, in part because it can be caused by a multitude of different mechanisms. One aspect of speciation that has received much attention lately is the importance that variation among environments may play in driving speciation, i.e., ecological speciation. However, attention has been largely limited to the role of divergent natural selection as a consequence of such variation. Nonetheless, variation among environment may also result in flexible individual responses: phenotypic plasticity, adjustment of the environment, and selection of the environment (incl. habitat choice). Here I discuss and give examples of how these other consequences of variation among environments can generate ecology-driven speciation, including when they are interacting with each other or with natural selection. I propose that such a bottom-up approach to speciation may help us uncover neglected aspects of the speciation process, including when speciation may not occur.

Keywords Ecological speciation · Ecological performance · Local adaptation · Population divergence · Phenotypic plasticity · Habitat selection · Selection of the environment · Adjustment of the environment

11.1 Approaches Toward the Study of Speciation

The process of speciation is beautiful and awe inspiring. But it can also be bewildering and sometimes frustratingly complex. This is because speciation happens via many different routes. It can be the outcome of many different events and mechanisms that may play larger or smaller roles (see other chapters in this book), that may interact with one another in various ways, and all of this dynamically varying in space and time during the course of speciation. To some extent, speciation is

P. Edelaar (✉)
University Pablo de Olavide, Seville, Spain
e-mail: edelaar@upo.es

D. T. Tietze (ed.), *Bird Species*, Fascinating Life Sciences,
https://doi.org/10.1007/978-3-319-91689-7_11

inevitable, and any given species will give rise to other species sooner or later in one way or another, unless it goes extinct before this happens.

In order to try to make some sense of this rather chaotic process, several means by which to categorize the diversity in speciation have been proposed. An important dimension that has been used a lot is that of geography, so classifying to what extent speciation happened in isolated distribution ranges or not (see chapter: Bird distributions and their evolution). Another important dimension to classify speciation events has been that of biological mechanism: natural selection (on ecological characters), sexual selection (on mating characters), and genetic drift (neutral divergence in ecological and/or mating characters) (Schluter 2000; Dieckmann et al. 2004; Nosil 2012). The relative importance of natural selection, sexual selection, or neutral events in driving speciation remains an active area of research.

Nonetheless, the emphasis has shifted more recently to yet another way to subdivide the drivers of speciation, restoring the historical focus on the interaction between the organism and its environment (Sobel et al. 2010). For example, a mating trait may diverge between speciating populations so sexual selection is involved. But is perhaps the divergence in the mating trait still caused by adaptation to the environment, similar to the case for naturally selected ecological traits? An example of this could be when a population lives in an environment with a different predator community, and therefore it evolves a sexual ornament that is rather inconspicuous toward their local predators yet still attractive to potential mates. In that case, the sexual ornament may become so different from other populations that individuals from our population are now unable to attract individuals from other populations as mates and speciation occurred. Hence, even though speciation was driven by sexual selection, the adaptation of the sexual ornament to the local environment is also an underlying component. In order to highlight this aspect, much recent research is addressing whether speciation is driven by ecology (adaptation to the environment) or not.

The importance of ecology for speciation, in birds or otherwise, has been recognized since the beginning of the study of speciation (reviewed in Nosil 2012; Sobel et al. 2010). It is probably involved in most cases of speciation one way or another (Sobel et al. 2010). Nonetheless, the term "ecological speciation" has been developed and proven useful in order to provide specific focus on the role of ecology in speciation. In his book on ecological speciation, Nosil (2012) defines ecological speciation as "the process, by which barriers to gene flow evolve between populations as a result of ecologically based divergent selection between environments." In this chapter, I will broaden this term a bit, for reasons that I will explain below. I will consider speciation to be ecological, if variation among environments induces divergent selective pressures on populations which somehow results into reproductive barriers.

But before going into detail, we first need to consider another aspect of the speciation process, because this is relevant to how I will approach the topic. I already addressed that speciation is complex and potentially driven by many different elements in variable ways. However, it usually also takes relatively much time, so it is exceptional (and unheard of in birds) to witness it from start to finish.

(Independent of the difficulty of defining when it has finished, see chapter on species definitions.) This is why we often resort to dynamical interpretations of static patterns: "What mechanism and situation has likely caused this outcome?" Next, we do this for a number of populations at different stages along the speciation events and extrapolate these insights in order to recreate a continuum: "How does each stage likely combine into a full speciation event?" This creates a methodological/ philosophical weakness in the study of speciation of birds: There has never been an application of the experimental approach to investigate speciation from beginning to end. Nonetheless, by comparing across many different systems and stages along the speciation process, progress has been made. For example, certain routes toward speciation can be judged to be unimportant in birds: Unlike in plants, speciation by hybridization or by genome duplication is rare if not absent in birds.

However, relying on this interpolating comparative approach also comes with risks. If we use patterns to infer processes, then we are using prior knowledge and beliefs to analyze and interpret the data at hand. This may cause us to see what we expect to see and to overlook certain alternative interpretations. Likewise, if we want to study particular elements of the speciation process, then we may pick study systems that a priori seem more suitable for this or that are more likely to include these elements. And this may cause us to confirm our initial knowledge or beliefs, and to overlook systems, in which completely different mechanisms are operating.

In this chapter, I will therefore continue the tradition of subdividing the speciation process into discrete categories, in this case applied only to ecological speciation. But I hope to do it in such a way that it promotes attention for drivers that are typically not taken into account when interpreting patterns or when selecting research questions/hypotheses and study systems, be it birds or other organisms. In a way, by approaching the topic of how birds interact with their environments (which could lead to speciation) in a logical manner and from the bottom up, the relevant research questions arise almost by themselves.

11.2 Four Ways to Increase Ecological Performance: Which May Each Drive Speciation

How well organisms interact with their environment is perhaps the most important aspect of life itself. Even if it is not the most poetic way of viewing things, most of what we see a bird doing in its everyday life ultimately serves only a single purpose: to leave as many descendants (including other relatives) as possible at the end of its life. For this, birds need to be good at what they do, for example, finding food or attracting mates. However, since the environment is not the same everywhere or all the time and there are so many other species with whom to interact (as food items, predators, diseases, etc.), there is no single constant solution to the challenge of being best in leaving descendants. Different environments require different solutions. This is one of the reasons of why new bird species arise and therefore why there are so many of them.

So what types of solutions are open to birds in order to be as good as possible at what they do? One useful way of thinking about this is to realize that when a population with certain characteristics is faced with a certain set of environments (including other species), there are *only four distinct ways by which ecological performance may increase* (Edelaar et al. 2017) and therefore by which reproductive success may increase. The first and most well-known mechanism is *natural selection*, meaning that individuals with unfavorable characteristics have a larger probability to die and a smaller probability to reproduce. (I therefore include sexual selection—competition over fertilizations—under this mechanism, which is not unusual: Sexual selection is often seen as a subset of natural selection.) If the favored characteristics have a heritable basis, then natural selection is a great way to adapt populations to their local environments (e.g., next generations of a bird evolving a denser plumage in colder environments).

However, natural selection cannot improve the performance of locally maladapted individuals directly. Therefore, three other ways to improve performance may evolve to help such maladapted individuals, as individually flexible responses. While these other three ways have evolved due to past natural selection, it seems useful to distinguish them from natural selection, because the way they operate is completely different and their effects (once evolved) are independent of current natural selection. The first one is *phenotypic plasticity*, whereby individuals change their characteristics such that they have a better match to the local environment (e.g., if the environment is too cold, fluff up your feathers or grow a denser plumage). However, given that ecological performance depends on the interaction between individual traits and environmental characteristics, alternatively individuals *may also change their environment such that this provides a better match to their phenotype*. This can be done by actively *adjusting the local environment*, i.e., manipulating and improving aspects of the current environment (e.g., building a warmer nest in a colder environment). Finally, individuals could *select environments* that have a better match to their phenotype (e.g., an individual with a thinner plumage feeds in sunnier places or at lower latitudes or altitudes). As far as I can tell, these are the only four ways by which ecological performance can increase: Either the phenotype or the environment is adjusted and this happens either via selection or via some sort of development (of the phenotype or of the environment), yielding four combinations.

The next question (and the topic of this chapter) is then, if these four distinct ways of improving ecological performance might promote the evolution of new bird species? To address this question, I will discuss them one by one. I will not perform an exhaustive review of all the relevant literature (partly because I believe it to be incomplete and potentially biased); instead, I will just use some examples that serve to illustrate the points made.

11.3 Ecological Speciation Driven by Natural Selection

There are many ways by which natural (incl. sexual) selection can cause ecological speciation. Adapting to specific aspects of one environment (e.g., food resources) may change traits (morphology, physiology, behavior, and life history) of a local population. This can influence how successful individuals are, if they happen to move into the environment of another population: such immigrants may simply have a lower probability to survive. This is known as ecologically based selection against immigrants (Nosil 2012). Alternatively, if they do manage to survive and reproduce with a member of the other population, they may produce offspring which have traits with intermediate values which are not adjusted to any of the available environments. Such mixed-pair offspring may then also have a lower probability to survive or reproduce, i.e., ecologically based selection against hybrids. In either case, there is a certain barrier to the reproductive mixing of populations adapted to different environments, and this can initiate, drive, or complete the speciation process. Such naturally selected differences between populations are likely to evolve when inhabiting different geographic ranges (i.e., in allopatry) but can also evolve within a single range (i.e., in sympatry), e.g., when specializing on different food items.

A good, if not the best, example of the link between divergent natural selection and speciation in birds are Darwin's finches. One group of these, called the ground finches (genus *Geospiza*), mainly feed on seeds. These seeds have strong seed coats that need to be cracked with their bills. Since small bills are not strong enough to crack large seeds, but large bills are too clumsy to handle small seeds, natural selection has been found to act against individuals that have bills which are not suitable for the type of seeds on offer. These finches live on the Galápagos Islands, and, on some islands, certain types of seeds (e.g., larger ones) are more common, and, on other islands, other types (e.g., smaller ones) are more available (Schluter and Grant 1984). Hence, populations of the same species living on different islands sometimes have evolved different bill sizes, adjusted to the local seed types.

But how is this morphological, naturally selected divergence related to speciation? There are several ways how morphological divergence is implicated in preventing homogenization of populations. First, any immigrant with a bill type that doesn't fit the local food resources has a lower survival. But on several islands, there are actually multiple types of seeds available and therefore multiple populations of distinct ground finches that are specialized on each type. How do these populations keep apart? One important aspect is that the finches seem to distinguish individuals as being from one population or the other by their morphology and prefer to mate with individuals that are of the same population (Fig. 11.1, Grant and Grant 2008). This was shown in an experiment where stuffed birds where used to measure the reaction of live individuals. It turned out that the sizes of the body, head, and bill affect whether an individual is recognized as belonging to the

Fig. 11.1 Medium Ground Finches (*G. fortis*) on Daphne Island. (**a**) An offspring of large-billed immigrants that mated with each other and (**b**) a typical smaller-billed resident (adult male) that doesn't mate with large-billed immigrants (Photos by B. R. Grant, adapted from Grant and Grant (2008), by permission from the Royal Society)

same population, both for territorial and mating interactions (Ratcliffe and Grant 1983). Most likely this is because chicks imprint on the morphology of their parents and later in life prefer to interact and mate with individuals that resemble their parents (Grant and Grant 2008). This imprinting behavior can be a strong facilitator of reproductive barriers among populations with different morphologies. Imprinting also occurs for male song: sons copy their father's song and daughters prefer a father-like song (see chapter on Song learning). While songs may diverge little by little among isolated populations on different islands just by chance (neutral evolution), the production of sounds also seems to be related to bill traits. For example, a large and heavy bill is not able to produce a rapid trill (Podos 2001). This means that when natural selection causes a divergence in bill size between populations living on distinct islands, then songs will change as well and, if birds from these populations meet again, they may avoid mating with each other, because their songs have become too different. And finally, if in spite of these barriers they do happen to mate and produce offspring, such offspring will have bills with intermediate sizes for which few seeds are available and almost certainly they will not obtain enough food for survival and reproduction. Hence, natural selection is involved in ecologically based selection against immigrants and hybrids but also in the change of mating traits, both directly (size) and indirectly (song production). Without divergent natural selection, most of these reproductive barriers would not exist, and only slow and accidental song divergence in allopatric populations would remain. Moreover, if allopatric populations that were not exposed to divergent natural selection came into secondary contact, they would be ecologically identical and therefore unlikely to coexist for a long time. Therefore, if it weren't for ecological speciation caused by natural selection, the total number of Darwin's finches on the Galápagos Islands as a whole and on any island in particular would be much reduced. Environmental variation drives biodiversity.

Above I discussed how natural selection may act against immigrants and hybrids directly and how it may influence mating traits. But we should also consider how ecology and selective pressures could affect the *expression* of mating traits. Many mating traits are not fixed (unlike Darwin finch size or song type) but depend on the ability of an individual to develop or maintain these traits. For example, individuals that have problems obtaining enough food may not be able to grow large or colorful plumage ornaments or produce a lot of song or the more difficult song notes. They may be more likely to lose fights when competing over matings or unable to impress their potential mates with behavioral displays or courtship feeding. Basically, their sexually selected traits are of lower quality. Hence, an immigrant or hybrid that is not adapted to the local environment could easily have a lower or even no reproductive success due to low-quality sexual traits and hence low sexual attractiveness, and in that way divergent selection leading to divergence in local ecological performance could further initiate, promote, or finalize the reproductive isolation among populations. This is because sexual selection and natural selection are acting in synergy and the total strength of divergent selection is increased, facilitating speciation (Van Doorn et al. 2009). Here, sexual selection is acting on mating traits, but these mating traits are actually representations of the degree of local adaptation of ecological traits. Indeed, mate preferences and condition-dependent sexual signals can even evolve because of their utility in avoiding mating with locally maladapted individuals, which would otherwise result in poorly adapted offspring, so sexual selection may arise, because it allows for adaptive ecological speciation (Van Doorn et al. 2009).

11.4 Ecological Speciation Driven by Phenotypic Plasticity

When individuals can adaptively change their phenotype, then to some extent, this may prevent natural selection from acting. This is because when locally maladapted individuals improve their ecological performance, natural selection has less variation in performance to act upon. And by reducing natural selection from acting, plasticity reduces the scope for genetic changes to build up between populations. If so, then while plasticity increases local ecological performance, it may decrease the probability of genetic population divergence and perhaps speciation (Price 2006).

Could plasticity also cause speciation? Let's do a thought experiment. Imagine a bird with an allopatric distribution, for example, with populations on a continent and on an isolated island and these populations look and behave very differently. They might therefore upon first sight be considered to be two different species. However, what if we were to translocate eggs from the island to the continent and vice versa and the chicks grew up to completely resemble their foster parent "species"? Would you still call them two species, if their difference completely depended on the environmental conditions during growth? And what if migrants would be so different that they would not succeed in mating? (I'll come back to those questions later.) There is not too much evidence that such a strong degree of plasticity is present in birds (in contrast to, e.g., plants or other sessile organisms), but it is possible that part of the divergence we see, especially for subspecies or locally adapted populations, is actually due to adaptive phenotypic plasticity.

There is actually a very nice example of how phenotypic plasticity can drive speciation: the African indigobirds (genus *Vidua*). These small passerines are brood parasites that lay their eggs in the nests of several estrildid species (waxbills, munias, and allies), but each species of indigobird uses a different host species (Sorenson et al. 2003). Similar to other brood parasites like cuckoos (family Cuculidae) or cowbirds (genus *Molothrus*), chicks are reared by foster parents of another species. This presents an interesting problem: How do the chicks know later on in life with whom to mate? In most brood parasites, species recognition is a hard-wired genetic trait. So whichever host rears it, a Common Cuckoo (*Cuculus canorus*) recognizes another Common Cuckoo for what it is. But in the indigobirds, chicks imprint upon their foster species. This means that males, when adult, will sing songs like their foster species (see also chapter on song learning). And that females, when adult, prefer males that sing songs like their foster species as well as look for nests of this foster species to parasitize. Although it is rare, sometimes females lay eggs in the nest of another host species, perhaps as a mistake or because they can't find a suitable nest of their host species in time. But this rare event has very interesting consequences. Chicks that are reared by this atypical host will now sing like (males) and prefer songs of (females) their alternate host. This means that such wrongly placed chicks will now seek each other out for mating, i.e., mating is assortative for the species of host parent. This then implies that, if eggs are placed in several nests of a new host that wasn't used by indigobirds yet, then all of the resulting indigobird chicks will form a reproductively isolated population or in fact a new species.

Sounds farfetched? Actually, the data supports this scenario for speciation in indigobirds (Sorenson et al. 2003). Imprinting on the host species appears to be a relatively new trait in this genus, and once it evolved, it resulted in more than ten largely sympatric species (Fig. 11.2). The rate of speciation in this group is also much higher than that in their hosts, strongly suggesting that the imprinting is a driver of speciation. In fact, we could see this as an example of adaptive radiation

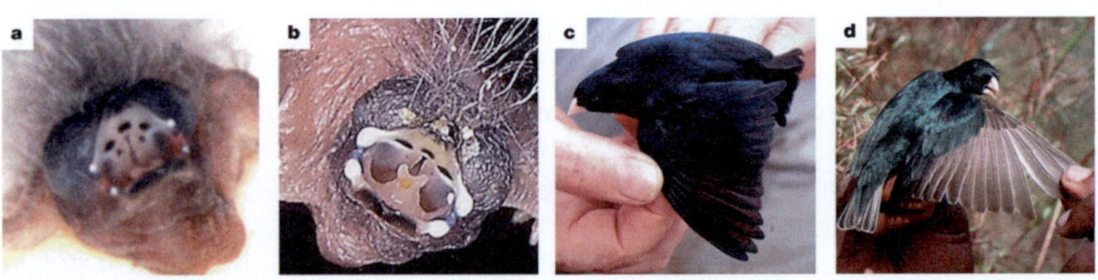

Fig. 11.2 The nestling mouth markings differ between (**a**) Cameroon Indigobird *Vidua camerunensis* and (**b**) Village Indigobird *V. chalybeata* to better mimic the young of their firefinch hosts, Black-bellied Firefinch *Lagonosticta rara* and Red-billed Firefinch *L. senegala*, respectively. Small differences between adult male (**c**) Village Indigobird *V. chalybeata* with dark plumage and wing and (**d**) Jambandu Indigobird *V. raricola* with greenish plumage and pale wing (Figure adapted from Sorenson et al. (2003), by permission from Nature Publishing Group)

Fig. 11.3 Adaptive radiation is the evolution of ecological and phenotypic diversity within a rapidly multiplying lineage (Schluter 2000). Depicted here are the Malagasy vangas (family Vangidae), their phylogenetic relations, and their foraging behavior (black gleaning, blue sallying, red probing). The vangas showed an explosive burst of speciation soon after colonization of Madagascar together with diversification into novel niches and the evolution of extraordinary ecomorphological diversity. However, over time the net rate of speciation declined, probably as the ecological opportunities became saturated. Adaptive radiations such as these show that the ecological conditions have a large effect on the final pattern of speciation. However, without additional information, it is not certain to what extent speciation is actually driven by ecology or whether ecological opportunity simply reduces the rate of extinction of species that evolved by other means. Other famous adaptive radiations in birds include the Darwin's finches from the Galápagos Islands, the honeycreepers from Hawaii, the crossbills from the northern hemisphere, or in fact the entire songbird clade after they managed to leave their area of origin, Australia (Figure from Reddy et al. 2012, by permission of the Royal Society)

(Fig. 11.3, Malagasy birds), where the key innovation of imprinting on the host allowed a rapid diversification to utilize a range of available resources (the different host species). While imprinting changes the phenotypes (male song as well as female host and song preferences) almost instantaneously via plasticity, adaptation via natural selection also does occur, because the indigobird chicks are reared alongside the chicks of the host and need to resemble host chicks in the markings of their mouth in order to be fed (Fig. 11.2). Hence, the assortative mating isn't just a by-product of imprinting: It actually is beneficial for females to prefer a male that sings the song of the same host as the one she was reared by, because it means that both male and female were accepted by the same host and therefore had correct mouth markings and, assuming that this trait is heritable, these markings are then passed on to their offspring and increases the survival of their offspring.

If there exists an unutilized host species and we would experimentally place a large number of indigobird eggs in their nests, we could create a new species of indigobird within the time span of one generation (!). Nonetheless, this new species would be indistinguishable from the original species in morphological traits as well as genetically. Indeed, indigobird species are very similar to each other, since there is little need to look different (Fig. 11.2). Nonetheless, over time neutral genetic differences will build up gradually, and indeed the different indigobirds are genetically differentiated, even when sympatric, confirming the presence of reproductive isolation among them. Do you remember my hypothetical example presented earlier of two allopatric populations looking like different species due to plasticity only, yet that would not interbreed? Did you consider those to be different species? Well, one could argue that here we indeed have a case where reproductive isolation is greatly if not fully determined by phenotypic plasticity, if we consider that the host species is the environment with which the developing indigobird chick interacts in such a way that it changes its phenotype (the male songs and female preferences), yet where the different populations are formally considered to be distinct species (Sorenson et al. 2003).

The different host species are the different environments that cause phenotypic divergence and are the different resources that indigobird species specialize upon and allow them to coexist. So ecology is key in mechanistically driving speciation, even if divergent natural selection plays virtually no role, in support of my broader treatment of ecological speciation. However, not all cases of reproductive isolation between populations living in different environments that are due to plastic traits are support for this. The learning of song or the development of sexual preferences (song, morphology) and habitat preferences or other behaviors via imprinting is quite common among birds (see chapter Song and the example above on ecological speciation due to natural selection in Darwin's finches involving imprinting). Such learned differences in signals, preferences, and behaviors are often important reproductive barriers. But the important distinction to be made is why these learned traits have diverged between populations. In many cases, this may just be coincidental, neutral change, also known as cultural drift (similar to genetic drift), which would therefore not qualify as ecological speciation. In other cases, these learned traits may have been selected to enhance performance in local environmental settings, which would be ecological speciation due to natural selection (but acting on a nongenetically inherited trait). Only when individuals develop traits in response to their local environment which cause reproductive isolation could we think about ecological speciation driven by plasticity. Additional ways how plasticity may initiate speciation are discussed in more detail toward the end of this chapter in the section on feedbacks.

11.5 Ecological Speciation Driven by Adjustment of the Environment

Another way that birds could improve their local performance is by making adjustments to the local environment such that it matches better with their individual phenotypic characteristics. When confronted with ecologically divergent environments, this could result in divergence in the adjustments to the environment, and hypothetically this could lead to reproductive isolation among individuals making these adjustments. For example, if environmental aspects of male territories are adaptively adjusted to the specific phenotype of each male individual, then it is likely that females would prefer to settle and mate with males that have phenotypes similar to their own because of the favorable aspects of those males' territories, thereby creating positive assortative mating for phenotype.

However, as far as I know, it seems that adjusting the local environment is not a driver of speciation in birds. While it is possible that we just haven't recognized this route toward speciation for neglect of the mechanism, there may also be good reasons why this isn't the case (which then still helps us to understand nature). First, birds are among the most mobile of animals, and therefore any mismatch between phenotypes and environments is more likely to be resolved via selection of the environment (see next section) than via adjustment of the environment. The same is to some extent also true for phenotypic plasticity, which is much more pronounced in organisms with low mobility that are forced to deal with the local environment as it is. Second, environmental variation is often spatial, so if the individual that made the change to its local environment moves to another environment (e.g., an immigrant into an allopatric population), it is unable to take the adjustment with it and therefore the adjustment itself cannot play much of a role in reproductive isolation. This is in contrast to plasticity, where the change in the phenotype is much more likely to move and remain as a fixed part of the individual.

Even though birds are not expected to make many adjustments to their local environment, a conspicuous exception is the construction of nests, probably because eggs and altricious chicks do not move as easily as adult birds, yet are in need of a specific environment. However, I know of no data that show that the nest is adapted to (within-species) individual characteristics, e.g., that individuals with a lower capacity to brood chicks will build a warmer nest. Also, I know of no evidence that variation among nests (i.e., divergence in adjusted environments) causes reproductive isolation. This may be because it is unlikely that nest characteristics are going to affect the viability or reproductive success of immigrant individuals or their hybrid offspring. (Note that the decorated, sexually selected nest-like bowers of bowerbirds (family Ptilonorhynchidae) are not ways to increase ecological performance, so these are not relevant for our discussion on ecological speciation, but they do show that adjustment of the environment can matter for speciation.)

In conclusion, there is no evidence to date that adjusting the local environment in order to increase ecological performance plays a role in speciation in birds. It is possible that this is because of the specifics of the biology of birds. But in theory it could work, so keeping this possibility in mind might help us to uncover such cases.

11.6 Ecological Speciation Driven by Selection of the Environment

A particularly efficient and in fact ubiquitous way to increase ecological performance is to select the environment in which a phenotype has to perform: If an individual selects an environment that combines well with the phenotypic characters it already has (and cannot change via plasticity), then this improved phenotype-environment match can increase survival and/or reproduction (Edelaar et al. 2008, 2017), e.g., due to greater food uptake. Selection of the environment can vary from very small temporal and spatial scales to large ones. In the context of speciation, the largest effects are probably going to come from larger scale selection of the environment, since this provides longer-term, more consistent effects. Large-scale selection of the environment is better known as habitat choice, and this is rampant among birds: Any species of bird typically uses only a subset of the habitats available within its reach. For example, a swallow may be hawking for insects over forest canopy or a pond but will never be found searching the forest floor or dive to the bottom of the pond to look for insects. This is not because there are no insects to be found there—other birds do utilize this resource. It is in part because its phenotype (e.g., type of wings, legs, and bill) simply is not suitable to move around and capture prey in those habitats, even if it would try. Likewise, forest skulkers or diving birds also stick to their own habitats, because alternative habitats just do not combine well with their phenotypes. This habitat segregation reduces competition over the same resources and thereby allows for more species to occur in a certain area, i.e., species coexistence. Indeed, habitat is the commonest ecological difference between sympatric sister species in birds (Schluter 2000). Habitat segregation is one reason why structurally more diverse environments have a greater biodiversity, because one species cannot dominate all the resources available. But could habitat choice also drive speciation? For this we need to know when habitat selection evolves and to what extent it may create reproductive barriers.

Let us look at the last topic first. Two aspects of habitat choice seem to be important here: Being in different environments reduces the probability of simply meeting one another, and being in different environments may change how natural selection acts (on which traits and its strength and direction). The power of this first aspect should not be underestimated: If you never meet a potential partner, it will never become an actual partner. This is why allopatry can be such a powerful setting for speciation, because it already avoids the creation of mixed pairs just because of spatial constraints. But such spatial constraints can also be self-imposed, by habitat selection.

A good example of the capacity of habitat choice to drive speciation is found in the Red Crossbill *Loxia curvirostra*, a fascinating songbird occurring in many coniferous forests across the northern hemisphere. Perhaps the most striking aspect about this bird is that the tips of its mandibles are crossing (Fig. 11.4), which allows it to separate the scales of closed cones of coniferous trees and thereby to obtain the now unprotected seeds (Benkman and Lindholm 1991). However, depending on the species or variety of conifer, sizes and shapes of cones come in many ways, cone scales may be thick and woody or thin and more paperlike, and even the sizes of seeds vary (Fig. 11.4). And this means that a single type of bill is simply not able to optimally exploit all this variation in this food resource. For example, a crossbill with a small and thin bill might be very efficient in prying out the small seeds from between the thin and weak scales of a small cone type but might be unable to separate the strong scales of a large cone with thick woody scales. In return, such a strong cone could be handled by a crossbill with a large and thick bill, but such a large bill would be quite clumsy and inefficient when dealing with small cones with thin scales and small seeds (Fig. 11.4). One might therefore expect that crossbills specialize even more, to feed on the seeds of only specific types of cones. And indeed, in those times of the year when food is scarce and foraging efficiency matters most, this is what happens (Benkman 1993). In North America, there are a number of conifers that hold their seeds in closed cones during the time of the year when food is scarce, and each of these conifers is used by crossbills which not only have the right bill size and shape to deal with these cones but also the appropriate internal bill structure to remove the cover from the seeds, given the size of the seeds (Fig. 11.4). This use of a particular conifer is a kind of selection of the environment, matching the type of cone to interact with the bill a crossbill has and cannot change (there is no plasticity in most bill traits). Of course, the resulting degree of phenotype-environment (bill-cone and seed) match will still affect food intake and therefore survival, i.e., natural selection is still expected to act. Indeed, in a resident population of crossbills where only a single type of conifer was available, it was found that natural selection acted on bill traits: Birds with bill traits that made them to obtain their food slower (known from observations in captivity) also had a lower survival rate in the wild (Benkman 2003).

This means that each type of cone will select for a specific type of crossbill that is able to feed on it. And this would easily lead to divergence among crossbills, if these conifers occurred in an allopatric setting, just as in the example with Darwin's finches on different islands. However, this is not quite the case—several different suitable conifers can be found in close vicinity of each other, including on the same mountain range (north and south slopes or at different altitudes) or even in large tracts of mixed forest. Moreover, since many conifers produce poor cone crops over large areas in some years, crossbills are highly mobile and may wander hundreds and even thousands of kilometers in search of suitable feeding areas, which brings them into contact with many different conifers. This means that the scope for mixing among diverging crossbill populations is very large. In such a setting, it seems unlikely that strong natural selection on bill traits could create divergence and speciation by itself. However, it would be naïve to assume that a crossbill would choose a type of conifer to feed upon randomly and that subsequently food intake affects survival (i.e., that natural selection acts). Of course we should consider that a crossbill can assess its food intake upon a particular cone type and, if it scores low, that the crossbill will move on in search of a better one. Several studies have shown that this occurs at a small scale: Predation on seeds of individual trees by crossbills is selective, and trees with "easy" cones are attacked most heavily (Benkman and Parchman 2009). However, it almost certainly also occurs at a large scale: The different specialist crossbill populations are typically found in association with only a subset of conifers, including the type they specialize upon (Benkman 1993). And it must be very easy, and highly rewarding, for a crossbill in search of food to not stop and sample at every tree it encounters, but to recognize the species of conifer from kilometers away when flying above the forest canopy and to skip over patches of the type of conifers that it knows to be unsuitable from previous experience.

The best example of how such habitat choice drives speciation probably comes from a small area in the south of Idaho (Cassia County, USA). Here, on two nearby mountain tops, a resident population of crossbills is locally adapted to the local conifer: The crossbills have a larger bill in order to deal with the larger and more strongly defended cones (Fig. 11.4). However, the two mountains are part of the Rocky Mountains complex which also holds various other populations of crossbills (Fig. 11.4). And these other types of crossbills, when in search of food, do arrive in the area regularly and sometimes in large numbers. Therefore, mixed resident-immigrant breeding pairs could easily arise and erase the adaptive divergence in crossbills. Nonetheless, these mixed pairs are extremely rare, because when the breeding season starts, the immigrants have disappeared. Where have they gone? Have they all died? Indeed, it is known that the immigrant crossbills have bills that are not very suitable for feeding on the local cones (Fig. 11.4). However, given what I wrote above about choosing certain conifers and given the distances that these immigrant crossbills clearly can move, it seems likely that instead of accepting this low food intake rate as a given fact and let natural selection act on them (i.e., most immigrants will not reproduce or even starve to death), they instead choose to move on again and emigrate to other areas: selection of the environment acts. Although actual movements are very hard to quantify in this small bird, this inference is supported by the observation that, if the disappearance of immigrants was only due to selective mortality, it would be due to one of the strongest natural selection pressures measured to date (Benkman 2017). Hence, the more likely explanation, which corresponds with all we know about the biology and behavior of crossbills mentioned above, is that large-scale habitat choice is the main factor reducing the inflow of immigrant crossbills into this locally adapted crossbill population, as a kind of self-imposed allopatric setting (Edelaar and Bolnick 2012). The resulting reproductive isolation is so strong that the crossbills are not only morphologically diverged but also vocally and genetically distinct, and therefore this population has recently been described and accepted as a new species, the Cassia Crossbill *Loxia sinesciuris*.

In crossbills, large-scale selection of the environment therefore seems to be a strong contributor to speciation (see also Edelaar et al. 2012). However, selection of the environment can operate in multiple ways. For example, for the indigobird were I discussed plasticity as a driver for speciation, selection of the environment is also involved: Each species of indigobird chooses the nest of a different estrildid species for its chicks to grow up into, and this can be seen as a matching of the environment (the host type during rearing) to the phenotype (the pattern of mouth markings in chicks). Another way by which selection of the environment can drive speciation is by choosing not where but *when* to breed. Even when individuals occur and breed in

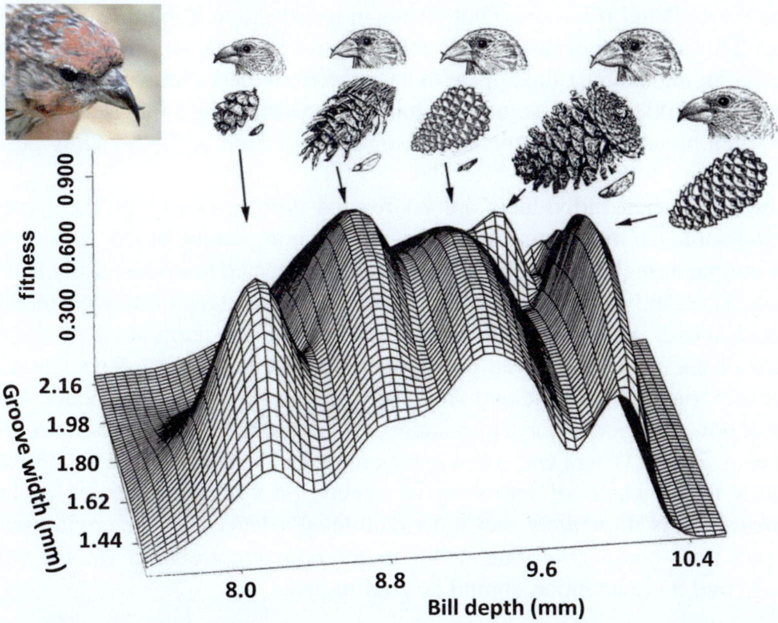

Fig. 11.4 The top-left panel shows a young male Red Crossbill *Loxia curvirostra* with his crossing bill tips—an apparent maldeformation which in fact greatly increases his ability to separate the scales of conifer cones, giving access to its seeds. The main panel shows for five different types of conifers (cones and seeds depicted) how only specific combinations of bill traits (bill depth and groove width) give a high feeding success (here expressed as relative fitness). A deeper bill increases biting force but requires more energy and reduces handling ability of small seeds, whereas the width of the husking groove inside the bill must match with the size of the seed, in order to facilitate removal of the protective seed coat. Each type of conifer is therefore utilized by a specialized crossbill population which differs in bill traits and overall size from other crossbill populations (note the marked increase in size from left to right and the more subtle variation in bill shape). At times, these different crossbills can be found in the same areas (e.g., the Rocky Mountains), but the largest crossbill on the right is restricted to just two mountain tops in the south of Idaho, because only there the specific conifer it specializes upon occurs (Figure adapted from Benkman 2003, by permission from John Wiley and Sons)

the same area, if they are reproductively active at different times, then mating will be greatly reduced. A good example of this is storm petrels of genus *Oceanodroma*. It was known that on a few islands on the Madeiran archipelago in the Atlantic, storm petrels bred twice a year: once during the hot season and once during the cool season. Only recently (Friesen et al. 2007) was it determined that these are actually not the same individuals: One petrel always breeds at the same time of the year and its chicks also breed at that time. In fact, it turned out that the timing is so conservative, that morphological and genetic differences are apparent between the two populations, and that they behave as two reproductively isolated species, resulting in the description of Monteiro's Storm Petrel (*O. monteiroi*). This new species overlaps fully in geographic distribution with the previously known Band-rumped Storm Petrel (*O. castro*) but differs in its selection of the time of year when to breed. This selection of time is probably not a neutral trait, since the species differ in size (including the bill) and appear to have different diets even when co-occurring (Bolton et al. 2008), so presumably Monteiro's Storm Petrel has a higher performance when breeding during the hot season and selection of the breeding period is likely adaptive.

In general, when individuals are confronted with a variety of environments, preferring some environments over others can give important benefits. Selection of the environment (incl. habitat choice) is therefore expected to evolve easily and to be common. There are various ways by which this can lead to reproductive isolation, as illustrated above. Yet it is insufficiently recognized that individual flexibility in selection of the environment can be an independent contributor to speciation. This is especially true for allopatric populations living in different environments, where it is unclear how members of each population would respond to alternate environments (Sobel et al. 2010). Where and when to become reproductively active are among the first steps in the chain of reproductive events, prior to mate choice or hybrid performance. It is thus likely that individual-level selection of the environment in space and time could contribute a lot to reproductive isolation (in the past or currently) and more attention should be paid to this.

11.7 Feedbacks Between Plasticity, Adjusting the Environment, Selection of the Environment, and Natural Selection

In the introduction, I already highlighted the often complex nature of speciation and that it can be the outcome of many different events and mechanisms that may interact with one another in various ways in space and time. Indeed, while above I discussed how the three different types of individual flexible responses to environmental variation could directly affect reproductive isolation, these responses often do not operate independently of each other. Nor do they operate independently of natural selection. On the one hand, they have evolved due to prior natural selection, which favored the evolution of solutions for locally maladapted individuals which allowed them to increase their ecological performance and thereby avoid natural selection. But on the other hand, these individual responses influence the continued operation of natural selection.

I already mentioned that plasticity can reduce natural selection, since maladaptation is reduced by plasticity. The same is of course also true for the other two individual responses (adjustment of the environment and selection of the environment). This individual flexibility to enhance ecological performance could reduce or even remove natural selection and thereby prevent adaptive genetic divergence of populations, which could reduce the probability of speciation. But this flexibility should also increase the probability to survive in a new environment. It may therefore increase the probability that a population manages to establish and survive when exposed to a new environment, e.g., after a natural range expansion, when accidentally blown to an oceanic island, when introduced by humans to a new geographical range (biological invasion), after dramatic change in the landscape (e.g., natural disasters, urbanization), and so on. In such populations that are exposed to a different environment and initially survive due to flexible individual responses, subsequent heritable changes due to natural selection could still occur. This then could cause populations to further diverge and potentially speciate. If so, then one could argue that it was the individual ability to change phenotypes or environments which allowed populations to speciate and therefore that this has been a key element of the final outcome.

These subsequent heritable changes that increase reproductive isolation could act on the same traits as the individual flexible response. For example, plasticity or habitat choice might become genetically fixed (genetic accommodation) and even evolve to values beyond those reached by individual flexibility. Upon secondary contact, such large and fixed differences between population might then be sufficient to keep them reproductively isolated, e.g., because one population occurs in the tree tops of coniferous forest and the other in the undergrowth of deciduous forest. But the heritable changes may also involve other traits, whose selective regimes have changed, because the individual flexible responses have changed the selective settings. For example, a population that shifted to use the caterpillars found in the tree tops of coniferous forest may subsequently evolve a range of additional traits that further increase ecological performance: shorter legs with stronger feet, long bills to probe between needles, improved physiology to digest caterpillars and to neutralize chemicals produced by conifers, breeding later in the year when caterpillars are most abundant, and so on. This is not only expected based on logic: Of course we have many examples of coadapted traits in birds. For example, woodpeckers have strong feet with curved nails to cling to bark, strong tail feathers to lean against trunks, a pointy bill to hammer into wood, a specialized skull that absorbs the hard blows which would cause brain damage in any other bird, an extra-long tongue with a harpoon-like tip to extract larvae from their tunnels and they climb along tree trunks and thick branches, hammer dead or damaged wood, hammer nesting cavities inside tree trunks, and so on. These morphological and behavioral traits work well together but would work poorly when combined with traits of other species (e.g., duck-like feet on a woodpecker). Therefore, any individual flexible response in a trait, be it via plasticity or environmental changes, could trigger the evolutionary change in other morphological, physiological or behavioral traits that on the one hand further increase ecological performance but on the other hand might also increase reproductive isolation from other populations. Indeed, the more traits that are different between populations, the greater the probability of a single or combined effect of traits on preventing hybridization (Nosil 2012).

In that context, and given their importance for reproductive isolation, it is worth stressing that mating traits might also secondarily change in response to the change in a first trait. This is because mating traits by definition always involve the interaction between at least two individuals and therefore they do not operate in isolation: There is a signaler and a receiver. We could divide this interaction into three steps: The expression (emission) of mating traits depends on the phenotype of the signaler, the transmission of mating traits may depend on the environment of the signaler and the receiver, and the reception depends again on the phenotype of the receiver. Therefore any change in phenotype or environment could influence the success of a mating trait. Assuming that mating traits are generally initially successful, these changes probably cause a decrease in reproductive success, and therefore sexual selection will favor traits that respond appropriately to these changes in phenotype or environment, independent of whether they were caused by plasticity, adjustment of the environment, selection of the environment, or natural selection.

There are several examples of how mating traits are influenced by phenotypic or environmental variation, which could cause reproductive isolation. Starting from the emitting individual's perspective, I already discussed how bill size might affect the ability of Darwin's finches to produce certain vocalizations. This is also found in other bird groups, e.g., antbirds (family Thamnophilidae; Seddon 2005) and woodcreepers (in family Furnariidae; Palacios and Tubaro 2000). The same is true for body size: Just like humans or musical instruments, smaller birds generally produce higher-pitched vocalizations (Seddon 2005). So two populations that evolved different body or bill sizes may subsequently produce songs that are so different that they are not recognized anymore by the other population. Such changes in song might also evolve due to differences in environments, affecting transmission. Sound travels differently through open space than through densely vegetated areas, e.g., low-frequency sounds travel best through dense foliage, and songs containing short notes repeated at longer intervals are less affected by reverberations caused by leaves. In order to have a song reach the intended audience, birds may therefore change the volume of their song, the pitch of their notes, the degree of modulation (pitch variation) within each note, the spacing between notes, and so on depending on their habitat. Therefore, a shift in habitat choice could result in a shift in song characteristics, as, for example, observed in antbirds living at different heights above the ground in the forest (Seddon 2005) or in a set of bird species from the Amazon that are specialized in bamboo patches (Tobias et al. 2010). This change in song can be a flexible response at the individual level. For example, birds that live in noisy places like cities are known to change their songs such that they can still be heard by others. But it can also be a trait that is fixed in the population as in the antbirds (Seddon 2005). A shift in habitat can also be important for plumage characteristics (or any sexually selected trait). Sexual selection on plumage traits normally operates on the showy aspect: larger color patches, brighter colors, larger feathers, and so on. However, survival selection often operates against the expression of sexual ornaments, because they may attract predators or because they are very costly to produce or maintain, resulting in sexual ornamentation that is a balance between the two selective forces. Hence, depending on the color of the background and the amount and color of ambient light (e.g., open sky versus filtered light reaching the forest floor), selection may favor more or larger patches, at different positions, of different colors, and so on. In leaf warblers (Phylloscopidae), species that live in darker forests generally have more pale patches in their plumage (Marchetti 1993). And a comparison across 20 pairs of closely related bird species found that species from more closed habitats (forests, etc.) generally had more orange and red colors in their plumage, because this is more visible in those habitats (McNaught and Owens 2002). Depending on the abundance of predators, food, or crucial nutrients, sexual ornaments may also change. For example, female House Finches *Haemorhous mexicanus* prefer intense red patches in males, but males need to be able to extract enough carotenoids from their food to produce such red feathers. Hence, when the environment is poorer in carotenoids, House Finch males concentrate their carotenoids into smaller intense red patches, instead of producing large but pale red patches. Environmental variation could even result in a shift in relative importance between plumage, vocalizations, or behavioral displays as sexual signals, which could cause diverged populations to not mate with each other anymore. For example, in finches, the more colorful species tend to sing less elaborate (Badyaev et al. 2002), and in general, nocturnal owls are very vocal but do not make much use of plumage or display behavior to attract the other sex or repel same-sex individuals. Finally, the traits of the receiver may also affect the evolution of sexual signals. For example, some families of birds (e.g., songbirds, parrots, and gulls) have evolved the capacity to see ultraviolet radiation (Horth 2007). Only in these groups would we expect to find sexual signals that involve the ability to reflect or absorb UV.

Overall then, the presence of feedbacks, especially toward natural selection, can magnify the impacts of initial flexible individual responses which increase ecological performance on population divergence and speciation. However, these flexible responses are expected to occur mainly initially during population divergence as they already operate at the individual level. If subsequently feedback into natural selection increases population divergence in the same and other (including mating) traits, then these initial changes due to flexible responses most likely leave little-to-no signature after the process. Hence, when we look at pairs of birds that have completed the speciation process, we may see large and fixed genetic difference between them (at the phenotypic level or at the genomic level) which strongly influence reproductive isolation. However, these final differences may cause us to overlook the initial differences that initiated the speciation process, and in some cases, these initial steps may simply not be visible anymore, e.g., due to genetic accommodation (i.e., the pattern misrepresents the process, as I suggested in the introduction). Hence, the capacity of individuals to flexibly and adaptively change their appearance or their local environment in response to local ecological settings may play a much more important role in the divergence and speciation of birds than we currently realize. However, it remains a methodological challenge to show this convincingly. But perhaps we should also be more receptive to the evidence when it presents itself. To a large extent, it is not sufficiently recognized that and how these individual flexible responses can drive speciation, so we may not even think about these alternative mechanisms when we look at a pattern (i.e., data from a certain study). By drawing attention to the existence of these flexible individual response to improve ecological performance and their distinction from one another and from natural selection and by presenting real and hypothetical examples of how they may drive speciation, I hope that future studies will be in a better position to interpret patterns or to test specifically for their operation and their importance in the speciation process. Hopefully this makes the process of speciation less bewildering and complex but no less beautiful and awe inspiring.

References

Badyaev AV, Hill GE, Weckworth BV (2002) Species divergence in sexually selected traits: increase in song elaboration is related to decrease in plumage ornamentation in Finches. Evolution 56:412–419

Benkman CW (1993) Adaptation to single resources and the evolution of Crossbill (*Loxia*) diversity. Ecol Monogr 63:305–325

Benkman CW (2003) Divergent selection drives the adaptive radiation of Crossbills. Evolution 57:1176–1181

Benkman CW (2017) Matching habitat choice in nomadic Crossbills appears most pronounced when food is most limiting. Evolution 71:778–785

Benkman CW, Lindholm AK (1991) An experimental analysis of the advantages and evolution of a morphological novelty. Nature 349:519–520

Benkman CW, Parchman TL (2009) Coevolution between Crossbills and black pine: the importance of competitors, forest area and resource stability. J Evol Biol 22:942–953

Bolton M, Smith AL, Gomez-Diaz E, Friesen VL, Medeiros R, Bried J, Roscales JL, Furness RW (2008) Monteiro's Storm Petrel *Oceanodroma monteiroi*: a new species from the Azores. Ibis 150:717–727

Dieckmann U, Doebeli M, Metz JAJ, Tautz D (2004) Adaptive speciation. Cambridge University Press, Cambridge

Edelaar P, Bolnick DI (2012) Non-random gene flow: an underappreciated force in ecology and evolution. Trends Ecol Evol 27:659–665

Edelaar P, Siepielski AM, Clobert J (2008) Matching habitat choice causes directed gene flow: a neglected dimension in evolution and ecology. Evolution 62:2462–2472

Edelaar P, Alonso D, Lagerveld S, Senar JC, Björklund M (2012) Population differentiation and restricted gene flow in Spanish Crossbills: not isolation-by-distance but isolation-by-ecology. J Evol Biol 25:417–430

Edelaar P, Jovani R, Gomez-Mestre I (2017) Should I change or should I go? Phenotypic plasticity and matching habitat choice in the adaptation to environmental heterogeneity. Am Nat 190:506–520

Friesen VL, Smith AL, Gomez-Diaz E, Bolton M, Furness RW, Gonzalez-Solís J, Monteiro LR (2007) Sympatric speciation by allochrony in a seabird. PNAS 104(47):18589–18594

Grant PR, Grant BR (2008) Pedigrees, assortative mating and speciation in Darwin's Finches. Proc R Soc B 275(1635):661–668

Horth L (2007) Sensory genes and mate choice: evidence that duplications, mutations, and adaptive evolution alter variation in mating cue genes and their receptors. Genomics 90:159–175

Marchetti K (1993) Dark habitats and bright birds illustrate the role of the environment in species divergence. Nature 362:149–152

McNaught MK, Owens IPF (2002) Interspecific variation in plumage colour among birds: species recognition or light environment? J Evol Biol 15:505–514

Nosil P (2012) Ecological speciation. Oxford University Press, Oxford

Palacios MG, Tubaro PL (2000) Does beak size affect acoustic frequencies in woodcreepers? Condor 102:553–560
Podos J (2001) Correlated evolution of morphology and vocal signal structure in Darwin's Finches. Nature 409:185–188
Price TD (2006) Phenotypic plasticity, sexual selection and the evolution of colour patterns. J Exp Biol 209:2368–2376
Ratcliffe LM, Grant PR (1983) Species recognition by Darwin's Finches (*Geospiza*, Gould). I. Discrimination by morphological cues. Anim Behav 31:1139–1153
Reddy S, Driskell A, Rabosky DL, Hackett SJ, Schulenberg TS (2012) Diversification and the adaptive radiation of the vangas of Madagascar. Proc R Soc B 279:2062–2071
Schluter D (2000) The ecology of adaptive radiation. Oxford University Press, Oxford
Schluter D, Grant PR (1984) Determinants of morphological patterns in communities of Darwin's Finches. Am Nat 123:175–196
Seddon N (2005) Ecological adaptation and species recognition drives vocal evolution in Neotropical suboscine birds. Evolution 59:200–215
Sobel JM et al (2010) The biology of speciation. Evolution 64:295–315
Sorenson MD, Sefc KM, Payne RB (2003) Speciation by host switch in brood parasitic Indigobirds. Nature 424:928–931
Tobias JA, Aben J, Brumfield RT, Derryberry EP, Halfwerk W, Slabbekoorn H, Seddon N (2010) Song divergence by sensory drive in Amazonian birds. Evolution 64:2820–2839
Van Doorn GS, Edelaar P, Weissing FJ (2009) On the origin of species by natural and sexual selection. Science 326:1704–1707

Chapter 12
Climate Change Impacts on Bird Species

Sven Trautmann

Abstract Climate change has been shown to be an increasingly important driver of changes in the distribution, abundance and life history of bird species, causing changes of biodiversity and community compositions, e.g., measured by climate change indicators. Especially changes in the distributional ranges of species have been demonstrated in many cases, with some species being driven towards their altitudinal or latitudinal limits leading to shrinking populations. But also changes of phenology, genetics, and population sizes have been proposed to be caused by climate change. Although there is recent evidence only for regional climate-driven extinction events for birds, climate change can be considered among the major risk factors that might lead to the complete extinction of bird species. Together with (and sometimes contradicting) land-use change and demographic effects, climate change is shown to be a risk factor especially for cold-dwelling, restricted-range, and slowly adapting species. However, indirectly, by means of climate change mitigation measures modifying land-use patterns, also widespread generalist species are becoming increasingly threatened. Despite demonstrated niche conservatism in several cases, also adaptation of species to climate change is taking place, changing their multidimensional niche spaces. Birds are not sufficiently tracking climate change, nor do northern and southern range limits shift at the same pace. Thus, altered migration phenology and distance have been proven, often proposed to result from phenotypic adaptation. In a few cases, however, genetic predisposition and/or microevolution appear to shape the impact climate change has on bird species. There is broad consensus that climate change will lead to a reshuffling of communities and altered selection pressures both within and among species. The extent of the projected changes largely differs among studies owing, e.g., to the uncertainty inherent in climate change predictions. Nevertheless, niche space is expected to undergo future changes for many species, which, in turn, can have beneficial or detrimental effects on the survival of the respective species or communities and may

S. Trautmann (✉)
Dachverband Deutscher Avifaunisten e.V., Münster, Germany
e-mail: trautmann@dda-web.de

D. T. Tietze (ed.), *Bird Species*, Fascinating Life Sciences,
https://doi.org/10.1007/978-3-319-91689-7_12

also lead to changes in the effectiveness of bird protection which in turn should be adjusted to climate change impacts.

Keywords Climate change · Climatic niche · Climate change indicator · Distribution · Phenology · Adaptation · Microevolution · Species richness · Leading/trailing range margins · Demographic changes · Warm-dwelling/cold-dwelling species · Niche conservatism · Stepping-stones

12.1 Introduction

Past climatic changes have regularly triggered large-scale alterations of plant and animal communities. Single large-scale extinction events occurred at least five times in Earth's history, the sixth mass extinction being suspected to be under way (Barnosky et al. 2011).

Non-analogue climatic regimes and communities stimulated both the extinction of existing and the evolution of new species. For birds, such stimuli comprise the Cretaceous-Paleogene mass extinction (Longrich et al. 2011) but also ice ages and the creation of new weather systems like the monsoon in Southeast Asia (Weir and Schluter 2004; Koch and Barnosky 2006; Päckert et al. 2012).

Since the twentieth century, life on Earth has been faced with an anthropogenic increase in Earth surface temperature of 0.74 °C, with mean temperature increases exceeding over two degrees in certain areas of the temperate and arctic climate zones (Pacifici et al. 2015; Houghton et al. 2001). Future climatic projections predict a continued rise in Earth surface temperature by 0.3–4.8 °C (relative to 1986–2005 conditions) until the end of the twenty-first century (IPCC 2013).

Ongoing climatic change markedly differs from preceding changes, because the latter supposedly did not occur at such small temporal scales and were not accompanied by changes in land use which act—at least partly—independently of them.

Accepting climate change as a scientific fact, this inevitably leads us to the question of how those unprecedented changes might influence the future of bird species.

12.2 Birds and Climate Change: Is There an Impact?

When focusing on climate change impacts, it becomes apparent that a vast amount of knowledge already exists showing various types and magnitudes of impacts on bird species.

We especially find phenologic, genetic, population level and biogeographic changes.

Regarding **phenology**, changes in the arrival (and partly also departure) timing of migratory species, in some cases diminishing migratory activity, alteration of migratory distances, and changes in breeding behavior related to earlier arrival and timing of breeding as well as increased length of the breeding season and numbers of broods have been found, but no strong effects on clutch sizes could be demonstrated (Doswald et al. 2009; Visser et al. 2009; Both and Visser 2001; Smallegange et al. 2010; Both et al. 2004; Møller et al. 2010; Parmesan and Yohe 2003; Winkler et al. 2002).

For most of those effects, it remains unclear whether phenotypical plasticity or rather **genetic changes** cause them (e.g., Gienapp et al. 2007). In some cases, it has been hypothesized that climate change might have led to altered selection pressures, e.g., resulting in changed coloration (Karell et al. 2011), smaller body sizes (Salewski et al. 2010), and the change of migration routes or migratory activity within few generations (Bradshaw and Holzapfel 2006; Plummer et al. 2015). Especially, selection for dispersal and reproductive traits is thought to increase toward range edges (Oliver and Morecroft 2014).

Population changes reveal themselves as a result of altered biotic and abiotic conditions and can be more or less pronounced, depending on the adaptability of species. Changes of species abundance have been increasingly related to climate change effects (Jiguet et al. 2010; Renwick et al. 2012; Tayleur et al. 2015; Huang et al. 2017).

Biogeographic changes are caused by the aforementioned factors, first in the form of spatial abundance shifts, local extinction, and colonization events and then, ultimately, as areal shifts, resulting in either gain or loss (in Europe at the northern and southern range margins, respectively) of species' ranges (e.g., Thomas and Lennon 1999; Hitch and Leberg 2007). In Europe, predominantly range extensions could be demonstrated, whereas range loss is much less commonly reported. Yet, climate change-driven global or continental extinction events for bird species remain to be unambiguously proven, although together with weather extremes it already seems to be a primary or secondary threat of extinction for around 10% of bird taxa (Szabo et al. 2012) and local extinctions have already been demonstrated in both northern temperate and tropical parts of the Americas (Wiens 2016).

Within the last two decades, range shifts have increasingly been debated as a proof of measurable climate change impact on bird species and communities.

In northern temperate regions, mostly, shifts at the northern or leading range margins toward the northeast or northwest have been reported (e.g., Lehikoinen and Virkkala 2016) and seem to be constricted by winter temperatures (Parmesan 2006), whereas evidence for corresponding shifts of the southern or trailing range margins remains scarcer (e.g., Gillings et al. 2015; Tayleur et al. 2016). Furthermore, range shifts in the tropics (Walther et al. 2002; Şekercioğlu et al. 2012) or southern temperate areas (Hughes 2003) have less frequently been reported.

But even for proven shifts, not in all cases the directions of the shifts fit with the expectations (e.g., western or eastern range shifts, shifts downslope or toward tropical latitudes, (Bateman et al. 2016; Huang et al. 2017; Lenoir and Svenning 2014; Gibson-Reinemer and Rahel 2015), and sometimes alleged shifts are disproven (Taheri et al. 2016). Also, altitudinal shifts caused by climate change are not unambiguously supported by the current literature (Tingley et al. 2012).

In the light of such inconsistent results, sometimes it has even been disputed, whether apparent range shifts are actually attributable to climate change or rather to land-use or demographic changes (Beale et al. 2008, 2009; Peterson et al. 2009; Bateman et al. 2016).

Thus, although climate change and its impacts are widely accepted as scientific facts, it should neither be considered as an isolated driver of changes in bird populations or communities nor can its effects always be distinguished from other drivers such as land use.

12.2.1 Climate Change Indicators

Despite all those blurred views, however, we still find a generalizable effect of climate change on a large spatial and temporal scale, especially at the community level (Araújo et al. 2009). Therefore, climate change indicators have been invented in order to measure effects that might be masked by the large spectrum of reactions of single species or populations.

As some species are more, others less dependent on climate, this individual relationship has to be captured in one common framework in order to reveal generalizable effects. Several attempts have been made to do so.

One of them is to examine changes of species richness as a measure of the overall climatic influence on birds (e.g., H-Acevedo and Currie 2003; Lemoine et al. 2007).

More frequently, the climatic associations of single species were estimated using species distribution modeling (see Chap. 9) and then aggregated to community level, e.g., resulting in the **climate impact indicator** (Gregory et al. 2009). This approach was recently modified to allow for different reactions of the same species to climate change, depending on the region in which the species occurs (Stephens et al. 2016). Changes in the local climatic suitability for bird species can be calculated using regression models of occurrence/abundance of species in a given area at a time and the climatic circumstances in the same area and time period. Those trends in climatic suitability can then be linked to the climatic changes in the same time period and area.

Apart from changes of suitability, a simpler measure of the climatic dependence or niche of a species can be used by calculating the average temperature throughout the range of their occurrence. This value can be used to distinguish warm-dwelling and cold-dwelling species. Furthermore, weighing the abundance (relative to other species) of a species in a given year by its temperature value and averaging over the whole community of species, a **community temperature index** (DeVictor et al. 2008, cf. Fig. 12.1) can be calculated, representing the "fever curve" of the bird community. This indicator is not just sensitive to changes in the distribution of species but also to changes in abundance and thus unveils the changes happening within ranges rather than focusing only on range margins.

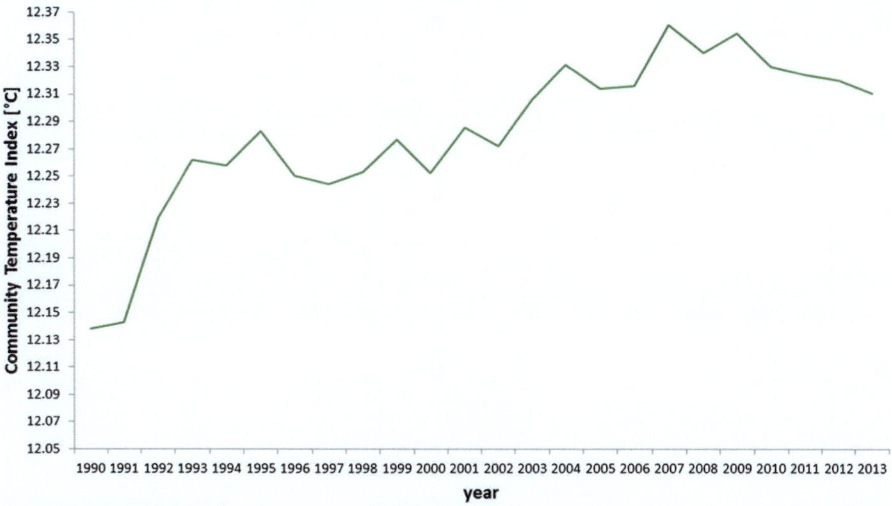

Fig. 12.1 The community temperature index in Germany is the "fever curve" of the bird community. It contains 88 common bird species and shows the relative changes in the proportion of warm- vs. cold-dwelling species by averaging the population trend indices, weighed by the temperature niche of the single species within their European distributional range. Changes do not seem to be tremendous between 1990 and 2013 but reflect very well the changes in mean April–July temperatures in Germany with increases at the beginning of the 1990s and 2000s and a decrease since 2009

All those approaches have added tremendously to our perception of climate change effects on birds. Thereby, it became known that bird species richness is changing, especially due to the immigration and increase of new species but regionally also due to the decline and/or loss of species. Additionally, proposed range shifts reflected very well changes among potential "winners" and "losers" of climate change on continental levels or even across continents (Gregory et al. 2009; Stephens et al. 2016). But also, smaller-scale changes in the climatic impact could be tracked, and the reshuffling of communities was demonstrated (DeVictor et al. 2008; Princé and Zuckerberg 2015).

Nevertheless, still some doubts remained about whether community-based indicators can disentangle climate and land-use change effects. Some effects that seem to be caused by climate may in fact be due to land-use changes acting only on one part of a community that has distinct properties of its climatic niche like, e.g., changes in temperate forest birds, which tend to be rather cold-dwelling (Clavero et al. 2011). On the other hand, differential effects (e.g., regarding sensitivity to land-use change) on subsets of communities might even cover climate change effects (Reif et al. 2010).

Thus, more recent community-based indicators aim to correct for effects of habitat specialization of communities to rule out such effects (Kampichler et al. 2012).

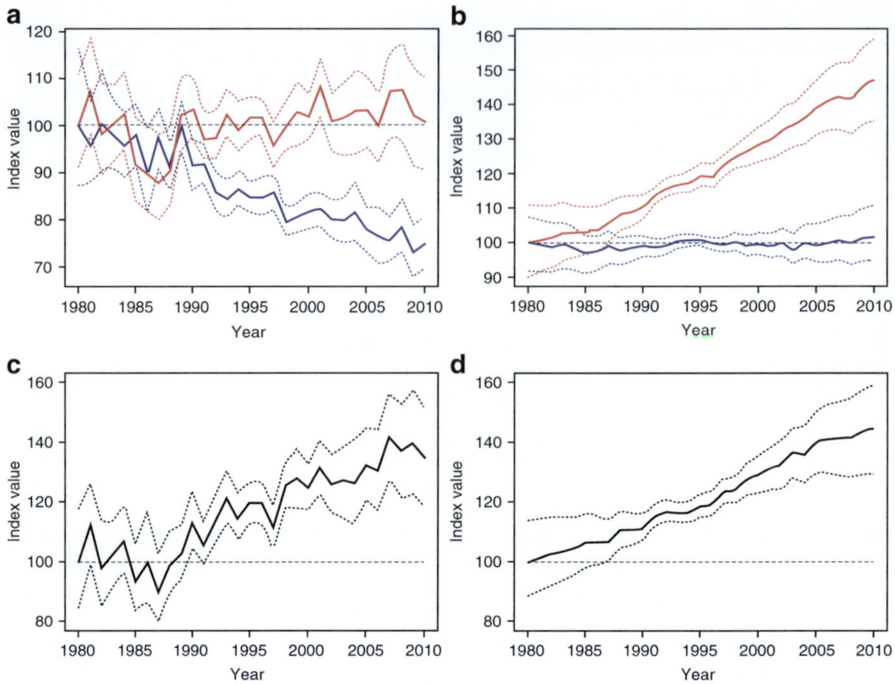

Fig. 12.2 The climate impact indicator (CII) connects changes in the climatic suitability for bird species in defined areas with their population changes among 1980 and 2010. For Europe, in (**a**) the average population changes are shown for species for whom climatic suitability increased (CST+) in red and for species for whom climatic suitability decreased (CST−) in blue (red and blue dashed lines depict 90% confidence intervals; the dashed horizontal line depicts the expectation if climatic suitability played no role). In (**b**), the same is shown for the USA. Annual values of the ratio of the CST+ index to the CST− index form the climate impact indicator (CII), which is shown for Europe in (**c**) and for the USA in (**d**). On both continents, since 1980, a consistently increasing climatic impact for bird populations can be demonstrated. Figure taken from Stephens et al. (2016)

Ultimately, results of species- or community-based indicator systems can provide insights into the climate change impacts which can yield either positive or negative consequences for birds (Fig. 12.2).

12.3 What Are the Consequences of Climate Change for Birds?

Climate change-induced alterations of bird phenology and distribution have been identified as important threats for single bird species. Less prominently, species might also profit from climatic change. Thus, a number of species in the climate impact indicator have been labeled as potential "winners" and others as "losers" of climate change. However, in the new edition of the CII (Stephens et al. 2016), both are somewhat intermingled, as the same species might in one part of its range be subject to positive yet in another part to negative climatic impacts. And in a number of species, no unambiguous climate change associations are apparent at all (e.g., Huang et al. 2017).

Positive aspects of climate change comprise increasing abundance and/or ranges for sedentary, general, and warm-dwelling species (Kampichler et al. 2012; Davey et al. 2012; Devictor et al. 2008). Also, changed competition structures might favor some species over others (Ahola et al. 2007; Wittwer et al. 2015). And especially, wintering conditions have already improving for a number of species (Princé and Zuckerberg 2015; Tellería et al. 2016).

Nevertheless, in most publications, negative impacts are expected to prevail for the majority of species.

Phenological mismatches have been hypothesized and regionally been proven (e.g., Both et al. 2010). Also, unequal climate change in breeding, passage and wintering areas and less flexibility to adapt to changes (especially for long-distance migrants, compare Gomez et al. 2016; Potvin et al. 2016) lead to a higher threat potential for migrants (Hüppop and Winkel 2006; Jones and Cresswell 2010). Changes of wintering or passage conditions also have additional detrimental effects on bird populations (Hüppop and Winkel 2006; Lehikoinen et al. 2013). And during the breeding season, enhanced extreme weather events are causing negative effects for bird species (Van de Pol et al. 2010).

However, at the community level, threats from climate change seem to arise especially from the changes in wintering and breeding distributions of species and the accompanying reshuffling of bird communities (La Sorte and Thompson 2007). Especially, asynchronous changes lead to altered competition and facilitation structures and make communities more susceptible to other detrimental (often anthropogenic) effects.

Ranges can expand or shrink due to climate change, depending on the adaptability of species to new conditions (Massimino et al. 2015). For generalist species or species which are able to shift their habitat preferences quickly, new habitats become available through climatic change, whereas other species might lose ranges, if habitat quality deteriorates due to climate change (Davey et al. 2012).

Although land-use change can be more important for habitat specialists (e.g., Virkkala et al. 2005), climate change might locally even have overtaken land use as the most important threat factor (compare Lemoine et al. 2007). However, the highest threat potential does probably not arise from climatic or land-use changes per se but rather from their interaction (Oliver and Morecroft 2014).

Not only is climate change known to drive changes in land-use patterns by changing environmental conditions for agriculture or forestry.

Also, climate change mitigation measures have the potential to reduce the impact of climatic change by up to 20% for birds (Warren et al. 2013) and in some cases are already promising regarding positive impacts on bird communities (Lindbladh et al. 2014). However, several mitigation measures show adverse effects on bird populations and diversity (Danielsen et al. 2009; Immerzeel et al. 2014).

So, it is estimated that one of the largest negative effects for birds at community level in Germany can currently be observed for the cultivation of energy crops. Especially, the increased cultivation of rapeseed for biofuel (E10) and winter wheat, rye, and especially corn for biogas production has led to higher agricultural intensification. Large-scale monocultures and the simultaneous loss of both extensive grassland and fallow land have probably led to a "biodiversity disaster" and might continuously add to an accelerated loss of farmland bird abundance and diversity (Flade 2012; Sauerbrei et al. 2014). Such negative effects could be minimized, if monocultures were avoided and habitats were enriched when cultivating bioenergy crops or grassland (Casado et al. 2014).

12.4 Projections of Potential Climate Change Impacts: What Else Is Waiting for Us?

Regarding future projections of climate change impacts, very often mismatches and uncertainty occur.

Interestingly, perceived climate change impacts and future projections very often differ markedly. While currently no climate-driven global extinctions of bird species can be proven, vast species losses are predicted for the future.

Dramatic distribution shifts of on average up to 550 km northward in Europe have been hypothesized (Huntley et al. 2008). Including land use into such projections leads to less pronounced shift projections of 335 km, at least under the expectation of *niche conservatism* (Barbet-Massin et al. 2012). Nevertheless, on a global perspective, vast land-use change acting jointly with climate change might very well lead to an even higher threat for biodiversity (Jetz et al. 2007; Mantyka-Pringle et al. 2015).

In the USA, 53% of species are projected to lose more than half of their current range (Langham et al. 2015). Especially, rare species with restricted ranges might be affected by such changes (Ohlemüller et al. 2008).

Furthermore, climate change impacts are proposed to be more severe for cold-dwelling species, restricted-range species, and species using seasonal habitats (Both et al. 2010). In contrast, warm-dwelling, generalist, and adaptable species are expected to profit from forecast climatic changes (Davey et al. 2012).

Additionally, although projections do not show associations to conservation prioritization (Langham et al. 2015), it has been proposed that protected areas might be able to at least partly buffer climate change impacts due to offering habitat niches and allowing for range shifts by providing refuges or stepping-stones for bird populations and communities (Hole et al. 2009).

Another mismatch occurs in the relationship between effects of climate and land-use change.

There are hypotheses stating that climate change might become an even more important threat factor than land-use change (Boit et al. 2016). However, there still exists considerable debate over this issue and especially for the tropics, but also for temperate regions, rather land-use change is expected to remain more important than climate change, whereas in arctic, boreal, and alpine environments, a larger effect of climate change is expected (Jetz et al. 2007; Sala et al. 2000).

Despite the unresolved question of which factor might be more important, future change projections of bird distributions often do not consider both climate and land-use change and their interactions which leads to an even higher uncertainty. Also, ecological features of species which influence dispersal, adaptability, resilience, and persistence of species are often ignored (Jiguet et al. 2007; Møller et al. 2008). Studies that incorporate those factors mostly reveal tremendously different or even contrasting possible outcomes in comparison to pure climate-based projections (e.g., Urban et al. 2012).

Even when incorporating interaction effects and biological mechanisms into future projections, there still remains considerable uncertainty regarding the extent to which single species, geographical areas, and ecosystems might be influenced by climate change. Future projections remain very vague for several reasons. They especially depend on many assumptions, e.g., about the fundamental and realized niches of species but also about the equilibrium of species distributions and climate. They sometimes even have to extrapolate to non-analogue conditions. And, overall, not only is the future extent of climate change on a global level unknown, but also its potential effects on regional and local scales are yet poorly understood. But for all that uncertainty about which effects climate change and other factors might have on biodiversity, the footprint of climate change is already apparent, e.g., illustrated by current bird population trends mirroring forecast climatic changes (Jiguet et al. 2013). This scientific fact clearly calls us into action.

12.5 Do Niches and Interactions with Abiotic and Biotic Environment "Evolve"?

As outlined before, future projections of climatic impacts on bird species or communities still remain vague. Despite the uncertainties of climate change, land-use change, their interactions, species traits, and species distribution models, major causes for this are assumptions of niche conservatism and stability. Those can in many cases be questioned and thus represent an additional source of uncertainty. Although for many species a clear relation to climate has been found, projections in time and space are hampered by several factors.

Species have been shown to react to climate change, e.g., by adapting their migratory behavior (Schaefer et al. 2008). Several species have shifted their arrival dates, changed wintering and stopover areas, or their migratory routes (Lehikoinen et al. 2004, 2013; Maclean et al. 2008).

But even without such changes, species might react to climatic changes in unexpected ways, especially when they are faced with non-analogue conditions (i.e., abiotic or biotic conditions they have not yet faced with within their former range). They can, e.g., occur in new environments which through climate change might offer more food resources in winter or during the breeding season. And they obviously react to climate change in different ways according to regionally different environmental conditions or altered community structures and competition. Thus, no straightforward predictions can be made for the "evolution" of non-analogue communities, i.e., communities which bird species have not encountered before (Urban et al. 2012).

It can however be expected that some species will react to such new challenges using ecological innovations. Especially changes in relative abundances within communities will necessarily force species under pressure to adapt to new conditions. The more generalist or innovative (e.g., niche-switching) species might fill niches that become available during the competition with species that are incapable of quick and efficient adaptation to new conditions.

Not only the competition among species but also subspecies distributions and furthermore genetic diversity might alter due to climate change (Peterson and Holt 2003; Pearman et al. 2010; McQuillan and Rice 2015).

But also external drivers like pathogens might influence species' capacities for climate change adaptation through impacts on the population structures of species. One example of induced pathogens that might have an impact on birds in the face of climate change is represented by the Usutu virus. This virus has been introduced into Europe by single events and caused mass mortality events in Common Blackbirds *Turdus merula* in 2001 in Austria and 2011 in Germany (Weissenböck et al. 2003; Becker et al. 2012). Such events lead to the selection of immune birds and might also regionally change relative abundance structures within bird communities as the blackbird is among the most abundant bird species in Central Europe. It remains yet to be examined to which extent other bird species are affected by this virus and whether other species that are immune to it might profit from changed competition structures within the bird community.

Also, at medium or long temporal scales, such pathogens by means of selection might shape the population genetic structure of species affected by them which in the case of Usutu cannot yet be demonstrated.

It can be expected that on larger timescales, such effects will become more prominent. Large-scale subliminal effects might already be existent but become apparent only in some decades.

Climate change is held responsible for the creation, differentiation, and extinction of species on evolutionary timescales. Availability of new niche space, altered selection pressure, and the loss of established niches are hypothesized to have driven evolution over millions of years. However, it seems difficult to predict how those processes act on small timescales of decades or centuries. Thus, current climatic changes and subsequent reactions of species have to be observed carefully in order to disentangle phenotypic adaptation from microevolutionary processes which might lead to altered conditions for species and communities.

One approach to do so includes a comparison of cues for past niche evolution to population trends. By this means, Lavergne et al. (2013) showed that niche conservatism correlates with population declines.

Nevertheless, not only the question of whether species adapt to climate change will be a matter of survival but also how quickly this can happen. If birds are not able to keep pace with climate change (DeVictor et al. 2008) or evolve new niches (Quintero and Wiens 2013), they will be subject to population declines and might ultimately be threatened with extinction.

The inherent uncertainties of those dangers and of climate change lead us to the question of how to combat climate change impacts.

12.6 Conservation Implications

Climate change has been shown to be an increasingly important threat factor for both bird species and communities. As it acts on large spatial and comparatively short temporal scales, it is a factor that (if only regionally) has the potential to quickly outperform land-use change as the current major threat factor for biodiversity (Eglington and Pearce-Higgins 2012). Major problems posed for conservation are that in contrast to land-use change, climate change is hard to revert on a short timescale but also difficult to detect as it is often intermingled with the first. And, even more dangerous appears the current lack of proof for negative impact of climate change on bird species. This may sound paradox, but despite a large body of research on the topic and numerous projections of future biodiversity loss, ultimately for no bird species, anthropogenic climate change has been unambiguously proven as the major driver of its extinction. Instead, the new appearances of many species outside of their previous ranges seem to overstress the positive effects of climate change. Also, the high uncertainty of future projections complicates climate change adaptation (see, e.g., Bagchi et al. 2013). Such projections are not only weakened by the inherent uncertainties of climate and land-use change predictions but also by the unpredictable responses of species.

In different parts of their ranges, species can react differently to changes, as shown by Stephens et al. (2016). Species can additionally be in disequilibrium with their climatic niche (DeVictor et al. 2008), or they might have a considerably broader fundamental niche than expected by their current realized climatic niche and not react to changes at all. Also, it remains unknown to which extent species will be able to adapt to non-analogue conditions (of climate, land use, community structure). Most of those issues are subject to both spatial and temporal scale effects.

Thus, conservation policy might be hampered by disequilibrium among the proposed long-term and the observed short-term impacts of climate change.

How can conservation policy react to this, and which measures have to be taken in order to help species and biodiversity adapt to climate change?

Importantly, uncertainty has to be communicated. Although this is quite unpopular with stakeholders and policy-makers, there is no one-solution-for-all situation.

Whereas politics currently still focuses on climate change mitigation, this should not counteract conservation and the necessary adaptation measures that have to be taken. Climate change is a fact and cannot be stopped immediately. However, it is currently happening and already showing impacts which have to be dealt with.

The inherent uncertainty and regional and temporal disparity of climate change has to be included into management concepts. Not only concepts for single species or on small spatial scales have to be developed, but rather concerted action plans are needed. It has already been shown that, not only in protected areas, adaptive landscape management can mitigate climate change effects (Princé et al. 2015).

Furthermore, conservation policy has to focus on international scales. Not only national breeding grounds but also breeding grounds in Europe, stopover sites, and also wintering areas have to be considered, especially for long-distance migrants (Sanderson et al. 2015).

In all regards, management necessarily has to be flexible so that it can be adapted to changed conditions. It should be combined with a comprehensive network of protected areas that can be used as habitat corridors in order to warranty connectivity of bird populations (Trautmann et al. 2013). Nevertheless, connectivity is only valuable, if it does not come at the cost of losing high-quality areas (Hodgson et al. 2009). Furthermore, it has been shown that habitat mosaics and diversity have positive effects on the resilience of communities to climatic change (Jarzyna et al. 2015).

One of the major threats is related to the mismatch among today's protected areas and the future ranges of species for which they were designed. Protected areas are designated and managed for species and might become less effective, if species shift ranges due to climate change (Coetzee et al. 2009). But how can adaptive management work in a changing world? And how can the success of a protected area network like Natura 2000 in Europe be assessed when species which triggered the designation of a protected area disappear due to climate rather than land-use change?

Those questions are currently pending an answer. However, protected areas seem to mitigate climate change impacts both at species and community level (Gaüzere et al. 2016).

Managing areas proactively might be one solution, at least if the new occurrence of a species of high conservation concern and/or climatic risk can be predicted with high certainty. Helping colonizing species to find new habitats can be a task for which existing protected areas could be used (Thomas et al. 2012). Also, as projections show that new refugia outside the protected area network might arise (Coetzee et al. 2009), new areas could be managed for prospective species, and protected area networks could be expanded.

This can also take place in current protected areas but then has to be well-balanced with attempts to maintain populations of diminishing species.

For some species, both approaches might be valuable in different parts throughout their ranges and might thus give them enough time and space to adapt to climate change. For this task, protected area networks give us a well-suited toolset yet remain to be extended and often still lack sufficient management planning (or at least information about it, as shown for individual countries, but also indicated by Gamero et al. 2017 for Europe). Also, it seems clear that protected areas cannot act as the only tool for climate change adaptation but have to be embedded in an enriched landscape matrix which has to provide stepping-stones for whole communities rather than only for single species.

References

Ahola MP, Laaksonen T, Eeva T, Lehikoinen E (2007) Climate change can alter competitive relationships between resident and migratory birds. J Anim Ecol 76:1045–1052

Araújo MB, Thuiller W, Yoccoz NG (2009) Reopening the climate envelope reveals macroscale associations with climate in European birds. Proc Natl Acad Sci USA 106:E45–E46

Bagchi R, Crosby M, Huntley B, Hole DG, Butchart SH, Collingham Y, Kalra M, Rajkumar J, Rahmani A, Pandey M, Gurung H, Trai LT, Van Quang N, Willis SG (2013) Evaluating the effectiveness of conservation site networks under climate change: accounting for uncertainty. Glob Chang Biol 19:1236–1248

Barbet-Massin M, Walther BA, Thuiller W, Rahbek C, Jiguet F (2009) Potential impacts of climate change on the winter distribution of Afro-Palaearctic migrant passerines. Biol Lett 5 (2):248–251. https://doi.org/10.1098/rsbl.2008.0715

Barbet-Massin M, Thuiller W, Jiguet F (2012) The fate of European breeding birds under climate, land-use and dispersal scenarios. Glob Chang Biol 18:881–890

Barnosky AD, Matzke N, Tomiya S, Wogan GO, Swartz B, Quental TB, Marshall C, McGuire JL, Lindsey EL, Maguire KC, Mersey B, Ferrer EA (2011) Has the Earth's sixth mass extinction already arrived? Nature 471:51–57

Bateman BL, Pidgeon AM, Radeloff VC, Van Der Wal J, Thogmartin WE, Vavrus SJ, Heglund PJ (2016) The pace of past climate change vs. potential bird distributions and land use in the United States. Glob Chang Biol 22:1130–1144

Beale CM, Lennon JJ, Gimona A (2008) Opening the climate envelope reveals no macroscale associations with climate in European birds. Proc Natl Acad Sci USA 105:14908–14912

Beale CM, Lennon JJ, Gimona A (2009) European bird distributions still show few climate associations. Proc Natl Acad Sci USA 106:E41–E43

Becker N, Jöst H, Ziegler U, Eiden M, Höper D, Emmerich P, Fichet-Calvet E, Ehichioya DU, Czajka C, Gabriel M, Hoffmann B, Beer M, Tenner-Racz K, Racz P, Günther S, Wink M, Bosch S, Konrad A, Pfeffer M, Groschup MH, Schmidt-Chanasit J (2012) Epizootic emergence of Usutu virus in wild and captive birds in Germany. PLoS One 7:e32604

Boit A, Sakschewski B, Boysen L, Cano-Crespo A, Clement J, Garcia-Alaniz N, Kok K, Kolb M, Langerwisch F, Rammig A, Sachse R, van Eupen M, von Bloh W, Zemp DC, Thonicke K (2016) Large-scale impact of climate change vs. land-use change on future biome shifts in Latin America. Glob Chang Biol 22:3689–3701

Both C, Visser ME (2001) Adjustment to climate change is constrained by arrival date in a long-distance migrant bird. Nature 411:296–298

Both C, Artemyev AV, Blaauw B, Cowie RJ, Dekhuijzen AJ, Eeva T, Enemar A, Gustafsson L, Ivankina EV, Järvinen A, Metcalfe NB, Nyholm NEI, Potti J, Ravussin P-A, Sanz JJ, Silverin B, Slater FM, Sokolov LV, Török J, Winkel W, Wright J, Zang H, Visser ME (2004) Large-scale geographical variation confirms that climate change causes birds to lay earlier. Proc R Soc Lond B 271:1657–1662

Both C, Van Turnhout CAM, Bijlsma RG, Siepel H, Van Strien AJ, Foppen RPB (2010) Avian population consequences of climate change are most severe for long-distance migrants in seasonal habitats. Proc R Soc Lond B 277(1685):1259–1266. https://doi.org/10.1098/rspb.2009.1525

Bradshaw WE, Holzapfel CM (2006) Evolutionary response to rapid climate change. Science 312:1477–1478

Casado MR, Mead A, Burgess PJ, Howard DC, Butler SJ (2014) Predicting the impacts of bioenergy production on farmland birds. Sci Total Environ 476:7–19

Clavero M, Villero D, Brotons L (2011) Climate change or land use dynamics: do we know what climate change indicators indicate? PLoS One 6:e18581. https://doi.org/10.1371/journal.pone.0018581

Coetzee BW, Robertson MP, Erasmus BF, Van Rensburg BJ, Thuiller W (2009) Ensemble models predict important bird areas in southern Africa will become less effective for conserving endemic birds under climate change. Glob Ecol Biogeogr 18:701–710

Danielsen F, Beukema H, Burgess ND, Parish F, Brühl CA, Donald PF, Murdiyarso D, Phalan B, Reijnders L, Struebig M, Fitzherbert EB (2009) Biofuel plantations on forested lands: double jeopardy for biodiversity and climate. Conserv Biol 23:348–358. https://doi.org/10.1111/j.1523-1739.2008.01096.x

Davey CM, Chamberlain DE, Newson SE, Noble DG, Johnston A (2012) Rise of the generalists: evidence for climate driven homogenization in avian communities. Glob Ecol Biogeogr 21:568–578

Devictor V, Julliard R, Couvet D, Jiguet F (2008) Birds are tracking climate warming, but not fast enough. Proc R Soc Lond B 275:2743–2748

Doswald N, Willis SG, Collingham YC, Pain DJ, Green RE, Huntley B (2009) Potential impacts of climatic change on the breeding and non-breeding ranges and migration distance of European *Sylvia* warblers. J Biogeogr 36:1194–1208

Eglington S, Pearce-Higgins JW (2012) Disentangling the relative importance of changes in climate and land use intensity in driving recent bird population trends. PLoS One 7:e30407

Flade M (2012) Von der Energiewende zum Biodiversitäts-Desaster–zur Lage des Vogelschutzes in Deutschland. Vogelwelt 133:149–158

Gamero A, Brotons L, Brunner A, Foppen R, Fornasari L, Gregory RD, Herrando S, Horak D, Jiguet F, Kmecl P, Lehikoinen A, Lindström A, Paquet J-Y, Reif J, Sirkiä PM, Skorpilova J, van Strien A, Szep T, Telensky T, Teufelbauer N, Trautmann S, van Turnhout CAM, Vermouzek Z, Vikström T, Vorisek P (2017) Tracking progress toward EU biodiversity strategy targets: EU policy effects in preserving its common farmland birds. Conserv Lett 10:395–402

Gaüzere P, Jiguet F, Devictor V (2016) Can protected areas mitigate the impacts of climate change on bird's species and communities? Divers Distrib 22:625–637

Gibson-Reinemer DK, Rahel FJ (2015) Inconsistent range shifts within species highlight idiosyncratic responses to climate warming. PLoS One 10:e0132103

Gienapp P, Leimu R, Merilä J (2007) Responses to climate change in avian migration time—microevolution versus phenotypic plasticity. Clim Res 35:25–35

Gillings S, Balmer DE, Fuller RJ (2015) Directionality of recent bird distribution shifts and climate change in Great Britain. Glob Chang Biol 21:2155–2168

Gomez C, Tenorio EA, Montoya P, Cadena CD (2016) Niche-tracking migrants and niche-switching residents: evolution of climatic niches in New World warblers (Parulidae). Proc R Soc Lond B 283:20152458

Gregory RD, Willis SG, Jiguet F, Vorisek P, Klvanova A, van Strien A, Huntley B, Collingham YC, Couvet D, Green RE (2009) An indicator of the impact of climatic change on European bird populations. PLoS One 4:e4678. https://doi.org/10.1371/journal.pone.0004678

H-Acevedo D, Currie DJ (2003) Does climate determine broad-scale patterns of species richness? A test of the causal link by natural experiment. Glob Ecol Biogeogr 12:461–473

Hitch AT, Leberg PL (2007) Breeding distributions of North American bird species moving north as a result of climate change. Conserv Biol 21:534–539

Hodgson JA, Thomas CD, Wintle BA, Moilanen A (2009) Climate change, connectivity and conservation decision making: back to basics. J Appl Ecol 46:964–969

Hole DG, Willis SG, Pain DJ, Fishpool LD, Butchart SMH, Collingham YC, Rahbek C, Huntley B (2009) Projected impacts of climate change on a continent wide protected area network. Ecol Lett 12:1–12

Houghton JT, Ding Y, Griggs DJ, Noguer M, van der Linden PJ, Dai X, Maskell K, Johnson CA (eds) (2001) Climate change 2001: the scientific basis. Cambridge University Press, Cambridge

Huang Q, Sauer JR, Dubayah RO (2017) Multidirectional abundance shifts among North American birds and the relative influence of multifaceted climate factors. Glob Chang Biol 23:3610–3622

Hughes L (2003) Climate change and Australia: trends, projections and impacts. Austral Ecology 28:423–443. https://doi.org/10.1046/j.1442-9993.2003.01300.x
Huntley B, Green RE, Collingham YC, Willis SG (2007) A climatic atlas of European breeding birds. Lynx Edicions, Barcelona
Huntley B, Collingham YC, Willis SG, Green RE (2008) Potential impacts of climatic change on European breeding birds. PLoS One 3:e1439
Hüppop O, Winkel W (2006) Climate change and timing of spring migration in the long-distance migrant *Ficedula hypoleuca* in central Europe: the role of spatially different temperature changes along migration routes. J Ornithol 147:344–353
Immerzeel DJ, Verweij PITA, Hilst FLOOR, Faaij AP (2014) Biodiversity impacts of bioenergy crop production: a state-of-the-art review. Glob Change Biol Bioenergy 6:183–209
IPCC (2013) Summary for policymakers. In: Stocker TF, Qin D, Plattner G-K, Tignor M, Allen SK, Boschung J, Nauels A, Xia Y, Bex V, Midgley PM (eds) Climate change 2013: the physical science basis—contribution of working group I to the fifth assessment report of the intergovernmental panel on climate change. Cambridge University Press, Cambridge
Jarzyna MA, Porter WF, Maurer BA, Zuckerberg B, Finley AO (2015) Landscape fragmentation affects responses of avian communities to climate change. Glob Chang Biol 21(8):2942–2953. https://doi.org/10.1111/gcb.12885
Jetz W, Wilcove DS, Dobson AP (2007) Projected impacts of climate and land use change on the global diversity of birds. PLoS Biol 5:e157
Jiguet F, Gadot AS, Julliard R, Newson SE, Couvet D (2007) Climate envelope, life history traits and the resilience of birds facing global change. Glob Chang Biol 13:1672–1684
Jiguet F, Devictor V, Ottvall R, Van Turnhout C, Van der Jeugd H, Lindström A (2010) Bird population trends are linearly affected by climate change along species thermal ranges. Proc R Soc Lond B 277:3601–3608
Jiguet F, Barbet-Massin M, Devictor V, Jonzé N, Lindström A (2013) Current population trends mirror forecasted changes in climatic suitability for Swedish breeding birds. Bird Study 60:60–66
Jones T, Cresswell W (2010) The phenology mismatch hypothesis: are declines of migrant birds linked to uneven global climate change? J Anim Ecol 79:98–108
Kampichler C, van Turnhout CAM, Devictor V, van der Jeugd HP (2012) Large-scale changes in community composition: determining land use and climate change signals. PLoS One 7:e35272. https://doi.org/10.1371/journal.pone.0035272
Karell P, Ahola K, Karstinen T, Valkama J, Brommer JE (2011) Climate change drives microevolution in a wild bird. Nat Commun 2:208
Koch PL, Barnosky AD (2006) Late quaternary extinctions: state of the debate. Annu Rev Ecol Evol Syst 37:215–250
La Sorte FA, Thompson FR III (2007) Poleward shifts in winter ranges of North American birds. Ecology 88:1803–1812
Langham GM, Schuetz JG, Distler T, Soykan CU, Wilsey C (2015) Conservation status of North American birds in the face of future climate change. PLoS One 10:e0135350. https://doi.org/10.1371/journal.pone.0135350
Lavergne S, Evans MEK, Burfield IJ, Jiguet F, Thuiller W (2013) Are species' responses to global change predicted by past niche evolution? Philos Trans R Soc B 368:20120091
Lehikoinen A, Virkkala R (2016) North by north-west: climate change and directions of density shifts in birds. Glob Chang Biol 22:1121–1129
Lehikoinen E, Sparks TH, Zalakevicius M (2004) Arrival and departure dates. Birds and climate change. In: Møller AP, Fiedler W, Berthold P (eds) Advances in ecological research, vol 35. Elsevier Academic, London, pp 1–32
Lehikoinen A, Jaatinen K, Vähätalo AV, Clausen P, Crowe O, Deceuninck B, Hearn R, Holt CA, Hornman M, Keller V, Nilsson L, Tomankova I, Wahl J, Fox AD (2013) Rapid climate driven shifts in wintering distributions of three common waterbird species. Glob Chang Biol 19:2071–2081
Lemoine N, Schaefer H-C, Böhning-Gaese K (2007) Species richness of migratory birds is influenced by global climate change. Glob Ecol Biogeogr 16:55–64

Lenoir J, Svenning J-C (2014) Climate-related range shifts – a global multidimensional synthesis and new research directions. Ecography 37:1–14
Lindbladh M, Hedwall P-O, Wallin I, Felton AM, Böhlenius H,Felton A (2014) Short-rotation bioenergy stands as an alternative to spruce plantations: implications for bird biodiversity. Silva Fennica 48:1135. p. 18
Longrich NR, Tokaryk T, Field DJ (2011) Mass extinction of birds at the Cretaceous–Paleogene (K–Pg) boundary. Proc Natl Acad Sci USA 108:15253–15257
Maclean IMD, Austin GE, Rehfisch MM, Blew J, Crowe O, Delany S, Devos K, Deceuninck B, Günther K, Laursen K, van Roomen M, Wahl J (2008) Climate change causes rapid changes in the distribution and abundance of birds in winter. Glob Chang Biol 14:2489–2500
Mantyka-Pringle CS, Visconti P, Di Marco M, Martin TG, Rondinini C, Rhodes JR (2015) Climate change modifies risk of global biodiversity loss due to land-cover change. Biol Conserv 187:103–111
Massimino D, Johnston A, Pearce-Higgins JW (2015) The geographical range of British birds expands during 15 years of warming. Bird Study 62:523–534
McQuillan MA, Rice AM (2015) Differential effects of climate and species interactions on range limits at a hybrid zone: potential direct and indirect impacts of climate change. Ecol Evol 5:5120–5137
Møller AP, Rubolini D, Lehikoinen E (2008) Populations of migratory bird species that did not show a phenological response to climate change are declining. Proc Natl Acad Sci USA 105:16195–16200
Møller AP, Flensted-Jensen E, Klarborg K, Mardal W, Nielsen JT (2010) Climate change affects the duration of the reproductive season in birds. J Anim Ecol 79:777–784
Ohlemüller R, Anderson BJ, Araújo MB, Butchart SH, Kudrna O, Ridgely RS, Thomas CD (2008) The coincidence of climatic and species rarity: high risk to small-range species from climate change. Biol Lett 4:568–572
Oliver TH, Morecroft MD (2014) Interactions between climate change and land use change on biodiversity: attribution problems, risks, and opportunities. Wiley Interdiscip Rev Clim Chang 5:317–335
Pacifici M, Foden WB, Visconti P, Watson JEM, Butchart SHM, Kovacs KM, Scheffers BR, Hole DG, Martin TG, Akçakaya HR, Corlett RT, Huntley B, Bickford D, Carr JA, Hoffmann AA, Midgley GF, Pearce-Kelly P, Pearson RG, Williams SE, Willis SG, Young B, Rondinini C (2015) Assessing species vulnerability to climate change. Nat Clim Chang 5:215–224. https://doi.org/10.1038/NCLIMATE2448
Päckert M, Martens J, Sun Y-H, Severinghaus LL, Nazarenko AA, Ting J, Töpfer T, Tietze DT (2012) Horizontal and elevational phylogeographic patterns of Himalayan and Southeast Asian forest passerines (Aves: Passeriformes). J Biogeogr 39:556–573
Parmesan C (2006) Ecological and evolutionary responses to recent climate change. Annu Rev Ecol Evol Syst 37:637–669
Parmesan C, Yohe G (2003) A coherent fingerprint of climate change impacts across natural systems. Nature 421:37–42
Pearman PB, D'Amen M, Graham CH, Thuiller W, Zimmermann NE (2010) Within-taxon niche structure: niche conservatism, divergence and predicted effects of climate change. Ecography 33:990–1003
Peterson AT, Holt RD (2003) Niche differentiation in Mexican birds: using point occurrences to detect ecological innovation. Ecol Lett 6:774–782
Peterson AT, Barve N, Bini LM, Diniz-Filho JA, Jiménez-Valverde A, Lira-Noriega A, Lobo J, Maher S, de Marco P Jr, Martínez-Meyer E, Nakazawa Y, Soberón J (2009) The climate envelope may not be empty. Proc Natl Acad Sci USA 106:E47
Plummer KE, Siriwardena GM, Conway GJ, Risely K, Toms MP (2015) Is supplementary feeding in gardens a driver of evolutionary change in a migratory bird species? Glob Chang Biol 21 (12):4353–4363. https://doi.org/10.1111/gcb.13070

Potvin DA, Välimäki K, Lehikoinen A (2016) Differences in shifts of wintering and breeding ranges lead to changing migration distances in European birds. J Avian Biol 47:1–10
Princé K, Zuckerberg B (2015) Climate change in our backyards: the reshuffling of North America's winter bird communities. Glob Chang Biol 21:572–585
Princé K, Lorrillière R, Barbet-Massin M, Léger F, Jiguet F (2015) Forecasting the effects of land use scenarios on farmland birds reveal a potential mitigation of climate change impacts. PLoS One 10:e0117850. https://doi.org/10.1371/journal.pone.0117850
Quintero I, Wiens JJ (2013) Rates of projected climate change dramatically exceed past rates of climatic niche evolution among vertebrate species. Ecol Lett 16:1095–1103
Reif J, Šťastný K, Bejček V (2010) Contrasting effects of climatic and habitat changes on birds with northern range limits in central Europe as revealed by an analysis of breeding bird distribution in the Czech Republic. Acta Ornithoica 45:83–90
Renwick AR, Massimino D, Newson SE, Chamberlain DE, Pearce-Higgins JW, Johnston A (2012) Modelling changes in species' abundance in response to projected climate change. Divers Distrib 18:121–132
Sala OE, Chapin FS, Armesto JJ, Berlow E, Bloomfield J, Dirzo R, Huber-Sanwald E, Huenneke LF, Jackson RB, Kinzig A, Leemans R, Lodge DM, Mooney HA, Oesterheld M, Poff NL, Sykes MT, Walker BH, Walker M, Wall DH (2000) Global biodiversity scenarios for the year 2100. Science 287:1770–1774
Salewski V, Hochachka WM, Fiedler W (2010) Global warming and Bergmann's rule: do central European passerines adjust their body size to rising temperatures? Oecologia 162:247–260
Sanderson FJ, Pople RG, Ieronymidou C, Burfield IJ, Gregory RD, Willis SG, Howard C, Stephens PA, Beresford AE, Donald PF (2015) Assessing the performance of EU nature legislation in protecting target bird species in an era of climate change. Conserv Lett 9:172–180
Sauerbrei R, Ekschmitt K, Wolters V, Gottschalk TK (2014) Increased energy maize production reduces farmland bird diversity. Glob Change Biol Bioenergy 6:265–274
Schaefer HC, Jetz W, Böhning-Gaese K (2008) Impact of climate change on migratory birds: community reassembly versus adaptation. Glob Ecol Biogeogr 17:38–49
Şekercioğlu ÇH, Primack RB, Wormworth J (2012) The effects of climate change on tropical birds. Biol Conserv 148:1–18
Smallegange IM, Fiedler W, Köppen U, Geiter O, Bairlein F (2010) Tits on the move: exploring the impact of environmental change on blue tit and great tit migration distance. J Anim Ecol 79:350–357
Stephens PA, Mason LR, Green RE, Gregory RD, Sauer JR, Alison J, Aunins A, Brotons L, Butchart SHM, Campedelli T, Chodkiewicz T, Chylarecki P, Crowe O, Elts J, Escandell V, Foppen RPB, Heldbjerg H, Herrando S, Husby M, Jiguet F, Lehikoinen A, Lindstrom A, Noble DG, Paquet J, Reif J, Sattler T, Szep T, Teufelbauer N, Trautmann S, van Strien AJ, van Turnhout CAM, Vorisek P, Willis SG (2016) Consistent response of bird populations to climate change on two continents. Science 352:84–87
Szabo JK, Khwaja N, Garnett ST, Butchart SHM (2012) Global patterns and drivers of avian extinctions at the species and subspecies level. PLoS One 7:e47080
Taheri S, Naimi B, Araújo MB (2016) Did British breeding birds move north in the late 20th century? Clim Chang Res 3:5
Tayleur C, Caplat P, Massimino D, Johnston A, Jonzén N, Smith HG, Lindström A (2015) Swedish birds are tracking temperature but not rainfall: evidence from a decade of abundance changes. Glob Ecol Biogeogr 24:859–872
Tayleur CM, Devictor V, Gaüzere P, Jonzen N, Smith HG, Lindström A (2016) Regional variation in climate change winners and losers highlights the rapid loss of cold-dwelling species. Divers Distrib 22:468–480
Tellería JL, Fernández-López J, Fandos G (2016) Effect of climate change on Mediterranean winter ranges of two migratory passerines. PLoS One 11:e0146958
Thomas CD, Lennon JJ (1999) Birds extend their ranges northwards. Nature 399:213

Thomas CD, Gillingham PK, Bradbury RB, Roy DB, Anderson BJ, Baxter JM, Bourn NAD, Crick HQP, Findon RA, Fox R, Hodgson JA, Holt AR, Morecroft MD, O'Hanlon NJ, Oliver TH, Pearce-Higgins JW, Procter DA, Thomas JA, Walker KJ, Walmsley CA, Wilson RJ, Hill JK (2012) Protected areas facilitate species' range expansions. Proc Natl Acad Sci USA 109:14063–14068

Tingley MW, Koo MS, Moritz C, Rush AC, Beissinger SR (2012) The push and pull of climate change causes heterogeneous shifts in avian elevational ranges. Glob Chang Biol 18:3279–3290

Trautmann S, Laube I, Schwager M, Böhning-Gaese K (2013) Sind Vögel vom Klimawandel gefährdet? Modellierung des Einflusses des Klimawandels auf Vögel. In: Vohland K, Badeck F, Böhning-Gaese K, Ellwanger G, Hanspach J, Ibisch PL, Klotz S, Kreft S, Kühn I, Schröder E, Trautmann S, Cramer W (Hrsg) Schutzgebiete Deutschlands im Klimawandel – Risiken und Handlungsoptionen. Naturschutz und Biologische Vielfalt 127, pp 25–31

Urban MC, Tewksbury JJ, Sheldon KS (2012) On a collision course: competition and dispersal differences create no-analogue communities and cause extinctions during climate change. Proc R Soc Lond B 279:2072–2080

Van De Pol M, Ens BJ, Heg D, Brouwer L, Krol J, Maier M, Exo K-M, Oosterbeek K, Lok T, Eising CM, Koffijberg K (2010) Do changes in the frequency, magnitude and timing of extreme climatic events threaten the population viability of coastal birds? J Appl Ecol 47:720–730

Virkkala R, Luoto M, Heikkinen RK, Leikola N (2005) Distribution patterns of boreal marshland birds: modelling the relationships to land cover and climate. J Biogeogr 32:1957–1970

Visser ME, Perdeck AC, Balen V, Johan H, Both C (2009) Climate change leads to decreasing bird migration distances. Glob Chang Biol 15:1859–1865

Walther G-R, Post E, Convey P, Menzel A, Parmesan C, Beebee TJC, Fromentin J-M, Hoegh-Guldberg O, Bairlein F (2002) Ecological responses to recent climate change. Nature 416:389–395

Warren R, VanDerWal J, Price J, Welbergen JA, Atkinson I, Ramirez-Villegas J, Osborne TJ, Jarvis A, Shoo LP, Williams SE, Lowe J (2013) Quantifying the benefit of early climate change mitigation in avoiding biodiversity loss. Nat Clim Chang 3:678–682

Weir JT, Schluter D (2004) Ice sheets promote speciation in boreal birds. Proc R Soc Lond B 271:1881–1887

Weissenböck H, Kolodziejek J, Fragner K, Kuhn R, Pfeffer M, Nowotny N (2003) Usutu virus activity in Austria, 2001–2002. Microbes Infect 5:1132–1136

Wiens JJ (2016) Climate-related local extinctions are already widespread among plant and animal species. PLoS Biol 14:e2001104

Winkler DW, Dunn PO, McCulloch CE (2002) Predicting the effects of climate change on avian life-history traits. Proc Natl Acad Sci USA 99:13595–13599

Wittwer T, O'Hara RB, Caplat P, Hickler T, Smith HG (2015) Long-term population dynamics of a migrant bird suggests interaction of climate change and competition with resident species. Oikos 124:1151–1159

Chapter 13
Impact of Urbanization on Birds

Caroline Isaksson

Abstract Urban habitats and landscapes are markedly different from nonurban "natural" habitats. The major difference is the transformation of the land, from natural green areas to anthropogenic structures and impervious surfaces. To survive in the urban habitat, birds are forced to either accept or avoid the new conditions. In addition, the urban sprawl has led to a highly fragmented landscape, with islets of suitable bird habitat surrounded by highways and buildings that frequently act as barriers, even for mobile creatures such as birds. These altered conditions have changed the avifauna dramatically, with many species vanishing once an area is urbanized, thus resulting in a significant loss of local biodiversity. However, some species seem to thrive in the city, and these urban-dwelling species often show pronounced phenotypic differences (e.g., in behavior, physiology, and morphology) to their rural conspecifics. These phenotypic changes have been linked to specific urban selective drivers such as air pollution, artificial light at night, noise, different kinds of food, different predation pressures, and human disturbances. However, these drivers are often confounded, and it is hard to separate one urban factor as the main driver for the differentiation. Although the urban habitat is a large threat to biodiversity, it is also an exciting environment for studies of population divergence, evolutionary responses, and ultimately speciation in real time.

Keywords Anthropogenic · Environmental stress · Habitat fragmentation · Light at night · Pollution

13.1 A Brief History of Urbanization

Human exploitation of land dates back to our earliest settlements, with massive, yet local, destruction and deforestation (Diamond 2005). Already the Romans transformed the landscape to the treeless landscape of the Mediterranean that we

C. Isaksson (✉)
Department of Biology, Evolutionary Ecology, Lund University, Lund, Sweden
e-mail: Caroline.Isaksson@biol.lu.se

D. T. Tietze (ed.), *Bird Species*, Fascinating Life Sciences,
https://doi.org/10.1007/978-3-319-91689-7_13

are familiar with today, probably with devastating consequences for the wildlife at that time. Similar scenarios can be found throughout human history and across the world. However, it was not until the Anthropocene that the urban human societies started to grow significantly across the globe. Western industrialization started in the 1700s and with that urbanization and urban sprawl became a significant part of the landscape. Today, urbanization is a global phenomenon with implications for birds as well as for all other animals. Yet, developing countries are still in the early phases of industrial revolution; thus, the impact of global urbanization is expected to increase. Likewise, due to continued growth of the human population, existing urban areas in the western countries are also predicted to intensify and expand in the future.

Together with climate change, urbanization is considered one of the largest threats to wildlife, including the persistence of many bird species. The foremost threat is probably habitat loss and fragmentation, which forces rapid decisions about emigrating (if possible) to more suitable habitats or stay and cope with the new conditions (Marzluff and Ewing 2001; McKinney 2002). The new urban conditions are not only through the process of urbanization per se but also the fact that the existing or remaining "green" areas are often changed, through plantation of non-native plant species, managed lawns, and removal of the mid-story canopy (Luck and Smallbone 2010; Aronson et al. 2014).

Although, the total number of bird species declines once an area is urbanized, many bird species do seem to flourish. In fact, birds are probably the loudest and most visible animal group in the urban habitat. The urban species often exploit anthropogenic resources such as the high abundance of novel food sources and artificial nesting holes, e.g., nest boxes and under roof tiles. In temperate regions, birds can also benefit from the warmer climate caused by the so-called "urban heat island" effect, which is caused by the heat-absorbing properties of the impervious surfaces and buildings together with the scattering effects of air pollution, trapping heat irradiation within the atmosphere of the city. However, in warmer or tropical regions, the urban heating effect can be devastating for birds, leading to heat stress and dehydration (Grimm et al. 2008).

While all anthropogenic change of the landscape (including farmlands and deforestation in sparsely populated areas) is challenging for birds, I will here primarily focus on the anthropogenic disturbance (or urbanization) that is associated with densification of humans and human activities, i.e., cities. Nowadays, up to 54% of the world's human population live in cities (United Nations 2014). The more humans present in an area, the greater need for buildings and infrastructure, increasing traffic-related air pollution, noise pollution, and artificial light at night. Thus, human population size is a relatively good indicator of city-level impact on birds (Fig. 13.1).

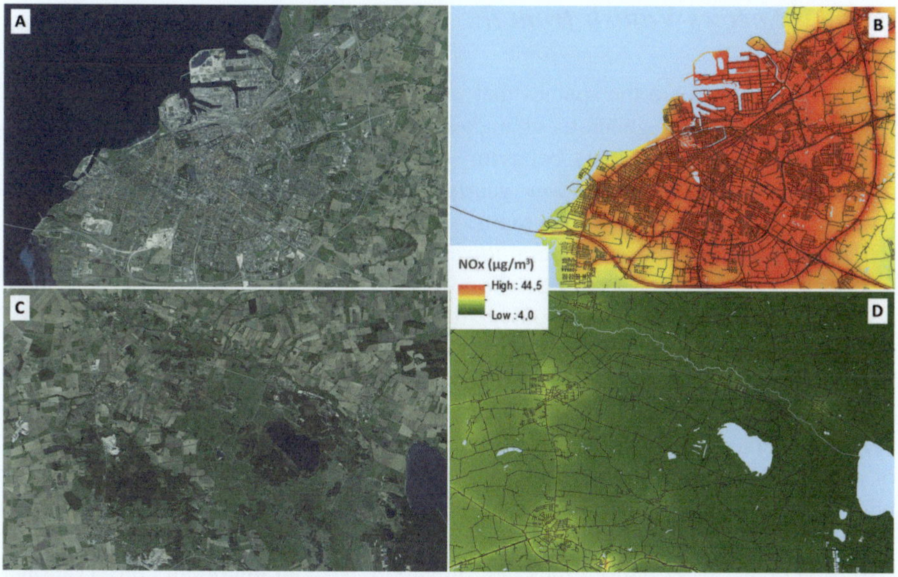

Fig. 13.1 Pollution map over a city and a nearby rural area. (**a**) Satellite image over Malmö, a city in southern Sweden with around 300,000 inhabitants; (**b**) pollution image (nitrogen oxides, NOx) over the same area in Malmö; (**c**) satellite image over a nearby rural area. The two larger lakes are Krankesjön and Vombsjön; (**d**) pollution map (NOx) over the rural area. The picture was produced by Emilie Stroh in 2018 by using ArcGIS (ESRI)

13.2 Birds and the City

Among the over 10,000 recognized bird species in the world, around 2000 (nearly 20%) occur in cities (Aronson et al. 2014). Although some bird species are cosmopolitan and thus occur in cities across the globe (e.g., Feral Pigeon *Columba livia*, House Sparrow *Passer domesticus*, Common Starling *Sturnus vulgaris*, Barn Swallow *Hirundo rustica*; Aronson et al. 2014), most urban species are representatives of the regional biogeographical species pool. Depending on their reliance on human resources, birds can be divided into three groups: urban avoiders, urban (suburban) adapters, and urban exploiters (Blair 1996, see further details below). Due to these different species-level responses to urbanization, once an area is urbanized, the species composition will change, with some species vanishing and others flourishing. In the sections below, I will start by reviewing the underlying drivers for these contrasting species-level responses and then continue with population divergence within a species as a result of urbanization and its implications for the species.

13.2.1 Species Vanish from the City

The urban avoiders are the species that immediately vanish when an area is urbanized. These species are generally characterized by ecological features such as having low natal dispersal, migratory behavior, fear toward humans (long flight-initiation distance), insectivory, and/or low yearly fecundity (Møller 2009 and references therein). This has, on a large geographic scale, led to a species-homogenizing effect (Clergeau et al. 2006) and a general lower species richness in the urban areas (e.g., Chace and Walsh 2006; MacGregor-Fors et al. 2012; Aronson et al. 2014). The main driver of this decline in avian biodiversity (i.e., that species vanish) is habitat loss, and there is a clear negative association between density of avian species and urban land cover (Chace and Walsh 2006; Aronson et al. 2014). However, cities that maintain native vegetation composition and structures will retain more native bird species than those that do not (Chace and Walsh 2006). Singapore is a good example of a megacity that has maintained a relatively high avian biodiversity. In a worldwide survey of bird species across 54 cities, 36 bird species were identified to be on the IUCN global Red List for threatened species, and 12 were recorded in Singapore (Aronson et al. 2014). Thus, cities like Singapore have an important conservational role for maintaining these species in the future—to defeat the species homogenization effect and to maintain suitable habitats for them. Since the human population continues to increase, so will urbanization; hence, the urban threat for birds is likely to be even more alarming in the future, and urban city planners and conservationists have an important task to maintain existing biodiversity.

The species that vanish immediately upon urbanization are relatively easy to identify. However, many species show a slower response to urbanization, with population decline over time as a result. To identify these species can be more difficult and requires long-term studies of populations and their dynamics (e.g., dispersal, fecundity, and survival). For studies of population dynamics, it will be important to (1) identify species for which the city act as an ecological trap, and (2) to establish, whether the urban population is a sink or source population. Both scenarios can result in impoverishment of the urban bird populations through lower lifetime fitness compared to birds in the surrounding nonurban habitats (Lepczyk et al. 2017). However, if the urban environment acts as an ecological trap, it can have implications for the persistence of the whole species. As the word *trap* indicates, an ecological trap acts by luring and attracting birds to a specific area. The city attracts many birds with its higher abundance of resources (i.e., food and nesting opportunities for cavity-nesting birds) and, in temperate regions, also with its milder winter climate compared to the surrounding nonurban habitats (Lepczyk et al. 2017). These factors make birds to evaluate the city habitat as a "high-quality" habitat, thus a preferred habitat compared to more natural habitats (Donovan and Thompson 2001). However, through the high rate of nest predation (e.g., by corvids and eastern gray squirrels *Sciurus carolinensis*), predation at feeding tables (e.g., by feral cats and dogs), poor nutritional value of the food sources, exposure to high pollution levels, and the high incidence of collisions with windows and cars reduce the overall fitness of the population in the urban habitat (e.g., Donovan and Thompson 2001; Erritzoe et al. 2003; Bonnington et al. 2015). Hence, if the ecological trap (city preference) is strong, the urban habitat will continue to attract rural birds to the city where they will suffer the negative consequences, ultimately reducing the species future chances of survival (Donovan and Thompson 2001). The second scenario, the source-sink scenario, can result in both positive and negative effects on the population depending on whether the urban population is a source or a sink. However, if the urban population is a sink (without being a trap), birds do not prefer the urban habitat over nonurban habitats. Instead, surplus individuals can be forced into the city due to lack of free niches in the natural habitat. However, since the urban habitat is not preferred, there will not be as severe effects on the species level as with the ecological trap (Dias 1996; Lepczyk et al. 2017). That argument, however, relies on the existence of natural habitats; thus, the continuation of habitat destruction and deforestation could be devastating also for source-sink dynamics across the urban/ rural landscape.

13.2.2 Species Flourish or Persist in the City

Although many species vanish from urbanized areas, some species thrive or persist in the city. The group of urban exploiters comprise species that seemingly flourish in the city. They have exploited the human resources to such a degree that they now depend on them to maintain current population densities. Many of the avian urban exploiters are invasive species, for example, Feral Pigeons, House Sparrows, and Common Mynas *Acridotheres tristis*. In fact, urban areas have greater abundance of birds (or biomass) per sampling unit than nonurban habitats (e.g., Chace and Walsh 2006; MacGregor-Fors et al. 2012). This is something that many raptor species have gained from, and some are now becoming increasingly common in urban areas, e.g., Great Horned Owl *Bubo virginianus*, Peregrine Falcon *Falco peregrinus*, and Black Sparrowhawks *Accipiter melanoleucus* (Chace and Walsh 2006; Kettel et al. 2017; Suri et al. 2017).

Similarly to above, when comparing city-dwelling species from the Western Palearctic with urban avoiders from the same region, the urban birds shared several ecological features (Møller 2009). Urban bird species are characterized by large breeding distributions, high propensity for dispersal, high rates of feeding innovation (novel ways of acquiring food), less fear toward humans (short flight-initiation distance), and a life history characterized by a high annual fecundity and high adult survival rate (Møller 2009). Urban species also have a larger size of the *bursa of Fabricius*, a specialized organ in birds that is part of the immune system, which suggests that they have the capacity to mount a stronger immune response compared to urban avoiders (Møller 2009). In addition, urban species have higher levels of dietary antioxidants (vitamin E and carotenoids) than the urban avoiders, which may help them to defeat oxidative pollution better, see below (Møller et al. 2010). Other studies have also concluded that urbanization favors omnivorous, granivorous, and cavity-nesting bird species but that it depends on geographical location and climate (Chace and Walsh 2006).

Although these features are common to species that are regularly seen in urban habitats, they are not indicative of whether a particular species is an exploiter or adapter species. Similar to the two scenarios above—the ecological trap and source-sink dynamic—to know whether a species is an adapter or an exploiter species can be very important for understanding species resilience to urbanization, the associated urban stressors, and also its dependence on humans. The urban adapter species are not dependent on human resources, but are happy to utilize them from time to time. Urban adapters include, for example, many of the small perching birds such as Great Tit *Parus major*, Blue Tit *Cyanistes caeruleus*, and House Finch *Haemorhous mexicanus* but also species such as the endemic New Zealand Tui *Prosthemadera novaeseelandiae*. A change in human behavior, the waste disposal system, or the urban landscape could potentially quite rapidly change the composition between urban adapter and exploiter species within an urban area, where adapters can outcompete the exploiters, if the resources decrease; however, the opposite scenario is probably more common.

As mentioned above, urban areas have overall lower species richness than nonurban habitats. However, a higher species richness is seen in Palearctic (European) cities as compared to cities in the Nearctic and Australasia (Aronson et al. 2014). This difference could be related to time since urbanization, i.e., the birds have had more time to adapt to the urban environment in the older cities. In the rest of this chapter, I will look into individual- and population-level responses to urbanization, which ultimately provide a platform for selection to act upon and for population divergence to appear, with potential implications for sexually selected traits and speciation.

13.2.3 Species Change

Urban environments provide an intriguing opportunity for studies of evolution in real time. Many, if not all, animal species show some phenotypic difference between the urban and nonurban counterparts. These phenotypic differences have been documented in all sorts of traits, from physiology and behavior to morphology, with potential effects on life-history traits and ultimately fitness. The drivers of these phenotypic changes are multiple and not always easy to disentangle. Likewise, the mechanistic underpinnings of the changes and whether they are caused by “non-genetic” phenotypic plasticity or genetic divergence (discussed below) are not always known.

Phenotypic plasticity here refers to the direct response of organisms to an environmental change like urbanization. In other words, a certain genotype will give rise to a different phenotype when in a urban compared to a rural environment. The term is broadly used to describe all phenotypic responses and includes, for example, physiological acclimatization and learning (West-Eberhard 1989). Some of these responses can change once and then persist throughout an individual's life or change continuously in response to environmental cues. Although I here refer to phentypic plasticity as a "non-genetic" trait, plasticity itself can be a inherited trait. How much genes are expressed and the timing of the expression are important determinants of a species' or individuals' plasticity or, in other words, potential to phenotypically change. To date, there is only one study that has compared transcriptomes of urban and rural birds (Watson et al. 2017). The differentially expressed genes revealed to be involved in gene regulation (such as DNA methylation), innate and adaptive immune responses, DNA repair, heavy metal detoxification, and fat metabolism. Furthermore, two species of the well-known Darwin's finches *Geospiza* sp. showed dramatic differences in DNA methylation patterns between urban and rural populations (McNew et al. 2017). Although the functional significance of these mechanistic differences is yet unknown, it is promising to find variation here for species capacity to rapidly change in response to urbanization (West-Eberhard 1989).

13.3 Urban Environment as a Barrier for Movement

A population needs to be reproductively isolated from other populations of the species to diverge genetically. For a long time, urban habitats were not considered a barrier for mobile species such as birds. However, habitat loss through intensification of urbanization or urban sprawl has been proven problematic for species with sedentary habits or limited vagility, i.e., the ability to move freely and migrate (reviewed in Delaney 2014). As a result, populations get trapped. Indeed, among the studies that have found strong genetic differentiation between urban and rural populations are those of relatively sedentary species such as the Wrentit *Chamaea fasciata* (Delaney et al. 2010). However, also the Song Sparrow *Melospiza melodia*, which is a more mobile species, shows a weak (yet significant) divergence (Unfried et al. 2013). In addition to dispersal ability, the strength of genetic structuring can also be associated with city characteristics and urbanization intensity (habitat fragmentation) (Delaney et al. 2010). The Great Tit, which is a fairly sedentary species, has shown mixed results in different cities. In Barcelona (Spain), the urban individuals show strong genetic divergence from those in the surrounding forest habitats (Björklund et al. 2010), whereas in Montpellier (France), there was a low yet significant genetic differentiation from individuals in the more rural sites (Perrier et al. 2017). However, when Great Tits were simultaneously screened from nine European cities and corresponding rural sites close to the city, the overall pattern was that Great Tits are very similar, genetically, across the whole of Europe and there was no divergence in relation to urbanization (Salmón 2017). However, the study could also reveal signals of selection on certain genetic regions, suggesting that the urban environment does pose a selective pressure on this common European species.

Furthermore, a newly colonized urban population can start to grow from a initially small or large gene pool. If the urban population starts to grow from a small number of individuals, this small population can be very different from the surrounding populations, the so-called founder effect, resulting in rapid population divergence (Delaney et al. 2010). Urban habitats may attract a certain kind of individuals such as those that are more explorative and bold (Riyahi et al. 2015; Audet et al. 2016; Charmantier et al. 2017; Senar et al. 2017). In the Great Tit, genes that are associated with exploration behaviors show the highest expression in the urban (from Barcelona) compared to the rural birds (Riyahi et al. 2015). Thus, colonization of urban habitats by a certain type of individuals rather than a subset of the population being trapped can more rapidly lead to genetic divergence but also loss of genetic diversity.

Studying genetic diversity is another common approach to explore the strength of a genetic barrier. Both isolated populations and populations showing founder effects are expected to show relatively low genetic diversity compared to their conspecifics. Indeed, this has been shown in urban populations of a number of species such as Dark-eyed Junco *Junco hyemalis*, Common Kestrel *Falco tinnunculus*, and Common Blackbird *Turdus merula* (Rasner et al. 2004; Rutkowski et al. 2006; Evans et al. 2009a). In the Blackbird study, the structure of the genetic signals suggested that this species has colonized urban environments multiple times (i.e., several independent founder effects) rather than one colonization event followed by local adaptation and dispersal (a leapfrog colonization model) (Evans et al. 2009a). A low genetic diversity is a concern, since it can lead to many negative effects. One example is the negative effects of inbreeding, which increases the occurrence of malformations and genetic disorders. A lower genetic diversity also means that natural selection has a reduced variation to act on; consequently populations of low genetic diversity are more sensitive to environmental perturbations. Interestingly, the genetic diversity in the abovementioned Great Tit populations from Barcelona and Montpellier showed contrasting results. The genetic diversity was higher for city birds in Barcelona and lower for city birds in Montpellier compared to their respective rural populations, suggesting that different population dynamics are in place in this species in these two European cities (Björklund et al. 2010; Perrier et al. 2017).

13.4 The Urban Drivers

There are at least four drivers (or environmental stress factors) that are directly related to urbanization and that are general across all geographical zones, namely, chemical pollution, noise, artificial light at night (ALAN), and human presence (Fig. 13.2). These four factors have led researchers to suggest that urban environments are relatively homogenous in the selection pressures that they might enforce.

The chemical pollutants that urban areas have in common are generated by traffic—the combustion of fossil fuels give rise to especially high levels of nitrogen oxides (NOx) and soot (Fig. 13.1; Salmón et al. 2018). Many Chinese cities have particularly high levels of these pollutants (Grimm et al. 2008). But pollution levels of a standard city of Sweden, which is a much more sparsely populated region, are also high enough to have negative impacts on birds as well as humans (Fig. 13.1;

Fig. 13.2 An illustration of the multiple urban stressors/drivers. When an area is urbanized, the habitat is changed in many ways; apart from habitat fragmentation and the increased number of buildings and impervious surfaces which affect the climate of the area, there are also increased noise levels, traffic-related pollution, and artificial light at night. In addition, we (humans) and our pets disturb birds. We also provide birds with food which may not always be of the best quality. All these urban stressors likely generate selection pressures on urban birds. Illustration made by Anna Persson. From Isaksson (2015), Functional Ecology. Copyright © 2015 by John Wiley Sons, Inc. Reprinted by permission of John Wiley & Sons, Inc.

Salmón et al. 2018). In addition, many urban areas are also polluted with heavy metals; however the difference between urban and rural sites in heavy metal contamination seems to vary widely, depending on local industry and history (Binkowski and Meissner 2013; Manjula et al. 2015; Orlowski et al. 2014). For example, in India many of the polluting industries are located in rural areas such as metal fabrication industries and leather tanning, which result in no differences between urban and rural populations in heavy metal incorporation in feathers (Manjula et al. 2015). Thus, the urban exposure to heavy metals is less tightly linked to level of urbanization as compared to NOx and soot exposure.

To find a silent place within an urban landscape is difficult. Constantly, cars are horning, sirens are alarming, hammering from construction work, and airplanes are taking off. This is jointly called urban noise pollution. Birds' responses to noise will depend on its loudness, frequency, consistency, and duration. Similarly, darkness is a rare thing in cities. Satellite images of the globe at night clearly illustrate the lit urban hotspots worldwide (Fig. 13.3). Europe, USA (mid- to eastern parts), and Asia, especially Japan and India, are lit hotspots on the map, whereas northern South

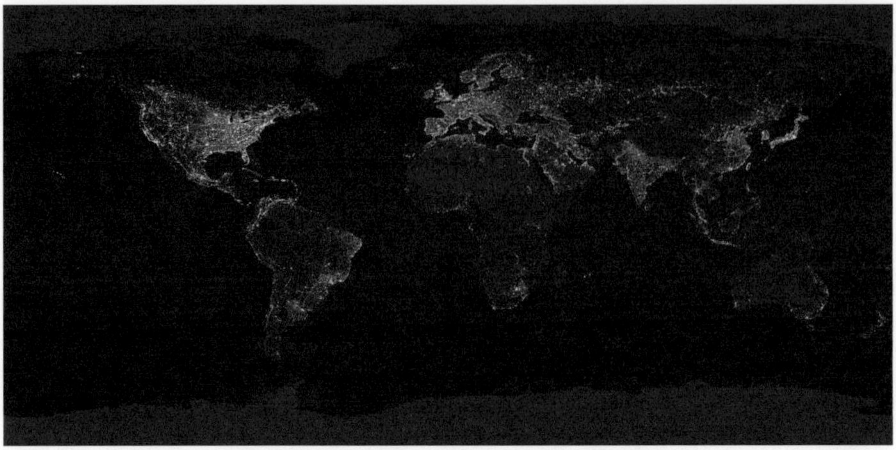

Fig. 13.3 A global map showing artificial light at night

America, central Africa, Russia, and Australia are still experiencing limited ALAN pollution (Fig. 13.3). All three pollution sources (chemical, noise and light pollution) are linked to phenotypic changes, which will be exemplified below, and also to costs in terms of reduced health, number of surviving offspring, and/or survival.

The last general urban driver is human presence. Birds perceive humans as a threat, and this threat is something that urban birds need to handle, since they interact and are exposed to humans constantly (Senar et al. 2017). Encounters with humans are likely to stress birds, in particular during the breeding season. Human behaviors toward birds are also likely to differ depending on species, for example, small birds are generally accepted to stay close, whereas larger species are vigorously scared away and even hunted in the city. Also human behavior toward birds are context-specific; we want them close when we choose to for example actively feed them (Fig. 13.4), but we do not want them too close while eating outside ourselves.

Apart from the abovementioned drivers, there are a few other factors that have been highlighted, namely, food abundance, pathogens, and predation (Fig. 13.2). However, these seem to be more affected by regional conditions, geography, and culture compared to the abovementioned factors. For example, food availability for urban birds can vary greatly across the globe. Supplementary feeding of birds is very common in western countries but more or less absent in eastern countries (Fuller et al. 2008). Thus, western urban birds seem to have plenty of food, whereas eastern urban birds may not (at least not from intentional feeding). Also, people living in sparsely populated farmlands can supplement birds heavily; thus food access is not directly linear to urbanization per se. However, poorer food quality, of both the natural and the anthropogenic sources, and reduced diversity of food items (e.g., grain and insect species) are also likely to be general across cities (e.g., Isaksson and Andersson 2007; Pollock et al. 2017).

Fig. 13.4 Anthropogenic food sources for birds. The abundance of food in the urban environment is one of the key factors that attract birds into the cities. However, the food that we provide them with is not always the most nutritional food

Pathogen abundance and diversity show also great geographic variation and even variation across cities from the same region (e.g., Evans et al. 2009b; Delgado-V and French 2011). Climate and presence of water bodies are the two main factors that affect pathogen abundance and diversity in the city. However, transmission of disease may be more rapid in urban areas due to the higher densities of birds and gathering of birds at feeding tables.

Predation pressures may ease or increase depending on the characteristics of the city and the region. On the one hand, there are presumably fewer natural predators (e.g., large mammals and raptors) in the urban areas compared to natural nonurban areas. On the other hand, corvids, feral cats, and American minks *Neovison vison* are more common. Likewise, many smaller raptors such as Peregrine Falcons and Lesser Kestrels *Falco naumanni* perform very well nowadays in the urban setting (Chace and Walsh 2006; Kettel et al. 2017). Also the presence of artificial nests seems to attract predators significantly more in urban than in rural areas (Wilcove 1985; Jokimäki and Huhta 2000). However, in a comparative study of multiple urban species, it was shown that feathers of urban birds were more difficult to pluck (i.e., an anti-predation response), which was interpreted as reduced pressures from natural selection caused by predation in urban habitats (Møller 2009). Although these three last factors are more variable across cities, when present, they are likely to be strong drivers of population-level shifts in phenotypic traits.

13.5 Phenotypic Changes and Responses as a Result of Urban Life

With this chapter, I do not intend to cover all kinds of phenotypic changes or drivers. In fact, in many cases the causal relationship between a driver and a phenotypic change is not known or difficult to single out from the ocean of possible urban drivers. Thus, many studies of phenotypic changes are linked to urbanization per se rather than to single drivers. The phenotypic changes documented can be both a result of (1) population-level changes in the genetic pool (see above) or through (2) "nongenetic" responses of an individual (phenotypic plasticity). It should be noted that these two sources for population-level variation are not mutually exclusive. In fact, a genetic change can give rise to a change in phenotypic plasticity, and changes in phenotypic plasticity can lead to a new phenotype for natural and sexual selection to act upon, changing the genetic pool to the coming generations.

13.5.1 Physiology

13.5.1.1 Stress Physiology and Its Implications

The main physiological responses investigated in relation to urbanization or to single urban stressors is stress physiology (oxidative stress and corticosterone, commonly referred to as a stress hormone). Oxidative stress is the key target for toxicological research but also in relation to cost of life, since oxidative stress is part of the unavoidable aging process (Isaksson et al. 2011; Isaksson 2015). Environmental influences on oxidative stress can be multiple, e.g., pollution, radiation, disease, and food intake. However, the main factor in the urban environment is probably chemical pollution (such as NOx and soot, Fig. 13.1). Many of the urban air pollutants act as prooxidants, which will react with and cause damage to life-sustaining molecules such as proteins, lipids, and DNA, unless they are detoxified by the protective antioxidants. Oxidative damages are commonly used as biomarkers of poor health, leading to premature death. The first response to pollution or prooxidants is to increase the antioxidant responses, and this is also what birds, as well as humans, generally do in urban environments (Isaksson 2010, 2015; Salmón et al. 2018). However, the upregulation is not always sufficiently high to avoid oxidative damage and species' as well as individuals' capacity to deal with oxidative stress varies. For example, some sparrow species (Passeridae) have a poorer capacity to block generation of oxidative damages to proteins and lipids in the urban environment compared to the tit species (Paridae) living in the same urban environments (Isaksson et al. 2017; Salmón et al. 2018). Consequently, the physiological response and capacity vary across urban-dwelling species and, ultimately, the negative physiological effects related to urbanization.

Another aging biomarker that may be linked to oxidative stress is the shortening of telomeres (von Zglinicki 2002; Boonekamp et al. 2017). Telomeres are the outer protective ends of the chromosomes, which shorten throughout an individual's life and can shorten more rapidly if exposed to stress such as irradiation, malnutrition, or pollution. When the telomeres reach a critical length, the cell cannot function, ultimately leading to cellular death. Early life in an urban environment has proven particularly challenging for Great Tits. In a cross-fostering experiment of 2-day-old chicks, half broods were swapped between an urban city park and a forest. Twelve days later, a blood sample was taken and analyzed for telomere length. The Great Tits that grew up in the urban habitat (independent of population origin) had significantly shorter telomeres than the chicks that grew up in the forest (Salmón et al. 2016). These individuals were followed to the next season, and it was clear that only the individuals with relatively long telomeres were recaptured the following breeding season, indicating that individuals with short telomeres had not survived the winter and this effect was significantly stronger in the urban environment (Salmón et al. 2017). This suggests that telomere length matters and that urban environmental stress significantly affects survival. To date, all studies on oxidative stress and telomeres are linked to urbanization and not to a specific urban driver; however, it is clear that regardless of source, the urban environment is challenging for birds and their physiological responses do not always circumvent the negative effects.

Another pathway to remove toxic heavy metals has been proposed for birds, namely, their incorporation into feathers. Melanin-pigmented feathers seem to incorporate more heavy metals, specifically zinc and lead, compared to feathers that are paler (non-melanin-pigmented). This was shown in pigeons of different color morphs which suggests that the dark melanin-pigmented Feral Pigeons could benefit in urban areas since they detoxify the blood stream from heavy metals, thereby reducing the potentially negative effect from pollution (Chatelain et al. 2014). Indeed, several studies have shown that the dark morph is more common in cities across Europe (Obukhova 2007; Jacquin et al. 2013).

Hormones have also been of great interest in relation to how birds respond to urbanization (Bonier 2012), especially stress and reproductive hormones. This is because hormones trigger behavioral and other physiological responses, thus representing key targets for selection. Changes in hormones have been associated with resource availability, conspecific interactions, predation, night light (see below), and human disturbance (Bonier 2012). Baseline levels of corticosterone seem to depend highly on sex, life-history stage, and/or species (Bonier 2012). Thus, studies on avian stress hormones have so far not been able to reveal any consistent pattern in relation to urbanization. Regarding hormones that affect reproduction, gonadotropin-releasing hormone is stimulated by day length. Due to the artificial night lighting in urban habitats the day becomes longer than in areas lacking street lights. Indeed, in urban environments gonadotropin-relasing hormone and other reproductive hormones are more stimulated which is the likely mechanistic explanation for the advancement of the timing of mating behaviors and reproduction (Deviche and Davies 2014).

13.5.1.2 Nutritional Physiology and Its Implications

Birds are provided with a great deal of anthropogenic food. This food can be either provided on bird feeders (e.g., peanuts, bread, sugar-water, sunflower seeds), or birds can scavenge from for example garbage bins or restaurant terraces. It is clear that many of the gregarious urban exploiter and invasive species take advantage of this resource (e.g., Robb et al. 2008; Galbraith et al. 2017). Food abundance and reliability change many phenotypic characters such as fat storage and flight-initiation distance, i.e., the tameness of the birds (Liker et al. 2008; Andersson et al. 2015; Møller et al. 2015). However, in relation to nutritional physiology, it is only recently that it has gained interest among ornithologists (e.g., Isaksson 2015). This goes hand in hand with human nutrition and the many negative health effects documented from our increased fat and sugar intake. Regarding fat intake, it is not only the increased consumption that is negative but also the changed composition of fatty acids, which affects, for example, inflammatory responses and metabolic rate. For example, in urban Great Tits there were less functionally important polyunsaturated fatty acids in yolks compared to yolks from eggs laid in the forest (Toledo et al. 2016). In addition, adult Great Tits show seasonal differences in fatty acid composition which reflect the differences in availability of anthropogenic food sources (Andersson et al. 2015). Furthermore and interestingly, the proportions of fatty acids were frequently in opposite directions when comparing species from the two families—Paridae and Passeridae. These patterns suggest that sparrows and tits feed on different food sources across the urban-rural gradient (Isaksson et al. 2017). Although the diet items were not quantified for fatty acids in any of the studies, certain polyunsaturated fatty acids are essential, i.e., they need to be obtained through the diet. The impact of these dietary differences shown in the birds' physiology is still unknown, but given the large-scale feeding of birds, it screams for attention (see, e.g., Harrison et al. 2010; Plummer et al. 2013).

Dietary antioxidants have also received attention in relation to urbanization, especially carotenoids (Isaksson and Andersson 2007; Møller et al. 2010). Carotenoids are synthesized by plants; thus birds need to obtain carotenoids through their diet. During the breeding season, many passerines rely on caterpillars for raising their brood, and these leaf-eating caterpillars represent a rich source of carotenoids (Isaksson and Andersson 2007). In cities, caterpillars are generally of lower abundance due to the high pollution levels and lack of native tree species (e.g., Pollock et al. 2017; Isaksson unpublished). However, it has also been shown that urban trees of birch *Betula* sp. and oak *Quercus* sp. (native to Northern Europe) produce less carotenoids (Isaksson 2009), which affects the carotenoid concentration of caterpillars and, ultimately, the carotenoid availability for birds (Isaksson and Andersson 2007). Carotenoids are important nutrients for proper development, the immune system, night vision, and yellow-red plumage pigmentation. Lack of carotenoids has been shown to affect breeding success, through smaller clutch sizes and reduced fledging success (Blount et al. 2002; Ewen et al. 2009). Nutritional limitation for breeding females and developing chicks is probably one of the reasons for the often-lower breeding success in urban environments (e.g., Chamberlain et al. 2009; Charmantier et al. 2017; Meyrier

Fig. 13.5 Urban (pale yellow) and rural Great Tits *Parus major* (yellow). The dietary pigments, carotenoids, that make feathers yellow are limited for urban birds. Hence the paler plumage coloration of this urban Great Tit. Photo: Caroline Isaksson

et al. 2017; Pollock et al. 2017). Moreover, carotenoid pigmentation is often used as an indicator of individual quality (i.e., nutritional quality and/or immunocompetence) used during mate choice. Urban Great Tits and also Great Tits living close to a copper smelter have a paler yellow (carotenoid-based) plumage coloration compared to their rural conspecifics (Eeva et al. 1998; Isaksson et al. 2005; Fig. 13.5). A paler plumage coloration affects the males' attraction potential (i.e., sexual signal value), and females may want or need to evaluate other characters, which may reduce female choice based on pigmentation and instead enhance the "sexiness" of other characters, which could lead to population divergence and reproductive isolation in the long run.

13.5.2 Behavior

13.5.2.1 Behavioral Responses to Chemical Pollution

Most responses documented to chemical pollution are physiological; however, also behavioral responses have been shown. In China, Li and colleagues documented the time it took for homing pigeons to find their way home. This race was conducted in an area with considerable air pollution, and it was shown that pigeons were homed faster when the air pollution level was especially high (Li et al. 2016). By doing this, the birds escape the high pollution levels, which can have negative effects on their performance and health. Another novel behavior is the use of cigarette butts in nests.

Both House Sparrows and House Finches have been shown to include cigarette butts as nest materials. The nicotine appears to work as an effective insect repellent against ectoparasites and the more cigarette butts the less infested was the nest (Suárez-Rodríguez et al. 2013). However, this repellent seems to come with physiological costs to the parent birds which show increased genotoxic damage (Suárez-Rodríguez et al. 2017).

13.5.2.2 Behavioral Responses to Noise

The list of behavioral responses to noise is long and includes changes in, for example, (1) vocal communication, (2) avoidance responses, and (3) fight-flight responses (e.g., Ortega 2012; Nemeth et al. 2013; LaZerte et al. 2017). The key feature of urban background noise is the low frequency, hence masking songs in this frequency range. Thus, a shift to higher frequency songs should be favored in the city. This is exactly what has been found in Great Tits (among many other species) in noisy natural environments (e.g., Slabbekoorn and Peet 2003). The higher minimum frequency song of urban Great Tits could also be experimentally induced, suggesting a highly plastic response (Halfwerk and Slabbekoorn 2009). However, a recent study challenges that plasticity is the cause for song differences across noisy habitats (Zollinger et al. 2017). The study by Zollinger and colleagues found that Great Tits sang consistently on pitch and with the same mean minimum frequencies in all noise conditions. This suggests that the observed changes between urban and forest populations may not be the result of individuals' plastic response, but instead be the outcome of slower, population-wide changes through selection (Zollinger et al. 2017). These population-level shifts in song could be driven by sexual selection, however, it could also be driven by body size and/or beak morphology (i.e., a smaller body and beak would lead to higher frequency songs). Urban environments have different food items available which repeatedly has been shown to affect beak morphology (e.g., Badyaev et al. 2008; Bosse et al. 2017). For example, urban House Finches were shown to feed on larger, harder foods than their counterparts in natural Sonoran Desert habitats (e.g., sunflower seeds *versus* cacti and grass seeds, respectively) (Badyaev et al. 2008). This led to a selection for larger bills in the urban population, which ultimately affected courtship song. This was suggested to give rise to a novel trade-off between bill size and song characteristics in urban environments. Possibly, these novel trade-offs between morphology and song in different environments could result in a nonoptimal song and mask the effect of noise on song.

Moreover, also species characters can influence the strength of vocal responses to urbanization. These species characters can, for example, be degree of vocal communication within species that use vocal communication to attract a mate, to defend a territory, or to warn for predators. Species also vary in their hearing capacity—some bird species hear certain frequencies and amplitudes better than other species, sometimes even within the ultrasonic range (Ortega 2012). Some species have solved the masking issue of their song and calls by changes in their daily rhythm. This is the case for European Robins *Erithacus rubecula*. The urban Robins reduce their acoustic interferences by singing during night, and the effect of daily noise was indeed a stronger driver of this change in behaviour than the night-light pollution through changed sleep patterns (see also below) (Fuller et al. 2007).

13.5.2.3 Behavioral Responses to ALAN

The effects on the navigation and orientation system of nocturnally migrating birds are the most well-known negative effects caused by ALAN. The migrating birds are attracted to urban ALAN, hence they appear more frequently in urban lit areas during autumn migration than during other seasons (La Sorte et al. 2017). Apart from the general ALAN, light installations are very popular nowadays; unfortunately, the light beams from installations can "trap" birds, i.e., birds are attracted to the beam and, while in the beam, they get disorientated and fly around in circles within the beam—they get trapped. This was shown for the light installation put up in New York as memorial tribute to the 9/11 victims. Over a billion birds were affected during a few days count repeated over multiple years (van Doren et al. 2017). Migrating birds were in 20 times higher densities in the light beams compared to the nearby surroundings. Nowadays, the city of New York turn of the light installations when the bird densities get too high within the beam. Another sensitive group for ALAN are seabird fledglings. During their first flight to the sea, they can get disorientated by ALAN and end up grounded at lit highways and roads (Rodriguez et al. 2014). By turning off road light during fledging it reduces the number of birds that ground on the road (dead or alive) (Rodriguez et al. 2014).

Reproductive timing and mating behaviors are also affected by ALAN (e.g., Kempenaers et al. 2010; Dominoni et al. 2013a). A study of city and forest Common Blackbirds revealed that when exposed to ALAN in captivity, the reproductive system developed one month earlier than in individuals that were kept with dark nights. However, even more interesting was the fact the bird that originated from the city responded stronger to ALAN treatment compared to the forest birds, i.e., they had an even earlier start than the forest birds (Dominoni et al. 2013b).

Furthermore, correlational studies of ALAN have for long suggested that birds start their dawn singing earlier in urban lit areas; however, recently a large-scale experiment using different light colors could not confirm this for the 14 species investigated (Da Silva et al. 2017). In another experimental study, Blue Tits responded to ALAN by advancing their daily activity onset and more so for red and white light than for green light (de Jong et al. 2017). Similarly, experimental lighting progressively advanced the dawn singing of Robins (Da Silva et al. 2016). Thus, at least for some bird species, urban influences through ALAN can be mitigated by changing the spectral characteristics and intensity of outdoor lighting.

Another concern with ALAN is the effect it might have on sleep. Anyone who has tried to sleep with the lights on knows how disruptive that can be. The sleep hormone, melatonin, is affected by light and a decline in melatonin with ALAN has been shown in Great Tits and Blackbirds (Dominoni et al. 2013a; de Jong et al.

2016) and female Great Tits spend a greater proportion of the night awake (Raap et al. 2015).

13.6 Concluding Remarks

Urbanization has led to an immense change of the avifauna. Species have fled and vanished in response to urbanization but also flourished and changed. It is clear that urbanization is a huge threat to biodiversity and the existence of many bird species and urbanization is not expected to slow down in any close future, rather the opposite (United Nations 2014). Thus, conservationists and city planners have an important task for the future. Their actions can in fact have great positive effects on the bird community, if the urban green space areas are managed well through plantation of native flora and enhanced complexity and if they enhance the urban green space or limit construction in key areas (e.g., Chamberlain et al. 2007; Aronson et al. 2014; Kang et al. 2015). Many political actions are also taken to reduce the impact of the different pollution sources, e.g., electric cars and LED lights that can be turned off or dimmed during sensitive periods, which will probably also show a positive effect on urban-dwelling species. However, much remains unknown about urban bird species resilience to urbanization and how plastic these species can be in their stress resistance responses to multiple stressors, before they reach a threshold with a population crash as a result (Watson et al. 2015). Future studies will entail, whether urbanization will be an opportunity for species radiation or if it will continue to be a habitat of species eradication and homogenization.

References

Andersson MN, Wang H-L, Nord A et al (2015) Composition of physiologically important fatty acids in great tits differs between urban and rural populations on a seasonal basis. Front Ecol Evol 3:93. https://doi.org/10.3389/fevo.2015.00093

Aronson MFJ, La Sorte FA, Nilon CH et al (2014) A global analysis of the impacts of urbanization on bird and plant diversity reveals key anthropogenic drivers. Proc R Soc Lond B 281:20133330. https://doi.org/10.1098/rspb.2013.3330

Audet J-N, Ducatez S, Lefebvre L (2016) The town bird and the country bird: problem solving and immunocompetence vary with urbanization. Behav Ecol 27:637–644. https://doi.org/10.1093/beheco/arv201

Badyaev AV, Young RL, Oh KP et al (2008) Evolution on a local scale: developmental, functional, and genetic bases of divergence in bill form and associated changes in song structure between adjacent habitats. Evolution 62:1951–1964. https://doi.org/10.1111/j.1558-5646.2008.00428.x

Binkowski LJ, Meissner W (2013) Levels of metals in blood samples from Mallards (*Anas platyrhynchos*) from urban areas in Poland. Environ Pollut 178:336–342

Björklund M, Ruiz I, Senar JC (2010) Genetic differentiation in the urban habitat: the great tits (*Parus major*) of the parks of Barcelona city. Biol J Linn Soc 99:9–19. https://doi.org/10.1111/j.1095-8312.2009.01335.x

Blair RB (1996) Land use and avian species diversity along an urban gradient. Ecol Appl 6:506–519. https://doi.org/10.2307/2269387
Blount JD, Surai PF, Nager RG et al (2002) Carotenoids and egg quality in the lesser black-backed gull *Larus fuscus*: a supplemental feeding study of maternal effects. Proc R Soc Lond B 269:29–36. https://doi.org/10.1098/rspb.2001.1840
Bonier F (2012) Hormones in the city: endocrine ecology of urban birds. Horm Behav 61:763–772. https://doi.org/10.1016/j.yhbeh.2012.03.016
Bonnington C, Gaston KJ, Evans KL (2015) Ecological traps and behavioural adjustments of urban songbirds to fine-scale spatial variation in predator activity. Anim Conserv 18:529–538. https://doi.org/10.1111/acv.12206
Boonekamp JJ, Bauch C, Mulder E et al (2017) Does oxidative stress shorten telomeres? Biol Lett 13:20170164. https://doi.org/10.1098/rsbl.2017.0164
Bosse M, Spurgin LG, Laine VN et al (2017) Recent natural selection causes adaptive evolution of an avian polygenic trait. Science 358:365–368. https://doi.org/10.1126/science.aal3298
Chace JF, Walsh JJ (2006) Urban effects on native avifauna: a review. Landsc Urban Plan 74:46–69. https://doi.org/10.1016/j.landurbplan.2004.08.007
Chamberlain DE, Gough S, Vaughan H et al (2007) Determinants of bird species richness in public green spaces. Bird Study 54:87–97. https://doi.org/10.1080/00063650709461460
Chamberlain DE, Cannon AR, Toms MP et al (2009) Avian productivity in urban landscapes: a review and meta-analysis. Ibis 151:1–18. https://doi.org/10.1111/j.1474-919X.2008.00899.x
Charmantier A, Demeyrier V, Lambrechts M et al (2017) Urbanization is associated with divergence in pace-of-life in great tits. Front Ecol Evol 5:53. https://doi.org/10.3389/fevo.2017.00053
Chatelain M, Gasparini J, Jacquin L et al (2014) The adaptive function of melanin-based plumage coloration to trace metals. Biol Lett 10:20140164. https://doi.org/10.1098/rsbl.2014.0164
Clergeau P, Croci S, Jokimäki J et al (2006) Avifauna homogenization by urbanization. Analysis at different European latitudes. Biol Conserv 127:336–344. https://doi.org/10.1016/j.biocon.2005.06.035
Da Silva A, Valcu M, Kempenaers B (2016) Behavioural plasticity in the onset of dawn song under intermittent experimental night lighting. Anim Behav 117:155–165
Da Silva A, de Jong M, van Grunsven RHA et al (2017) Experimental illumination of a forest: no effects of lights of different colours on the onset of the dawn chorus in songbirds. R Soc Open Sci 4:160638. https://doi.org/10.1098/rsos.160638
de Jong M, Jeninga L, Ouyang JQ et al (2016) Dose-dependent responses of avian daily rhythms to artificial light at night. Physiol Behav 155:172–179. https://doi.org/10.1016/j.physbeh.2015.12.012
de Jong M, Caro SP, Gienapp P et al (2017) Early birds by light at night: effects of light color and intensity on daily activity patterns in blue tits. J Biol Rhythm 32:323–333. https://doi.org/10.1177/0748730417719168
Delaney KS (2014) Landscape genetics of urban bird populations. In: Gil D, Brumm H (eds) Avian urban ecology. Oxford University Press, Oxford, pp 143–154
Delaney KS, Riley SPD, Fisher RN (2010) A rapid, strong, and convergent genetic response to urban habitat fragmentation in four divergent and widespread vertebrates. PLoS One 5:e12767. https://doi.org/10.1371/journal.pone.0012767
Delgado-V CA, French K (2011) Parasite-bird interactions in urban areas: current evidence and emerging questions. Landsc Urban Plan 105:5–14. https://doi.org/10.1016/j.landurbplan.2011.12.019
Deviche P, Davies S (2014) Reproductive phenology of urban birds: environmental cues and mechanisms. In: Gil D, Brumm H (eds) Avian urban ecology. Oxford University Press, Oxford, pp 98–115
Diamond J (2005) Collapse: how societies choose to fail or succeed. Viking Press, New York
Dias P (1996) Sources and sinks in population biology. Trends Ecol Evol 11:326–330

Dominoni DM, Goymann W, Helm B et al (2013a) Urban-like night illumination reduces melatonin release in European blackbirds (*Turdus merula*): implications of city life for biological time-keeping of songbirds. Front Zool 10:60 http://www.frontiersinzoology.com/content/10/1/60
Dominoni DM, Quetting M, Partecke J (2013b) Artificial light at night advances avian reproductive physiology. Proc R Soc Lond B 280:20123017. https://doi.org/10.1098/rspb.2012.3017
Donovan TM, Thompson FR III (2001) Modeling the ecological trap hypothesis: a habitat a demographic analysis for migrant songbirds. Ecol Appl 11:871–882. https://doi.org/10.1890/1051-0761(2001)011[0871:mtetha]2.0.co;2
Eeva T, Lehikoinen E, Ronka M (1998) Air pollution fades the plumage of the great tit. Funct Ecol 12:607–612. https://doi.org/10.1046/j.1365-2435.1998.00221.x
Erritzoe J, Mazgajski T, Rejt L (2003) Bird casualties on European roads – a review. Acta Ornithol 38:77–93. https://doi.org/10.3161/068.038.0204
Evans KL, Frantz AC, Simeoni M et al (2009a) Independent colonization of multiple urban centres by a formerly forest specialist bird species. Proc R Soc Lond B 276:2403–2410. https://doi.org/10.1098/rspb.2008.1712
Evans KL, Gaston KJ, Sharp SP et al (2009b) Effects of urbanisation on disease prevalence and age structure in blackbird *Turdus merula* populations. Oikos 118:774–782. https://doi.org/10.1111/j.1600-0706.2008.17226.x
Ewen JG, Thorogood R, Brekke P et al (2009) Maternally invested carotenoids compensate costly ectoparasitism in the hihi. PNAS 106:12798–12802. https://doi.org/10.1073/pnas.0902575106
Fuller RA, Warren PH, Gaston KJ (2007) Daytime noise predicts nocturnal singing in urban robins. Biol Lett 3:368–370. https://doi.org/10.1098/rsbl.2007.0134
Fuller RA, Warren PH, Armsworth PR et al (2008) Garden bird feeding predicts the structure of urban avian assemblages. Divers Distrib 14:131–137. https://doi.org/10.1111/j.1472-4642.2007.00439.x
Galbraith JA, Jones DN, Beggs JR et al (2017) Urban bird feeders dominated by a few species and individuals. Front Ecol Evol 5:81. https://doi.org/10.3389/fevo.2017.00081
Grimm NB, Faeth SH, Golubiewski NE et al (2008) Global change and the ecology of cities. Science 319:756–760
Halfwerk W, Slabbekoorn H (2009) A behavioural mechanism explaining noise-dependent frequency use in urban birdsong. Anim Behav 78:1301–1307. https://doi.org/10.1016/j.anbehav.2009.09.015
Harrison TJ, Smith JA, Martin GR et al (2010) Does food supplementation really enhance productivity of breeding birds? Oecologia 164:311–320. https://doi.org/10.1007/s00442-010-1645-x
Isaksson C (2009) The life-cycle of carotenoids: from plants to birds. Ardea 97:125–128
Isaksson C (2010) Pollution and its impact on wild animals: a meta-analysis on oxidative stress physiology. EcoHealth 7:342–350. https://doi.org/10.1007/s10393-010-0345-7
Isaksson C (2015) Urbanisation, oxidative stress and inflammation: a question of evolving, acclimatizing or coping with urban environmental stress. Funct Ecol 29:913–923. https://doi.org/10.1111/1365-2435.12477
Isaksson C, Andersson S (2007) Carotenoid diet and nestling provisioning in urban and rural great tits, *Parus major*. J Avian Biol 38:564–572. https://doi.org/10.1111/j.2007.0908-8857.04030.x
Isaksson C, Örnborg J, Stephensen E et al (2005) Plasma glutathione and carotenoid coloration as potential biomarkers of environmental stress in great tits. EcoHealth 2:138–146
Isaksson C, Sheldon BC, Uller T (2011) The challenges of integrating oxidative stress into life history biology. Bioscience 60:194–202. https://doi.org/10.1525/bio.2011.61.3.5
Isaksson C, Andersson MN, Nord A et al (2017) Species-dependent effects of the urban environment on fatty acid composition and oxidative stress in birds. Front Ecol Evol 5:44. https://doi.org/10.3389/fevo.2017.00044
Jacquin L, Récapet C, Prévot-Julliard A-C et al (2013) A potential role for parasites in the maintenance of color polymorphism in urban birds. Oecologia 173:1089–1099. https://doi.org/10.1007/s00442-013-2663-2

Jokimäki J, Huhta E (2000) Artificial nest predation and abundance of birds along an urban gradient. Condor 102:838–847. https://doi.org/10.2307/1370311

Kang W, Minor ES, Park C-R et al (2015) Effects of habitat structure, human disturbance, and habitat connectivity on urban forest bird communities. Urban Ecosyst 18:857–870. https://doi.org/10.1007/s11252-014-0433-5

Kempenaers B, Borgstroem P, Loes P et al (2010) Artificial night lighting affects dawn song, extra-pair siring success, and lay date in songbirds. Curr Biol 20:1735–1739. https://doi.org/10.1016/j.cub.2010.08.028

Kettel J, Gentle LK, Quinn JL et al (2017) The breeding performance of raptors in urban landscapes: a review and meta-analysis. J Ornithol 159(1):1–18. https://doi.org/10.1007/s10336-017-1497-9

La Sorte FA, Fink D, Buler JJ et al (2017) Seasonal associations with urban light pollution for nocturnally migrating bird populations. Glob Chang Biol 23:4609–4619. https://doi.org/10.1111/gcb.13792

LaZerte SE, Otter KA, Slabbekoorn H (2017) Mountain chickadees adjust songs, calls and chorus composition with increasing ambient and experimental anthropogenic noise. Urban Ecosyst 20:989–1000

Lepczyk CA, Aronson MFJ, Evans KL et al (2017) Biodiversity in the city: fundamental questions for understanding the ecology of urban green spaces for biodiversity conservation. Bioscience 67:799–807. https://doi.org/10.1093/biosci/bix079

Li Z, Courchamp F, Blumstein DT (2016) Pigeons home faster through polluted air. Sci Rep 6:18989. https://doi.org/10.1038/srep18989

Liker A, Papp Z, Bókony V et al (2008) Lean birds in the city: body size and condition of house sparrows along the urbanization gradient. J Anim Ecol 77:789–795. https://doi.org/10.1111/j.1365-2656.2008.01402.x

Luck GW, Smallbone LT (2010) Species diversity and urbanization: patterns, drivers and implications. In: Gaston KJ (ed) Urban ecology. Cambridge University Press, Cambridge

MacGregor-Fors I, Morales-Pérez L, Schondube JE (2012) From forest to cities: effects of urbanization on tropical birds. In: Lepczyk CA, Warren PS (eds) Urban bird ecology and conservation, Studies in avian biology (no 45). University of California Press, Berkeley, pp 33–38

Manjula M, Mohanraj R, Devi MP (2015) Biomonitoring of heavy metals in feathers of eleven common bird species in urban and rural environments of Tiruchirappalli, India. Environ Monit Assess 187:267. https://doi.org/10.1007/s10661-015-4502-x

Marzluff JM, Ewing K (2001) Restoration of fragmented landscapes for the conservation of birds: a general framework and specific recommendations for urbanizing landscapes. Restor Ecol 9:280–292. https://doi.org/10.1046/j.1526-100x.2001.009003280.x

McKinney ML (2002) Urbanization, biodiversity, and conservation. Bioscience 52:883–890. https://doi.org/10.1641/0006-3568(2002)052[0883:ubac]2.0.co;2

McNew SM, Beck D, Sadler-Riggleman I et al (2017) Epigenetic variation between urban and rural populations of Darwin's finches. BMC Evol Biol 17:183. https://doi.org/10.1186/s12862-017-1025-9

Meyrier E, Jenni L, Bötsch Y et al (2017) Happy to breed in the city? Urban food resources limit reproductive output in Western Jackdaws. Ecol Evol 7:1363–1374. https://doi.org/10.1002/ece3.2733

Møller AP (2009) Successful city dwellers: a comparative study of the ecological characteristics of urban birds in the Western Palearctic. Oecologia 159:849–858. https://doi.org/10.1007/s00442-008-1259-8

Møller AP, Erritzøe J, Karadas F (2010) Levels of antioxidants in rural and urban birds and their consequences. Oecologia 163:35–45. https://doi.org/10.1007/s00442-009-1525-4

Møller AP, Tryjanowski P, Díaz M et al (2015) Urban habitats and feeders both contribute to flight initiation distance reduction in birds. Behav Ecol 26:861–865. https://doi.org/10.1093/beheco/arv024

Nemeth E, Pieretti N, Zollinger S-A et al (2013) Bird song and anthropogenic noise: vocal constraints may explain why birds sing higher-frequency songs in cities. Proc R Soc Lond B 280:20122798

Obukhova NY (2007) Polymorphism and phene geography of the blue rock pigeon in Europe. Russ J Genet 43:492–501. https://doi.org/10.1134/S1022795407050031

Orlowski G, Kasprzykowski Z, Dobicki W et al (2014) Residues of chromium, nickel, cadmium and lead in rook *Corvus frugilegus* eggshells from urban and rural areas of Poland. Sci Total Environ 490:1057–1064. https://doi.org/10.1016/j.scitotenv.2014.05.105

Ortega CP (2012) Effects of noise pollution on birds: a brief review of our knowledge. Ornithol Monogr 74:6–22. https://doi.org/10.1525/om.2012.74.1.6

Perrier C, del Campo AL, Szulkin M et al (2017) Great tits and the city: distribution of genomic diversity and gene-environment associations along an urbanization gradient. Evol Appl. https://doi.org/10.1111/eva.12580

Plummer KE, Bearhop S, Leech DI et al (2013) Winter food provisioning reduces future breeding performance in a wild bird. Sci Rep 3:2002. https://doi.org/10.1038/srep02002

Pollock CJ, Capilla-Lasheras P, McGill RAR et al (2017) Integrated behavioural and stable isotope data reveal altered diet linked to low breeding success in urban-dwelling blue tits (*Cyanistes caeruleus*). Sci Rep 7:5014. https://doi.org/10.1038/s41598-017-04575-y

Raap T, Pinxten R, Eens M (2015) Light pollution disrupts sleep in free-living animals. Sci Rep 5:13557. https://doi.org/10.1038/srep13557

Rasner CA, Yeh PJ, Eggert LS et al (2004) Genetic and morphological evolution following a founder event in the dark-eyed junco, *Junco hyemalis thurberi*. Mol Ecol 13:671–681. https://doi.org/10.1046/j.1365-294X.2004.02104.x

Riyahi S, Sánchez-Delgado M, Calafell F et al (2015) Combined epigenetic and intraspecific variation of the DRD4 and SERT genes influence novelty seeking behavior in great tit *Parus major*. Epigenetics 10:516–525. https://doi.org/10.1080/15592294.2015.1046027

Robb GN, McDonald RA, Chamberlain DE et al (2008) Food for thought: supplementary feeding as a driver of ecological change in avian populations. Front Ecol Environ 6:476–484. https://doi.org/10.1890/060152

Rodriguez A, Burgan G, Dann P et al (2014) Fatal attraction of short-tailed shearwaters to artificial lights. PLoS One 9:e110114. https://doi.org/10.1371/journal.pone.0110114

Rutkowski R, Rejt L, Szczuka A (2006) Analysis of microsatellite polymorphism and genetic differentiation in urban and rural kestrels *Falco tinnunculus* (L.). Pol J Ecol 54:473–480

Salmón P (2017) On urbanization and birds: insight from a molecular and physiological perspective. Thesis, Media-Tryck, Lund University, Lund

Salmón P, Nilsson J, Nord A et al (2016) Urban environment shortens telomere length in nestling great tits, *Parus major*. Biol Lett 12:20160155. https://doi.org/10.1098/rsbl.2016.0155

Salmón P, Nilsson J, Watson H et al (2017) Selective disappearance of great tits with short telomeres in urban areas. Proc R Soc Lond B 284:20171349. https://doi.org/10.1098/rspb.2017.1349

Salmón P, Stroh E, Herrera-Dueñas A et al (2018) Oxidative stress in birds along a NOx and urbanisation gradient: an interspecific approach. Sci Total Environ 622–623:635–643

Senar JC, Garamszegi LZ, Tilgar V et al (2017) Urban great tits (*Parus major*) show higher distress calling and pecking rates than rural birds across Europe. Front Ecol Evol 5:163. https://doi.org/10.3389/fevo.2017.00163

Slabbekoorn H, Peet M (2003) Birds sing at a higher pitch in urban noise – great tits hit the high notes to ensure that their mating calls are heard above the city's din. Nature 424:267

Suárez-Rodríguez M, López-Rull I, Macías Garcia C (2013) Incorporation of cigarette butts into nests reduces nest ectoparasite load in urban birds: new ingredients for an old recipe? Biol Lett 9:20120931. https://doi.org/10.1098/rsbl.2012.0931

Suárez-Rodríguez M, Montero-Montoya RD, Macías Garcia C (2017) Anthropogenic nest materials may increase breeding costs for urban birds. Front Ecol Evol 5:4. https://doi.org/10.3389/fevo.2017.00004

Suri J, Sumasgutner P, Hellard É et al (2017) Stability in prey abundance may buffer black sparrowhawks *Accipiter melanoleucus* from health impacts of urbanization. Ibis 159:38–54. https://doi.org/10.1111/ibi.12422

Toledo A, Andersson MN, Wang H-L et al (2016) Fatty acid profiles of great tit (*Parus major*) eggs differ between urban and rural habitats, but not between coniferous and deciduous forests. Sci Nat 103:55. https://doi.org/10.1007/s00114-016-1381-0

Unfried TM, Hauser L, Marzluff JM (2013) Effects of urbanization on song sparrow (*Melospiza melodia*) population connectivity. Conserv Genet 14:41–53

United Nations (2014) World urbanization prospects: the 2014 revisions, highlights. United Nations publication, ST/ESA/SER.A/352

van Doren BM, Horton KG, Dokter AM et al (2017) High-intensity urban light installation dramatically alters nocturnal bird migration. PNAS 114:11175–11180. https://doi.org/10.1073/pnas.1708574114

von Zglinicki T (2002) Oxidative stress shortens telomeres. Trends Biochem Sci 27:339–344. https://doi.org/10.1016/S0968-0004(02)02110-2

Watson H, Cohen AA, Isaksson C (2015) A theoretical model of the evolution of actuarial senescence under environmental stress. Exp Gerontol 71:80–88

Watson H, Videvall E, Andersson MN et al (2017) Transcriptome analysis of a wild bird reveals physiological responses to the urban environment. Sci Rep 7:44180. https://doi.org/10.1038/srep44180

West-Eberhard MJ (1989) Phenotypic plasticity and the origins of diversity. Annu Rev Ecol Syst 20:249–278

Wilcove DS (1985) Nest predation in forest tracts and the decline of migratory songbirds. Ecology 66:1211–1214

Zollinger SA, Slater PJB, Nemeth E et al (2017) Higher songs of city birds may not be an individual response to noise. Proc R Soc Lond B 284:20170602. https://doi.org/10.1098/rspb.2017.0602

Glossary

Adaptive radiation	The evolution of ecological and phenotypic diversity within a rapidly multiplying lineage
ALAN	Artificial light at night
Alignment	Base (or peptide) sequences of DNA/RNA (or protein) with corresponding bases (or peptides) from different individuals arranged in the same orientation
Allochrony	Diverge in timing without temporal overlap
Allopatry	Spatially non-overlapping distributions
Anthropogenic	Relating to or resulting from the influence of humans on nature such as pollution, artificial constructions, and noise
Assembly	Smaller sequence reads are combined to a larger sequence
Call	Simple and rather short and predominantly innate vocalization given in many different behavioral contexts
Candidate gene	Gene that is expected a priori to be involved in a certain trait; very often a gene with focal functions identified in model organisms that may have a similar function on the focal trait, here migration, in other animals
Carotenoids	A set of organic pigments, often producing yellow to red colors in bird feathers and skin that cannot be produced by birds themselves, but that are mainly produced by plants and algae
Chronotype	Characterization of consistent timing of an individual's behavior, relative to other individuals; an example is wake-time relative to sunrise

D. T. Tietze (ed.), *Bird Species*, Fascinating Life Sciences,
https://doi.org/10.1007/978-3-319-91689-7

Circadian	Internal rhythm with a period length of approximately 24 hours
Circannual	Internal rhythm with a period length of approximately 365 days
Codominant	Both alleles (not only a dominant one) affect the phenotype
Coverage	Average number of reads per locus in DNA sequencing
Cytosine-phosphate-Guanine (CpG) islands	Regions with a high frequency of cytosine and guanine
Density-dependent effect on diversification	The concept that speciation rate slows down over time as the number of species increases in particular clades, as a consequence of a saturation of opportunities for new speciation events
Diagnostic	In taxonomy, a marker or trait is said to be diagnostic, if it presents different character states in different taxa
Dialect	A regional variant of song differing in structural, frequency, or time parameters from other variants from adjacent regions
Dispersal	In ecological terms, describes the movement of an individual from one breeding site to another. At evolutionary scales, it refers to a species' colonization of a previously unoccupied area
Diversification rates	Refers to the net rate of speciation (the formation of new species) minus extinction (living species going extinct) rate and can be estimated from fossils and phylogenetic trees
Ecological opportunity	The availability of a wealth of resources through a relaxation of interspecific competition or the relaxation of selection on some ecologically important traits as a consequence of a key innovation, the invasion of a novel or unoccupied habitat. Ecological opportunity is considered one of the major regulators of the rate and extent of phenotypic differentiation and also speciation
Ecological speciation	Typically described as the process by which barriers to gene flow evolve between populations as a result of ecologically based divergent selection between environments. Here used in a broader sense, whenever variation among environments induces divergent selective pressures on populations and individuals which somehow results into reproductive barriers (i.e., not necessarily via divergent natural selection alone)
Ecological trap	Refers to scenarios in which rapid environmental change makes individuals prefer to settle in poor-quality

	habitats, i.e., they misjudge the environmental cues that help them to evaluate and identify high-quality habitats
Ecomorphology	The study of morphological adaptations of an organism in relation to their ecological context
Element	A single sound event, represented by a continuous black line in a sonagram. Fixed combination of notes (motifs or syllables) can be repeated in songs in a stereotyped sequence
Genetic accommodation	The evolution of a genetic coding to enable the constant expression of a trait whose expression was previously affected by the environment (i.e., from phenotypically plastic to phenotypically fixed)
Genetic drift	The change in the genetic composition of a population across generations that is due to stochastic sampling of the parental genetic variation availability; most pronounced in smaller populations
Gondwana	Also known as Gondwanaland, an ancient super continent, which existed from c. 600–530 to 160–30 million years ago. This supercontinent covered most of the landmasses in today's southern hemisphere, including Antarctica, South America, Africa, Madagascar, the Arabian Peninsula, the Indian subcontinent, and the Australian continent
Granivorous	Grain feeding birds
Heritability	Quantitative genetics analyses characterize the relative contributions of genetic and environmental variation to phenotypic variation. Heritability (h^2) is the response to selection of the trait in focus
High-throughput sequencing techniques	High-throughput parallel sequencing (revealing the nucleotide sequence of DNA) and parallel analysis has revolutionized our ability to analyze genomes. Current state-of-the-art technology allows sequencing and characterization of whole genomes as de novo reference for basically any study species at much lower costs
Imprinting	The development of a preference for a trait or an environment based on exposure during an earlier, sensitive period in life
Integrative taxonomy	The theory and practice of documenting and evaluating hypotheses about the taxonomic status of putative species by integrating multiple lines of evidence

Intergradation	The transfer of genetic material between individuals from two (groups of) populations with or without unique evolutionary histories; also used to describe a pattern (clinal variation), rather than a process
Isolation by time	Limitation in gene flow caused by two populations reproducing at different times
Key innovations	The acquisition of evolutionary novelty, e.g., a novel trait that creates an ecological opportunity, positively influences ability to utilize previously unexploited resources, and promotes species proliferation
K-Pg boundary	The abbreviation for Cretaceous–Paleogene boundary representing the end of the Cretaceous at approximately 65 million years ago and the beginning of the Cenozoic era we still live in today. This transitional period was associated with a mass extinction event that led to the disappearance of 75% of plants and animals (including all non-avian dinosaurs) on Earth during a short period and that was triggered by the impact of a massive asteroid at Chicxulub, Mexico
Ligation	Linkage of two ends of DNA or RNA molecules
Light-level geolocator	Miniature light-level archival logger that allows tracking of animal movement over large distances for a certain time period. The device uses a light sensor to store light-level data at certain time points. Latitude/longitude location data of the animal can be inferred based on day length (latitude) and the time of solar noon (longitude). The geolocator must be retrieved for data download
Lineage	A metapopulation extended in time
Migratory divide	Natural hybrid zones where birds from two neighboring populations (or (sub)species) with different migratory orientation meet and mate
Modification of the environment	Manipulating and improving aspects of the current environment
Monophyletic lineage	A group of evolutionary lineages that includes the most recent common ancestor of those lineages and all of its past and current descendants
Monophyly	The grouping together of an ancestor and all of its descendants in a clade; the situation where all organisms and populations are descended from the same common ancestor and so more closely related to each other than any of them are to any other such group

Natural selection	The systematic differences in reproduction between phenotypes (including mortality before reproduction), whether caused by interactions with members of the same species (i.e. including sexual selection) or with other biotic or abiotic aspects of the environment
Neutral evolution	The change in heritable traits across generations that is not due to selection
Niche	The set of interactions with the biotic and abiotic environment which can be used to describe and understand the presence of an organism.
Non-urban habitat	This is different in different studies and ranges from natural habitats such as forests, deserts, and wetlands to farmland habitats with little urban construction, yet human modification of the land
Note	A single sound event, represented by a continuous black line in a sonagram. Fixed combination of notes (motifs or syllables) can be repeated in songs in a stereotyped sequence
NO_x	Nitrogen oxides
Omnivorous	Birds that have a mixed/varied diet
Orientation cage experiment	Behavioral experiments to characterize the migratory direction of birds in captivity make use of the characteristic migratory restlessness behavior caged migrants exhibit once the migratory season approaches. Orientation cages allow recording and quantifying migratory behavior activity and the directedness of the bird's orientation during periods of migratory restlessness. Cage design can vary; the basic form is usually referred to as "Emlen funnel" (named after its inventors, Stephen T. and John T. Emlen), where the walls of a funnel-shaped cage are covered with scratch sensitive paper. Scratches that are left by the tested bird on the inclined walls of the cage are quantified to identify the bird's mean bearing
Period length	Time taken for a full cycle of a rhythmic behavior; an example is the time passing between waking up on consecutive days
Phenotypic plasticity	The ability of a single genotype to produce more than one phenotype, depending on the environment to which it is exposed
Photoperiod	Light fraction of the 24-hour day

Phylogeographic analysis	Analysis of the geographic distribution of (intraspecific) phylogenetic lineages, study of how genetic diversity within a species is distributed across its geographic range and apportioned phylogenetically within the species
Polymorphic	Occurrence of more than one form in the same individual or population of a species
Repertoire	The entirety of song types of an individual
Reproductive isolation	Mechanisms preventing interbreeding among individuals from different species due to mating incompatibilities caused by morphological, physiological, or genetic barriers
Selection of the environment	Comparing and choosing among a set of (potentially) available environments
Sexual ornament	A trait that (mainly) functions to improve sexual attractiveness to the other sex
Sexual selection	The systematic differences in reproduction between phenotypes due to competition over fertilizations
Sink-source dynamic	This can be used to explain the flow of individuals between an urban and a rural habitat. A sink population has a negative population growth and a source population has a positive population growth. The surplus individuals from a source population often disperse to a sink population
Sonagram	Spectrographic visualization of sound events with frequency plotted against time
Song	Complex and rather long vocalization often composed of many different combinations of notes and displayed in two major behavioral contexts: territorial defense and mate attraction. Variable song types of an individual can be distinguished by different structural parameters or different note combinations
Speciation	Evolutionary divergence of subsets of one ancestral species into two (or more) different species
Species criterion	A rule designed to delimit species
Species delimitation	The designation of the taxonomic (and often also geographic) boundaries of one or more taxa
Species property	A property of a species taxon that may be used to infer the evolutionary history of that taxon (e.g. diagnosability, reciprocal monophyly) or its interactions with other taxa (adaptive zone, reproductive isolation)
Stable-isotope signature	The chemical composition of stable isotopes in a certain tissue at a specific time is unique to the area where the bird resides during the time of food take

	up. Depending on the molt schedule of the bird species you study, feathers may be perfectly suited for stable-isotope analyses as for each species they grow during a clearly defined molt period. Once the feather is grown, its stable isotope becomes metabolically inert. If the bird undergoes complete molt in its wintering grounds, the feather coat it wears upon arrival on the breeding ground mirrors the isotope signature of its wintering grounds and thus can be used as proxy for wintering grounds and consequently migratory orientation
Sticky ends	DNA ends with an overhang of nucleotides
Subspecies	There are two major definitions: (1) a population that differs taxonomically from other populations, (2) a population possessing some properties of species (e.g. diagnosability or reciprocal monophyly), but lacking a defining property (usually reproductive isolation)
Sympatry	Co-occurrence of two distinct species (or subspecies) due to spatially overlapping distributions
Syntax	Sequential order of notes or motifs
Taxon (plural: taxa)	A named subspecies, species, genus, and so on for any taxonomic rank
Taxonomy	The science of biodiversity; the theory and practice of delimiting biological taxa
Urban adapter species	Birds that rely on natural resources, but are able to utilize human subsidies. These species are very common in sub-urban areas or in city parks, but survive better in natural environments
Urban avoider species	Birds that avoid urban areas. These are species that have special requirements such as large territories, certain foods or breeding grounds
Urban exploiter species	Birds that depend on human resources to maintain their population size, species that flourish in urban environments. These species are often considered as pest species or invasive species
Urban heat island	Metropolitan area that is significantly warmer than its surrounding rural areas due to human activities
Urban stressor (driver)	Anthropogenic introduced factors that are more present in urban areas and that affect birds, such as pollution, human disturbance, and increased temperature

Urbanization	This is a broad term and its definition varies with the study focus. Here I use urbanization for anthropogenic disturbance that is associated with densification of humans and human activities, i.e. cities
Vocalization	Any kind of tonal utterance used for communication among individuals of the same or even of different species

MIX
Papier aus verantwortungsvollen Quellen
Paper from responsible sources
FSC® C105338

If you have any concerns about our products,
you can contact us on
ProductSafety@springernature.com

In case Publisher is established outside the EU,
the EU authorized representative is:
Springer Nature Customer Service Center GmbH
Europaplatz 3, 69115 Heidelberg, Germany

Printed by Libri Plureos GmbH
in Hamburg, Germany